D0207655

Every Molecule Tells a Story

Every Molecule Tells a Story

Simon Cotton

Honorary Senior Lecturer in Chemistry
University of Birmingham, UK

Chemist Emeritus
Uppingham School, UK

CRC Press
Taylor & Francis Group
Boca Raton   London   New York

CRC Press is an imprint of the
Taylor & Francis Group, an **informa** business

CRC Press
Taylor & Francis Group
6000 Broken Sound Parkway NW, Suite 300
Boca Raton, FL 33487-2742

© 2012 by Taylor & Francis Group, LLC
CRC Press is an imprint of Taylor & Francis Group, an Informa business

No claim to original U.S. Government works

Printed in the United States of America on acid-free paper
Version Date: 20110705

International Standard Book Number: 978-1-4398-0773-6 (Hardback)

Library of Congress Cataloging-in-Publication Data

Cotton, Simon.
   Every molecule tells a story / Simon Cotton.
      p. cm.
   Includes bibliographical references and index.
   ISBN 978-1-4398-0773-6 (hardback)
   1. Chemical bonds. 2. Molecules. I. Title.

QD461.C655 2012
540--dc23   2011022328

Visit the Taylor & Francis Web site at
http://www.taylorandfrancis.com

and the CRC Press Web site at
http://www.crcpress.com

Dedication

The late Nikolaus Pevsner famously dedicated the Bedfordshire, Huntingdon, and Peterborough volume of his Buildings of England series to the "Inventor of the Iced Lolly."

As a teacher without easy access to a university library, I have felt there could be something to be said for a dedication to the Internet, were it not for Sir Timothy Berners Lee and his invention, this book would not have been possible.

Instead, however, I dedicate this book to Hilary, to Jill, to Judith, and to Lisa, for their encouragement, assistance, and inspiration.

Contents

Preface

Everything that exists is made of atoms, and most of those substances contain groups of two or more of these bonded together to form molecules. Chemistry is the science of molecules. From cooking to medicine, from engineering to art, it is everywhere.

What is this book? For a start, it is not a textbook, nor is it a collection of reviews. If I can put this into words, it is a celebration of molecules and of chemistry. Over 200 molecules that give you life, enrich your life, or may end your life if you are not careful. Though I cannot write as well or as clearly as Peter Atkins or John Emsley, I have tried to say why I think that these molecules matter, putting them into their context rather than seeing them as just substances of academic interest. I've possibly written more than I should have concerning some molecules that I have found interesting, topical, or concerning which it does not seem so easy to obtain information, but the reader will judge that. For many years, chemistry has been described as the central science, notably in America. Despite that, it has been an unfashionable area. I have been unfashionable for most of my life, so I am not worried about that. It is primarily addressed to chemistry students on either side of the school/university divide and to their teachers. I even nurse a vain hope that it might become something of a *vade mecum* to teachers and lecturers and, to this end, I have done my best to provide a fairly comprehensive bibliography.

This book has been over 3 years in the making, 2 years in the writing, and perhaps 18 years in its conception, when I picked up a copy of, I think, *Time* magazine in February 1993, read an article about chocolate and phenylethylamine, and decided that it would be a good vehicle to teach students about amine chemistry. That led to "Soundbite Molecules", then to "Molecule of the Month", and, eventually, to this book. But I can sympathise with the writer of Ecclesiastes 12:12 in his comment "of making many books there is no end; and much study is a weariness of the flesh".

Many people have discussed or helped with material, including authors who have sent copies of papers. Among the individuals I should thank are, from the Chemistry Department of the University of Cambridge, Peter Wothers, Judith Battison, and Mike Todd-Jones for their help in different ways, and Phillip Broadwith and Nina Notman of the Royal Society of Chemistry for giving me the opportunity to write and deliver "Chemistry in Its Element—Compounds" podcasts on the Chemistry World site. Paul May of the University of Bristol has, over the past decade, not just tolerated my attempts to write "Molecule of the Month" for its website, but also given me material assistance in producing them. Successive members of the editorial team of the RSC's *Education in Chemistry* have helped me on the way, especially Kathryn Roberts, who, in 1996, believed in my idea of "Soundbite Molecules" and gave it air time within the pages of the journal. Finally, I would particularly like to thank those who have most often advised, commented, answered queries, and procured pdf files for me—namely, James Anderson, Peter Bodily, John Emsley, Gordon Gribble, John Mann, and Geoff Rayner-Canham.

So here it is then, ladies and gentlemen; I hope that you find this book useful or even valuable, though I have come to share the view of Saint Thomas Aquinas that all I have written seems like so much straw.

Simon Cotton
Holy Saturday 2011

The Author

Simon Cotton obtained his BSc and PhD at Imperial College London, followed by research and teaching appointments at Queen Mary College, London, and the University of East Anglia. He subsequently taught chemistry in both state and independent schools for over 30 years, retiring from Uppingham School as Chemist Emeritus. In September 2011, he was appointed an Honorary Senior Lecturer in the Department of Chemistry at the University of Birmingham. He has lectured widely in the UK, and carries out research on the chemistry of iron, cobalt, scandium, and the lanthanide elements.

His "Soundbite Molecules" feature has run as a regular column in the magazine *Education in Chemistry* since 1996, reaching every secondary school in the UK. He has written nearly 40 "Molecules of the Month" columns, which are featured online at http://www.chm.bris.ac.uk/motm/motm.htm and recognised globally. Additionally, he has delivered over a dozen "Chemistry in Its Element" podcasts for the Royal Society of Chemistry's *Chemistry World* website at http://www.rsc.org/chemistryworld/.

In 2005 he shared the Royal Society of Chemistry Schools Education Award. He was editor of "Lanthanide and Actinide Compounds" for the *Dictionary of Organometallic Compounds* and the *Dictionary of Inorganic Compounds* (Chapman and Hall) between 1984 and 1997.

He wrote the account of lanthanide coordination chemistry for the second edition of *Comprehensive Coordination Chemistry* (Pergamon) and the accounts of lanthanide inorganic and coordination chemistry for the first and second editions of the *Encyclopedia of Inorganic Chemistry* (Wiley).

This is his sixth book; his previous books are

D. J. Cardin, S. A. Cotton, M. Green, and J. A. Labinger, *Organometallic Compounds of the Lanthanides, Actinides and Early Transition Metals,* London: Chapman & Hall, 1985.

S. A. Cotton, *Chemistry of Precious Metals,* London: Blackie, 1997.

S. A. Cotton, *Lanthanide and Actinide Chemistry,* Chichester, England: John Wiley & Sons, 2006.

S. A. Cotton, *Lanthanides and Actinides,* Basingstoke, England: Macmillan, 1991.

S. A. Cotton and F. A. Hart, *The Heavy Transition Elements,* Basingstoke, England: Macmillan, 1975.

1 Atmosphere and Water

NITROGEN (DINITROGEN, N$_2$)

$$N\equiv N \qquad\qquad (1.1)$$

Though nitrogen gas, first isolated in 1772 by the Scottish doctor Daniel Rutherford, makes up over three-quarters of the atmosphere (in dry air, 78.08% to be precise), it is so unreactive that people might question its value. The very strong triple bond in the nitrogen molecule (1.1) makes it very unreactive in a chemical laboratory, though one or two very hot metals like magnesium have enough energy to break the bond and combine. In contrast to the 450°C and 200 times atmospheric pressure needed for a reasonable yield in the Haber process, the nitrogenase enzyme in rhizobium bacteria, found in nodules on the roots of leguminous plants (e.g., pea, bean, vetch, clover, lentil, alfalfa, lupin, peanut) can "fix" nitrogen at ambient temperature and pressure. Some fixation also occurs in thunderstorms when nitrogen and oxygen react to form nitrogen oxides as a flash of lightning supplies the energy.

This lack of reactivity of N$_2$ is vital, however, because it dilutes the oxygen present in the atmosphere and ensures that forest fires and other combustion reactions do not get out of control. On January 27, 1967, three astronauts were killed as a fatal fire swept through the command module, mounted on a Saturn rocket, in a test run for the first Apollo mission at Cape Kennedy. The speed of the spreading fire was due to the use of a pure oxygen atmosphere in the cabin, and this was subsequently replaced by a 60% O$_2$ and 40% N$_2$ mixture. Humans function better breathing a nitrogen–oxygen mixture; pure oxygen leads to problems like fluid buildup in the lungs.

Large amounts of nitrogen are by-products of the manufacture of oxygen by fractional distillation of liquid air, and its chemical inertness finds particular application as an "inert atmosphere" in industry, car tyres, and fuel systems, as well as in packaged foods (e.g., potato crisps). Liquid N$_2$ (b.p. −196°C) is the most widely used cryogenic refrigerant. For many years it was believed that N$_2$ would not bind to a metal, unlike the O$_2$ molecule; however, since the isolation of $[Ru(NH_3)_5(N_2)]^{2+}$ in 1965 by Allen and Senoff, many dinitrogen complexes have been isolated. It is accepted that a metal–N$_2$ complex is involved in the fixation of nitrogen by the molybdenum- and iron-containing enzyme nitrogenase, and chemists are still trying to find a way of carrying this out under mild conditions in industry.

Even though it is much less abundant in the earth, the interchange of nitrogen between the atmosphere and living systems in the nitrogen cycle is vital because it is an essential element in all amino acids (and proteins); likewise, a wide range of nitrogen compounds have immense commercial importance, such as ammonia and nitric acid in making fertilisers, explosives, and dyestuffs (see pp. 38–39 and 119, 122). Nitrogen molecules are unreactive, yet conversely it is the immense strength of the

N≡N bond (bond energy 945 kJ mol^{-1}) which contributes to the energy release from nitrogen-based explosives like TNT and ammonium nitrate in which N_2 is formed as a product.

OXYGEN (DIOXYGEN, O_2)

$$O=O \tag{1.2}$$

Oxygen is the most abundant element in the earth and the second most abundant in the atmosphere (21.0% by volume, in dry air). It is absolutely essential for life; living organisms depend upon it for cellular respiration, and it owes its presence in the air to photosynthesis in green plants, in the simplified form:

$$6\ CO_2(g) + 6\ H_2O\ (g) \rightarrow C_6H_{12}O_6(s) + 6\ O_2(g)$$

The discovery and isolation of O_2 gas by Priestley and Scheele in 1773–1774 and its subsequent exploitation in the understanding of combustion by Lavoisier were a key step in the development of chemistry. Along with nitrogen, oxygen is produced industrially in excess of 100 million tons a year by the fractional distillation of liquid air, the separation relying upon the difference in the boiling points of nitrogen (77K; –196°C) and oxygen (90K;'–183°C). Liquid nitrogen and liquid oxygen also differ in their colours: colourless and pale blue respectively. Industrially, oxygen is used in large quantities to make steel and other chemicals, to support combustion and generation of high temperature (e.g., H_2/O_2 rockets, oxy-acetylene cutters), and in life support in health care (oxygen tents and masks, emphysema treatment, divers, aircraft, and spaceflight). Oxygen has a small but significant solubility in water that is essential to marine and aquatic life; its solubility decreases with increasing temperature, so a tropical fish tank requires an aerator.

The oxygen molecule (1.2) contains a strong O=O bond (bond energy 498 kJ mol^{-1}), though not as strong as that in the N_2 molecule. Breaking this bond is therefore quite difficult, so the high activation energy for reactions like the combustion of hydrocarbons prevents these compounds being spontaneously flammable at room temperature. However, the small size of the oxygen atom leads it to form strong bonds with elements like hydrogen and carbon (bond energies H–O 463 and C=O ~ 740 kJ mol^{-1}), contributing to the exothermic combustion of hydrocarbon fuels. The high charge/radius ratio of the O^{2-} ion leads to high lattice energies of ionic oxides, a factor favouring their formation. At very high pressure (20 Gpa) solid oxygen (O_2) turns to the ε-phase, which is dark red in colour. This contains O_8 units, in which four O_2 molecules are associated together by weak chemical bonds.

OZONE (TRIOXYGEN, O_3)

$$\tag{1.3}$$

Ozone is an allotrope of oxygen, which is less stable than O_2. It is a strong oxidising agent and harmful to life. In contrast to linear CO_2, ozone forms a bent molecule (1.3). The central oxygen has two more electrons than the carbon in CO_2; they have to be accommodated as a lone pair.

It is formed in the upper atmosphere when ultraviolet light generates oxygen atoms, which then react with O_2 and form an ozone molecule. The subsequent decomposition of ozone also absorbs ultraviolet light, so these reactions in the "ozone layer" 30 km above the earth filter out much harmful ultraviolet radiation from the sun. Research in the mid-1970s by Frank Sherwood Rowland and Mario Molina showed that breakdown of CFCs (chlorofluorocarbons used as refrigerants, solvents, and aerosol propellants) in the upper atmosphere led to reactive Cl• radicals, which destroyed ozone. Together with Paul J. Crutzen, they were awarded the 1995 Nobel Prize for Chemistry; their research led to the 1987 Montreal protocol phasing out CFCs so that the ozone layer could recover.

Ozone is also produced at ground level, especially in cities. Sunlight acts on mixtures of hydrocarbons and nitrogen oxides emitted from car exhausts to form peroxyacylnitrates (PANs; $RCOOONO_2$ with R = alkyl); these are the compounds which typically make your eyes water on a hot summer's day in city streets. They react further forming ground-level ozone, which, unlike the "ozone layer" in the stratosphere, is a bad thing. Because it is a strong oxidising agent, ozone is especially damaging to the human respiratory system and also damages organic materials like rubber. It has a sharp, chlorine-like smell, detectable at the 0.01 ppm level. Ozone is formed by the action of electrical discharges upon O_2 molecules, and the smell of ozone can sometimes be detected near electrical equipment like photocopiers.

Ozone has its uses commercially as an alternative to chlorine to disinfect water supplies and to kill bacteria in hospital operating theatres. It is also a useful reagent in chemical synthesis.

CARBON DIOXIDE, CO_2

$$O=C=O \tag{1.4}$$

For its abundance in the atmosphere, carbon dioxide (1.4) punches above its weight. Because it absorbs strongly in the infrared region, carbon dioxide, along with water, methane, and ozone, is a greenhouse gas. If not for a "natural greenhouse effect" reducing the loss of long-wavelength IR heat by radiation, the earth's surface temperature would be around $-20°C$, not exactly hospitable. During my lifetime, the concentration of CO_2 has moved from just over 0.030% to nearly 0.04%. Controversy lies in the extent to which currently elevated temperatures, "global warming", are due to an "enhanced greenhouse effect" from increases in these atmospheric greenhouse gases caused by human activity. As well as in combustion of fossil fuels and other carbon-based materials, CO_2 is emitted by respiration in plants and animals and, conversely, is removed from air by photosynthesis in green plants. Of the world's CO_2 emissions, 5% come from cement making because it is formed when calcium carbonate is decomposed.

Carbon dioxide is acidic, well demonstrated by dropping a piece of dry ice into a beaker or measuring cylinder of water containing an acid–base indicator.

Unlike the other gases in the atmosphere, CO_2 only exists as a liquid above 5.1 atm pressure. Thus, chilling CO_2 gas below $-78°C$ produces solid carbon dioxide; conversely, when solid CO_2 ("dry ice") warms up, it turns directly to a gas and sublimes. This property is put to good employment when dry ice is used to refrigerate frozen foods because the packaging is not wetted. Liquid carbon dioxide can be heard, if not seen, in some CO_2-containing fire extinguishers used for electrical fires. This "supercritical" CO_2 is used industrially as a nontoxic covalent solvent—not least to extract caffeine from coffee.

Carbon dioxide is slightly soluble in water, and the solubility increases significantly with increasing pressure; its pleasantly sharp taste was one of Joseph Priestley's many discoveries (1772). His invention of soda water was taken up by Jacob Schweppe, who founded a commercial empire. Carbonation of drinks is also employed in beers and sparkling wines. More importantly, much carbon dioxide is dissolved in this planet's oceans. Increasing levels of CO_2 in the environment have led to a slight decrease in the pH of seawater, and concern is felt about the effects of this ocean acidification upon coral reefs and shellfish.

The inability of carbon dioxide to support combustion in any but the hottest fires (e.g., burning magnesium) is put to good use in fire extinguishers. Levels of CO_2 much more than 5% in air are dangerous to health, and there are caves in France and Italy where high levels of dense CO_2 near the cave floor are alleged to be lethal to dogs. Disturbance of lake water leading to release of CO_2 (originating in volcanic activity) associated with two "killer lakes" in Cameroon is believed to have killed 37 people at Lake Monoun in 1984 and 1,700 more around Lake Nyos 2 years later.

NOBLE GASES

At a time when scientists were discovering a number of new gases, Henry Cavendish (1731–1810) made important contributions. The significance of one experiment was not understood at the time (1785). He removed the nitrogen from air by sparking it with an excess of oxygen, dissolving the NO_x in water, then removing the remaining oxygen with potassium sulfide. This left a small amount of gas—just under 1% of the original volume—that simply would not react with anything. At the time, this was inexplicable.

Much later, on 19 April 1894, Lord Rayleigh announced to the Royal Society that nitrogen obtained from the air had a slightly greater density than nitrogen prepared from a chemical reaction. In the audience was William Ramsay, who was prompted to carry out an experiment in which he repeatedly passed 23 litres of aerial nitrogen backward and forward over hot magnesium (which reacted with the nitrogen, forming Mg_3N_2) until he was left with 100 cm^3 of totally unreactive gas. This turned out to be monoatomic and to have an atomic weight of 39.9. He called it argon.

The following year, Ramsay examined the gas formed when the uranium-containing mineral cleveite reacted with sulfuric acid; removing nitrogen and oxygen, he was left with a gas whose spectrum contained the same absorption at 587.49 nm already

TABLE 1.1

Abundance of Gases in the Earth's Atmosphere

Gas	%	ppm
N_2	78.1	781,000
O_2	20.9	209,000
CO_2	0.0390	390
He	5.2×10^{-4}	5.2
Ne	1.82×10^{-3}	18.2
Ar	0.93	9,340
Kr	1.14×10^{-4}	1.14
Xe	8.6×10^{-6}	0.086
Rn	Variable	Variable

noted in the solar spectrum in 1865 by Lockyer and Janssen. Ramsay had thus isolated helium, and in 1898 he obtained neon, krypton, and xenon by fractional distillation of liquid air. Dorn identified radon in 1900.

At that time, no one could explain why all these gases were made up of single atoms; it was 20 years before it was postulated that a filled outer shell of electrons was needed for stability. Chemists tried unsuccessfully to form compounds with them; thus, in 1895, Henri Moissan tried unsuccessfully to make an argon fluoride, but it was not until 1962 that Neil Bartlett made the first xenon compound. Compounds stable at room temperature are now known for Xe, Rn, and Kr. $HArF$ is stable at $-233°C$.

Across the universe, the abundance of the noble gases decreases with increasing atomic number, as expected. Two factors modify this pattern on earth (Table 1.1); argon is by far the most abundant (0.93% of the atmosphere) because ^{40}Ar is generated by β-decay of ^{40}K in rocks. Helium is only the third most abundant noble gas; despite being generated by α-decay of U and Th minerals, it is too light to be retained by the earth's gravity (as with H_2). The abundance of radon varies, depending on the levels of radon-emitting minerals in a region, but in some regions it represents a very real cause of cancer. The recent (2011) isolation of XeO_2 adds to speculation that insertion of Xe into SiO_2 in the earth's crust may lead to atmospheric xenon depletion.

Despite their lack of reactivity, the noble gases have several uses. Because it has the lowest boiling point of any substance, liquid helium (b.p. 4.2K) is a vital coolant, especially for superconducting magnets (e.g., in magnetic resonance imaging spectrometers). On account of its low density and lack of flammability, it replaced hydrogen as a lifting gas for airships. Helium has a very low solubility in water (or blood) and is used in $He-O_2$ breathing mixtures in place of nitrogen because it eliminates the "bends".

Several noble gases give a coloured glow when an electric current is passed through them; they are used in advertising lighting—classically, in red "neon signs".

Xenon lamps are used in car headlights. Argon is used as a filler gas for traditional (filament-type) incandescent light bulbs, as well as in providing inert atmospheres when welding reactive metals. Element 118 (ununoctium), the seventh member of this group, was synthesised by Russian and American researchers led by Yuri Oganessian and reported in 2006. They used nuclear reactions such as

$$^{249}_{98}Cf + ^{48}_{20}Ca \rightarrow ^{294}_{118}Uuo + 3n$$

The atoms have lifetimes in the order of a millisecond, so chemical study is not feasible. Predictions suggest that this element would have a boiling point ca. 80°C; if so, it would not be an "inert" gas.

WATER

(1.5)

Over 70% of the surface of the earth is covered in water; the vast majority of it (97%) is in the oceans. We in the Western world are inclined to take it for granted, never having had to walk miles to get drinking water because we have chlorinated water on tap. It is quite conceivable that water supply may yet become a major international issue.

Water in itself is a remarkable substance. It is nontoxic; its boiling point of 100°C is much greater than that of other molecules of the same size, like methane (–162°C) or even ammonia (–33°C); and its liquid range coincides with ambient temperatures on much of the earth. Without these properties, human life on the planet would not be possible. Its solvent properties are also vital; its polar nature enables solvation of ions, whilst it can hydrogen-bond to many covalent molecules. It transports substances through organisms. Water vapour in the atmosphere contributes to the natural greenhouse effect, ensuring a steady ambient temperature conducive to the maintenance of life.

Because the electronegativity of oxygen is much greater than that of hydrogen, the water molecule is polar. Electron density in the O–H bond is strongly polarised toward oxygen, leaving the hydrogen with a very slight positive charge. The attraction between such a hydrogen and a lone pair of electrons belonging to another water molecule constitutes a significant intermolecular force: a hydrogen bond (Figure 1.1). This can also occur with molecules containing N–H and H–F bonds (but not C–H). The fact that a water molecule has two polar O–H bonds and two lone pairs means that it can participate in up to four hydrogen bonds to neighbouring molecules, an *optimum* number compared to NH_3 or HF, which therefore have weaker intermolecular forces and lower boiling points.

In fact, the number of hydrogen bonds per water molecule is slightly greater in ice than in liquid water, leading to a more open structure, a lower density for the solid form, and hence the unusual property of ice floating on water or ice expanding out of the ice-cube maker tray in the freezer compartment of the 'fridge. The strength

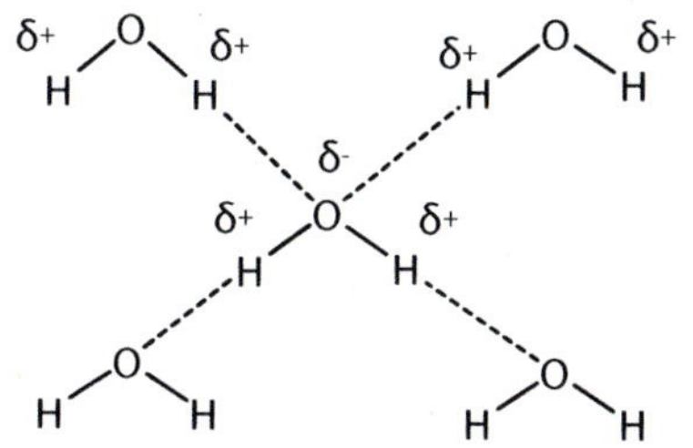

FIGURE 1.1 Hydrogen bond.

of hydrogen bonding also leads to the high surface tension of water, enabling pond skaters and needles to float on it, and to its capillary action (so that it can reach the top of high trees). Another anomalous property of water is that its greatest density occurs at 3.98°C, so in winter there is almost always liquid water at the bottom of lakes, enabling fish to survive.

In the late 1960s, scientists in the West became aware of experiments conducted by Boris Derjaguin in which water condensed in capillary tubes was found to have an anomalously high boiling point and density. Its infrared spectrum was different from that of ordinary water, and it was suggested that water molecules were associated in clusters. The word "polywater" was coined, and something of a media-fed frenzy began, with suggestions that this material was dangerous. It was subsequently discovered that this "polywater" contained hydrated silica and salts leached out of the glass or quartz tubes, but not before over 500 scientific papers had been published and some scientific reputations tarnished.

DEUTERIUM OXIDE (HEAVY WATER)

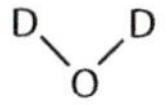

$$\begin{array}{c} D \diagdown \diagup D \\ O \end{array} \qquad (1.6)$$

For most practical purposes, substituting one isotope for another in a compound has no effect on bulk physical properties. In the case of such a light molecule as water, the difference between H_2O and D_2O leads to significantly different melting points (0°C and 3.81°C, respectively) and boiling points (100°C and 101.42°C, respectively); G. N. Lewis and his student, Ronald T. MacDonald, were therefore able to obtain pure deuterium oxide (1.6) by fractional distillation under reduced pressure in 1933. Molecular dimensions of the isolated molecules are very slightly different; for an isolated H_2O molecule, O–H is 0.9724 Å and H–O–H is 104.50°, whilst for an isolated D_2O molecule, O–D is 0.9687 Å and D–O–D is 104.35°. In the condensed liquid phase, the O–H bond in H_2O is 1.01 Å, whilst the O–D bond in D_2O is 0.98 Å, shorter by 0.03 Å; on the other hand, the intermolecular bond in D_2O is longer by 0.07 Å.

One significant difference between H_2O and D_2O is that deuterium oxide can act as a moderator in a nuclear reactor to produce neutrons of the right velocity to induce nuclear fission in ^{235}U. In the late 1930s and in the early years of World War II, this led to competition between the Germans and French for supplies of heavy water and, subsequently, to attacks on the Ryukan plant in Norway that produced it; the film *Heroes of Telemark*, starring Kirk Douglas and Richard Harris, was loosely based upon these events. Since the war, chemists have made use of the different nuclear properties of hydrogen and deuterium in a different way; a compound with labile N–H or O–H groups will exchange the hydrogens on shaking with heavy water, so the signal due to these hydrogens in the proton nuclear magnetic resonance spectrum of the compound will disappear, facilitating assignment of the spectrum.

$$R–O–H + D_2O \rightleftharpoons R–O–D + D–O–H$$

HYDROGEN PEROXIDE

Hydrogen and oxygen form one more binary compound, H_2O_2. Like water, hydrogen peroxide (1.7) is stable with respect to the elements:

$$H_2(g) + \tfrac{1}{2} O_2(g) \rightarrow H_2O(l) \; ; \; \Delta H_f = -286 \text{ kJ mol}^{-1} \; ; \; \Delta G_f = -237 \text{ kJ mol}^{-1}$$

$$H_2(g) + O_2(g) \rightarrow H_2O_2(l) \; ; \; \Delta H_f = -188 \text{ kJ mol}^{-1} \; ; \; \Delta G_f = -118 \text{ kJ mol}^{-1}$$

But hydrogen peroxide is thermodynamically unstable with respect to decomposition to water and oxygen:

$$H_2O_2(l) \rightarrow H_2O(l) + \tfrac{1}{2} O_2(g) \; ; \; \Delta H = -98 \text{ kJ mol}^{-1}$$

The activation energy for this decomposition is fairly high at 75 kJ mol^{-1}, but is greatly lowered by catalysts such as dust, MnO_2 ($E_a = 58$ kJ mol^{-1}), and, particularly, the enzyme catalase ($E_a = 23$ kJ mol^{-1}). Thus, solutions of H_2O_2 are normally kept in polythene or Teflon bottles and stabilisers added.

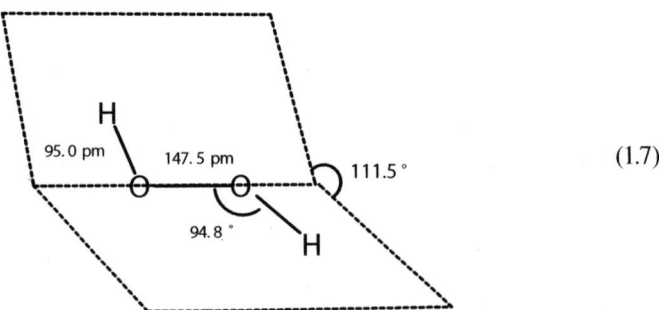

(1.7)

Hydrogen peroxide has an unusual skewed structure with a dihedral angle of 111.5° (gas phase), which minimises repulsion between the lone pairs and the O–H bond pairs. The dihedral angle is affected by hydrogen bonding; it is 90.2° in solid H_2O_2.

Small amounts of H_2O_2 exist in our bodies as by-products of cellular respiration in cells. This could be very damaging to the cells, were it not for the presence of very efficient catalase and peroxidase enzymes in our bodies, which rapidly decompose the H_2O_2 to water and oxygen. The African bombardier beetle, *Stenaptinus insignis*, stores a mixture of hydrogen peroxide and hydroquinones and, when needed, pumps this into a reaction chamber where it is mixed with catalysts (catalases and peroxidases) that decompose the peroxide to oxygen, which in turn oxidises the hydroquinones to quinones, very exothermically. This hot and irritating mixture is shot out of the beetle's abdomen at any lurking predator, usually ants.

Hydrogen peroxide has applications as an antiseptic and disinfectant, as well as a chlorine-free bleach familiarly used on hair by media figures like Marilyn Monroe (*Gentlemen Prefer Blondes*), Reese Witherspoon (*Legally Blonde*), Jean Harlow (*Platinum Blonde*), and Debbie Harry (of the band Blondie). The H_2O_2 oxidises the melanin pigment in the hair to a colourless substance. It will bleach teeth, too; as well as being used as a mouthwash and to clean contact lenses, it will whiten your skin if it is spilt on a finger. The biggest commercial use of H_2O_2 is for bleaching wood pulp or paper.

Concentrated hydrogen peroxide is unstable but has found use in fuels, including the Messerschmitt 163 "Komet" rocket aircraft of World War II, which was fuelled by the reaction between 80% H_2O_2 and a methanol/hydrazine mixture:

$$2\ H_2O_2(l) + N_2H_4(l) \rightarrow 4\ H_2O(g) + N_2(g)$$

No Allied aircraft could keep up with the 550 mph jet, but many Me 163s were lost in landing accidents, pilots either dying in the resulting explosions or being dissolved alive in 80% hydrogen peroxide. Two U-boats were built that ran on 95% H_2O_2 fuel; the steam that formed on its decomposition drove turbines. Hydrogen peroxide saw longer service as a propellant for torpedoes, but these caused problems. On August 12, 2000, the huge Russian submarine *Kursk* was lost off the Kola peninsula; its entire crew of 118 perished. It is now believed that hydrogen peroxide leaking from a torpedo decomposed violently, triggering explosions of kerosene fuel and, subsequently, of several torpedo warheads.

2 Carbohydrates and Artificial Sweeteners

INTRODUCTION

For many, the word "carbohydrate" means the food of survival, but for others it is synonymous with "sugar". All carbohydrates have a formula that is a multiple of CH_2O, though their structures are very different. "Sugar" means a sweet-tasting substance, usually a carbohydrate that is a monosaccharide or disaccharide; most often this is sucrose, lactose, or fructose, especially sucrose, which is table sugar. Along with spices, sugar originated in Asia and first entered Europe from the East in the wake of the Crusades, providing an alternative to traditional sweeteners like honey.

To meet increasing demand, sugar was grown in the West Indies and Brazil from the sixteenth century and fuelled the slave trade. The expansion of cultivation of sugar cane meant that sugar became more accessible to the poorer classes of society from 1700 onward, as did the discovery of sucrose in sugar beet, which provided a source of sugar that could be grown in temperate European climates. The detailed chemistry of sugars and initial understanding of the structures grew up in the late nineteenth century, led by Emil Fischer (1852–1919), who achieved the total synthesis of glucose from glycerol in 1890. He was awarded the Nobel Prize in Chemistry in 1902 for "the extraordinary services he has rendered by his work on sugar and purine syntheses".

Apart from sucrose, the most important sugar, though not a very sweet one, is **glucose,** which is found in many fruits and vegetables. Glucose, which takes its name from the Greek *glukus* (γλυκύς; sweet), was first isolated from raisins in 1747 by the German pharmacist Andreas Sigismund Marggraf (1709–1782). It is a key molecule in both photosynthesis and respiration. Glucose is also the substrate in fermentation to convert sugar into ethanol in making beer and wine. Present in human blood, glucose is the immediate source of energy through respiration in cells. A typical human has 5 to 6 grams of glucose in the blood stream—enough for the body's energy needs for about 15 minutes; thus, it has to be replaced continually from the body's source: glycogen in the liver. The body stores about a day's supply of glycogen—more than that and glucose is converted to, and stored as, fat.

Glucose exists as two optical isomers. Only the D-form (2.1) occurs naturally and only that form can be metabolised in the body, not the mirror-image L-isomer. The D-form rotates the plane of polarised light to the right (dexter), which is why another name for D-glucose is dextrose. It is the presence of all those OH groups in the glucose molecule that is responsible for the water solubility of glucose and simple carbohydrates; it enables the formation of many hydrogen bonds with water molecules.

$$\text{(2.1)}$$

Though the six-membered ring in glucose (and other carbohydrates) is not flat, structures are normally represented as Haworth projections, which are especially useful in showing the stereochemistry clearly, whether the groups are above or below the ring. In the solid state, glucose is only present as ring-shaped molecules, α-D-glucose and β-D-glucose; these two forms of D-glucose only differ in the orientation of the hydrogen and hydroxy groups at carbon 1. In solution, a small amount of an open-chain isomer (0.02%) is in equilibrium with α-D-glucose and β-D-glucose (Figure 2.1). The presence of the open-chain isomer in solution has two consequences: It permits interconversion between the α- and β-forms and also, because it has a terminal carbonyl group, it is an aldose, making glucose a reducing sugar.

Importantly, glucose molecules are used as building blocks in larger and important carbohydrate molecules. Thus, the disaccharide maltose is made from two glucose molecules joined through a condensation reaction; similarly, sucrose is formed when an α-D-glucose condenses with a β-D-fructose molecule.

Sucrose (2.2) is the standard against which all sweeteners are judged; table sugar is sucrose, not glucose. The first source of sugar to be discovered was sugar cane (a perennial grass of the genus *Saccharum*), which is native to Southeast Asia and the southern Pacific; someone made the discovery that when sugar cane was chewed, it tasted sweet. Its use spread to India and from there was brought west by invading Persians and thence to Europe by returning Crusaders ca. 1100. It became an important trade commodity and subsequently was widely cultivated, along with sugar beet; it can make up 10–20% of the dry weight of these plants. As with glucose, sucrose was first isolated in the eighteenth century by Marggraf, from sugar beet; water extracts sucrose from macerated beet or cane and, on crystallisation, "granulated sugar" is obtained. Sucrose is the main sugar in some fruits like the pineapple and apricot and also the sweet ingredient of flowers' nectar.

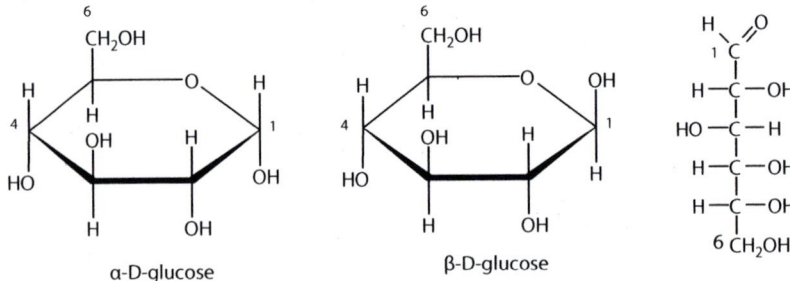

FIGURE 2.1   In solution, a small amount of an open-chain isomer (0.02%) is in equilibrium with α-D-glucose and β-D-glucose.

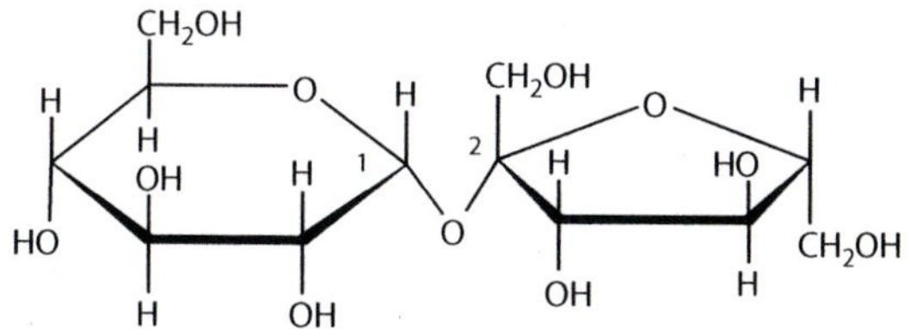

(2.2)

Using an invertase enzyme, honey bees convert sucrose into a mixture of fructose and glucose, and these are the principal sugars present in honey. Sucrose is optically active. If a beam of plane polarised light passes through sucrose solution, its plane of polarisation is rotated clockwise. On hydrolysis (e.g., with the enzyme sucrase), a mixture of glucose and fructose is formed, a solution that now rotates the beam of plane polarised light anticlockwise. This reversal of the direction of rotation is called inversion, so this mixture of glucose and fructose is known as "invert sugar":

$$\text{sucrose} \rightarrow \text{fructose} + \text{glucose}$$

When fruits are heated with sugar in jam making, some hydrolysis of sucrose occurs; that mixture of invert sugar is sweeter than the starting sucrose and is also less likely to crystallise in the jar.

Fructose (2.3) is the sweetest sugar known, some 50% sweeter than sucrose. The main sugar in honey, fructose is found in numerous fruits and is the major sugar in many, including pears, grapes, and apples. In the solid state and solution, it is largely present as a hexagonal ring, but when it binds to glucose to form sucrose, it forms a five-membered ring.

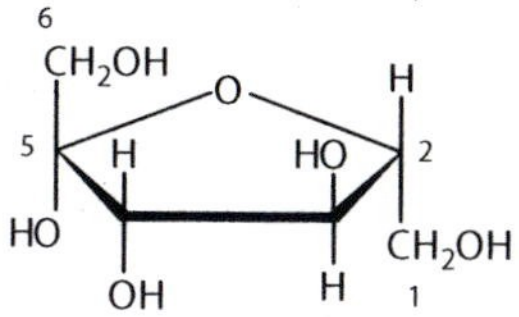

(2.3)

Maltose (2.4), malt sugar, is a disaccharide formed from two units of glucose joined by an α-1,4-glycosidic bond. It is formed by the enzymatic breakdown of starch in germinating cereals like barley. The maltose is then fermented by yeast during brewing.

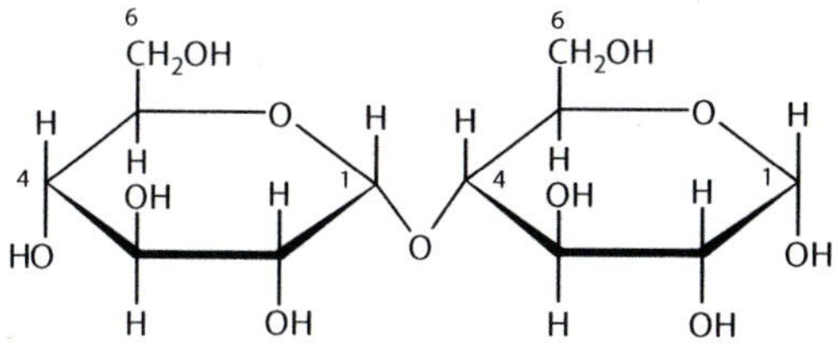

(2.4)

San Francisco sourdough bread is produced using a combination of yeasts—either *Candida milleri* or *Saccharomyces exiguus,* neither of which can metabolise maltose, but which metabolise the other sugars like glucose—and the bacterium *Lactobacillus sanfranciscensis,* which requires maltose, excreting lactic acid and ethanoic acid. The acids give the bread its characteristic tangy taste.

Lactose (2.5) is a disaccharide formed from the monosaccharides galactose and glucose, through a β-1,4-glycosidic linkage; galactose is a diastereoisomer of glucose, differing only in the configuration at carbon 4. Lactose is widely found in mammalian milk, making up 5% of cows' milk, with rather more lactose in human milk.

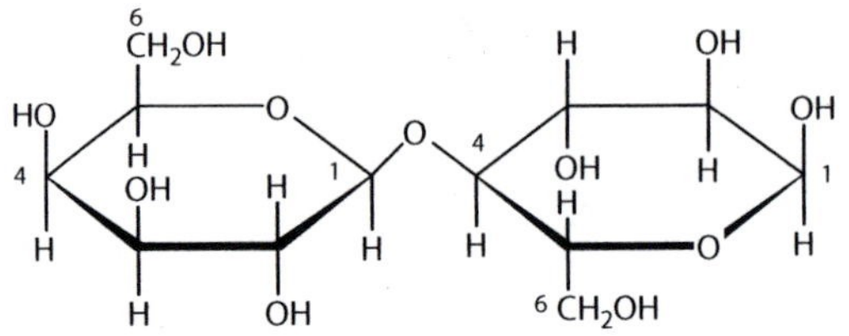

(2.5)

Lactose cannot be absorbed as such; it has to be hydrolysed by the lactase enzyme into galactose and glucose in the intestines. Some humans do not possess the lactase enzyme and thus cannot digest lactose and are said to be "lactose intolerant".

TREHALOSE

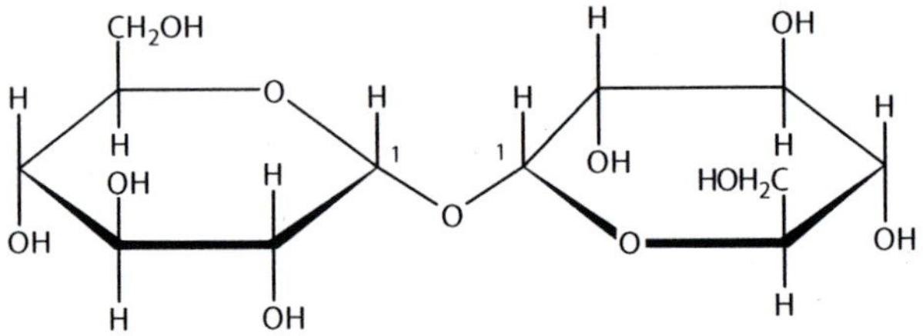

(2.6)

In the Old Testament of the Bible, the book of Exodus (16:11–36) recounts how the Israelites were fed in the wilderness by a sweet substance called manna: "And the house of Israel called the name thereof Manna: and it was like coriander seed, white; and the taste of it was like wafers made with honey" (verse 31).

Several substances have been identified with manna, such as the Kurdish manna produced on the leaves of oak trees by certain aphids, or the excretion of scale insects on leaves of the tamarisk in the Sinai desert. Others have linked it with the cocoons of the Iraqi parasitic beetle *Trehala manna.* **Trehalose** (2.6) causes the sweetness of many of these substances; it is a disaccharide made of two α-glucose molecules joined by an α,α-1,1-glycosidic linkage. It is very stable and hard to split into glucose, except by specialist trehalase enzymes. It has ca. 45% the sweetness of sucrose; because it is a nonreducing sugar, it is unreactive, so it does not participate in Maillard reactions with amino acids and proteins and thus will not cause browning. It is used by insects as an energy storage molecule in their blood-lymph system; they can break it back into glucose for energy for flight. It has the remarkable property of preserving life in creatures that can lie dormant for long periods of time under drought conditions (anhydrobiosis), and has attracted attention for possible applications as a food additive and stabiliser; to preserve cells, tissues, and organs; and in storing blood and antibodies. Polymers based on it are being investigated. Trehalose can now be mass produced from starch.

POLYSACCHARIDES

The polysaccharides starch (in potatoes, wheat, rice) and glycogen (in animals) have α-glucose monomers assembled together in slightly different ways. This enables plants and animals to store their "energy molecule", glucose, in an insoluble form. Nevertheless, we have enzymes that can hydrolyse the α-1,4-glycosidic links between the units to release glucose, which can then be transported and used. On the other hand, when β-glucose monomers are linked together, the insoluble cellulose is the result, and this is the structural material of the cell walls in green plants.

Starch is the key dietary polysaccharide; green plants use it as an energy store and it is abundant in foods such as potato, rice, wheat, maize, and cassava. It does not dissolve in water, but instead forms colloidal suspensions in which the colloidal particles are big enough to scatter light (which is why starch "solutions" look cloudy). It is easily broken down in the body to glucose when required; if you chew a piece of bread for a while, it tastes sweet because salivary amylase hydrolyses glycosidic links, releasing glucose. Two kinds of polymer are found in starch; both **amylase** (2.7) and **amylopectin** (2.8) have chains of glucose molecules joined by α-1,4-glycosidic linkages, whilst the branched amylopectin features additionally 1,6-α-glycosidic bonds at the branches.

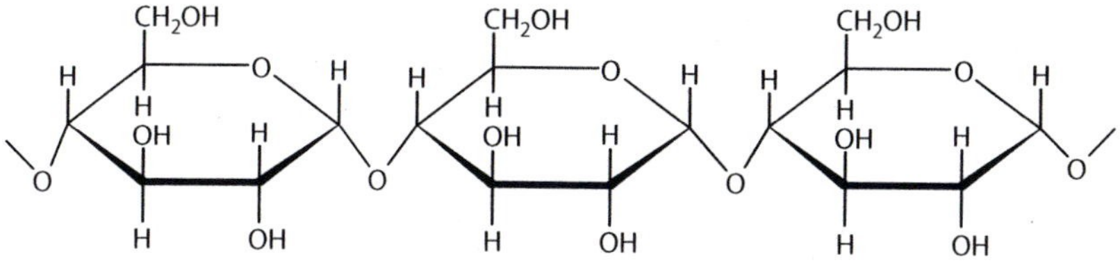

(2.7)

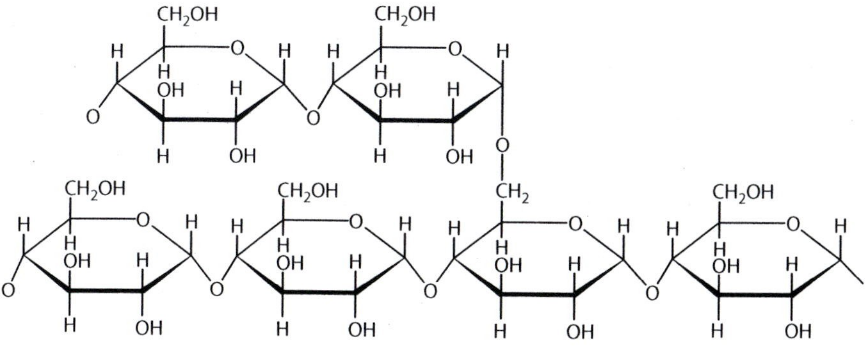

(2.8)

Glycogen is the storage polysaccharide in animal cells—a sort of animal starch, if you like. It has a similar structure to amylopectin, but with branches every 8–12 glucose units, in contrast to the branches every 30 units in amylopectin. This means that there are more points where ready glucose release can occur.

Cellulose (2.9) is the structural polysaccharide of which plant cell walls are made. It is the most common organic compound on earth, since about one-third of all plant matter is cellulose. It is the main component of cotton and can be converted into cellophane and rayon, as well as used to make paper and card. Wood consists of cellulose fibres embedded in a polymer called lignin.

$$(2.9)$$

Cellulose is made of thousands of glucose units, linked through β-1,4-glycosidic bonds (in contrast to starch and glycogen, which have α-1,4-glycosidic bonds). Alternate glucose molecules in the polysaccharide chain are flipped over compared with those on each side of it. This leads to essentially linear chains, with none of the coiling or branching seen in starch or glycogen. Neighbouring chains are parallel to each other and are linked by hydrogen bonds, which leads to strong microfibrils; this rigidity and strength are important to the role of cellulose in plant cell walls.

Humans and many other mammals do not have enzymes that can hydrolyse the β-1,4-glycosidic links in cellulose and thus cannot metabolise cellulose (grass and hay). Ruminants like cows and sheep have intestinal microbes that will do that.

Alternative sweeteners are sweet molecules that are effectively calorie free; since they are so much sweeter than sugar, only a small amount is needed to have the same sweetening power as a spoonful of sugar. Many people have a sweet tooth, but this can lead to being overweight and also to dental decay. Diabetics also want sugar-free sweeteners. So for the past century and more, scientists have sought out sweeteners that are calorie free. Alternative sweeteners are taken for granted nowadays, but were once looked down on as "unnatural"—unlike sugar—and there were moves to ban them in the early twentieth century. Most of these are synthetic molecules, but **stevia** is a natural material which is up to 300 times sweeter than sugar. *Stevia rebaudiana* (sweetleaf) is a herb originating in Paraguay whose leaves have a sweet taste due to two glycosides: stevioside (E960) (2.10) and rebaudioside A, whose structures are derived from a parent alcohol, steviol, where one hydrogen in an OH group is replaced by a glucose molecule and the other by linked glucoses.

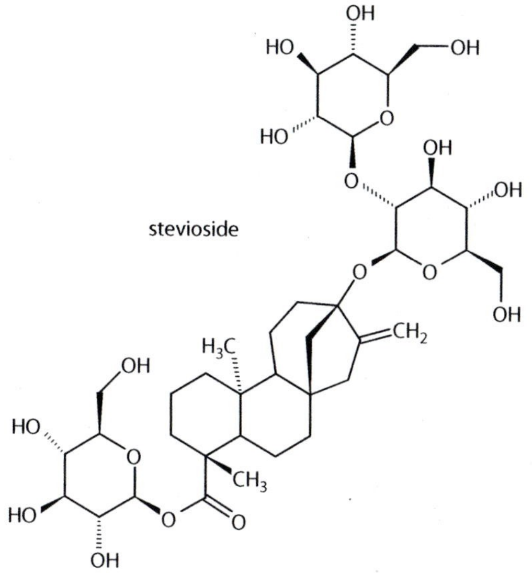

stevioside

$$(2.10)$$

These are now used to make sweeteners accepted throughout much of Asia and the United States, though there have been some complaints of a liquorice-like after-taste and of bitterness.

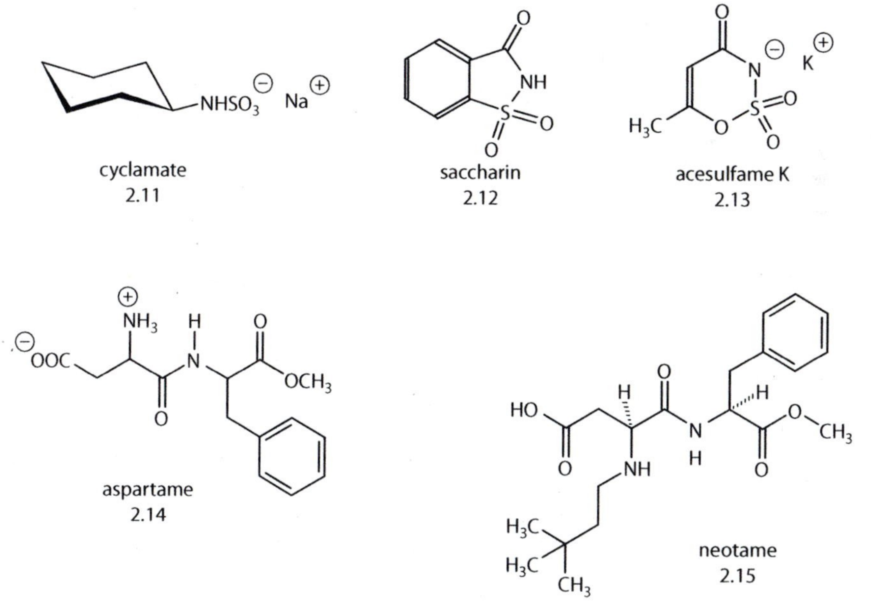

cyclamate
2.11

saccharin
2.12

acesulfame K
2.13

aspartame
2.14

neotame
2.15

$$(2.11–2.15)$$

Cyclamate (2.11; E 952) was another accidental discovery, recalling the days before modern health and safety practices. In 1937, a chemist at the University of

Illinois named Michael Sveda put down his cigarette on the lab bench, picked it up again and noticed a very sweet taste from the chemical—cyclamate—into which it had been placed. Cyclamate has about 25 times the sweetening power of sugar. It has most usually been used as a 10:1 mixture with saccharin, where each sweetener cancels out the aftertaste of the other. Cyclamate-based dietary products were first marketed in 1953 in the United States; however, it was banned in 1969 following test results claiming that rats fed huge doses of cyclamate could develop bladder tumours. It is used in many other parts of the world.

Saccharin (2.12; E954) is the longest established of the artificial sweeteners. It was first made by Ira Remsen and Constantin Fahlberg at Johns Hopkins University in 1879. The story goes that one day Fahlberg was eating dinner and noticed a sweet taste on his hand; he realised that it was due to a chemical he had handled in the laboratory that day. Saccharin made its public debut at the 1893 World's Fair in Chicago. At that time, saccharin, which is about 300 times sweeter than sucrose, was of particular help to diabetics as a sugar-free sweetener. It was very valuable during the two World Wars, when sugar was in short supply, and it was not until well into the post-World War II era that fashion trends changed toward slimness and the idea of calorie-free sweeteners gained ground. Americans are familiar with pink packets of Sweet'N Low—a granulated version of saccharin.

Acesulfame potassium (acesulfame K, 2.13; E950), about 200 times sweeter than sucrose, was another accidental discovery; the German chemist Karl Clauss had just handled acesulfame when he licked his fingers to pick up a filter paper. It is widely used in Europe under the name of Sunnett but has not found favour in the United States.

The most used artificial sweetener at present is probably **aspartame** (NutraSweet, Canderel; 2.14; E951), which is about 200 times sweeter than sucrose. This was first made in 1965 by a chemist named James Schlatter, who was trying to make new antiulcer drugs. Aspartame is the methyl ester of a dipeptide, based on the amino acids phenylalanine and aspartic acid; it undergoes rapid hydrolysis in the body into methanol and the two acids. About 1 person in 15,000 suffers from the rare genetic disorder phenylketonuria (PKU) and cannot metabolise phenylalanine properly and thus needs to avoid any sources of it (the reason for warnings that "this product contains phenylalanine" are found on all packs of aspartame). The small amounts of methanol formed do not pose a problem because the human body can readily metabolise it, and indeed higher levels are found in fruit. Aspartame is decomposed by heat, so it cannot be used in cooking.

Neotame (2.15; E961) was developed from aspartame and is another methyl ester of a dipeptide. It is around 10,000 times sweeter than sucrose (and around 50 times sweeter than aspartame), so only tiny doses are needed. It is metabolised rapidly in the body into methanol and the dipeptide. The very bulky 3,3-dimethylbutyl group attached to the aspartic acid section blocks peptidase enzymes from hydrolysing the peptide linkage between the phenylalanine and the substituted aspartic acid. This means that phenylalanine cannot be released and thus sufferers from PKU can use neotame safely. It received FDA approval in 2002.

Sucralose (2.16; E955) was discovered in 1975 when a postdoctoral researcher named Shashikant Phadnis was making chlorinated derivatives of sucrose. On the phone to his supervisor, Leslie Hough, he misheard an instruction to "test" a compound as "taste"; thus, he discovered the sweetness of sucralose (also known as Splenda), which is about 600 times sweeter than sucrose. One advantage of sucralose is that it is heat stable and thus can be used in baking recipes. The presence of the three chlorines in the molecule means that it is not metabolised, so it is a true zero-calorie sweetener.

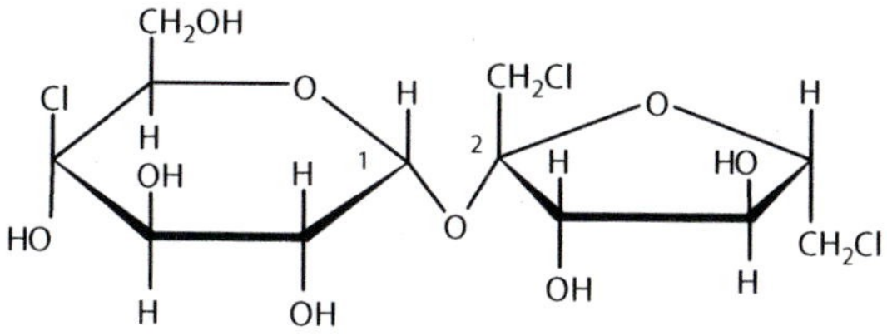

(2.16)

3 Hydrocarbons

Because twenty-first century civilisation depends so much on them, just mention the word "hydrocarbons" and many people think "fuel" without knowing much more about them. Hydrocarbons are easy to recognise and define; they are compounds containing only carbon and hydrogen atoms bonded together, but there is so much more to them than that. We take for granted combustion reactions like the following:

$$CH_4(g) + 2O_2(g) \rightarrow CO_2(g) + 2H_2O(g) \ \Delta H = -890.7 \text{ kJ mol}^{-1}$$

A good fuel must have more than a clean and a very exothermic combustion. The activation energy must be just right; otherwise, you have materials like the boron hydrides, many of which are spontaneously flammable in air (besides giving a solid combustion product). The products of combustion have very strong C=O and O–H bonds, so a lot of energy is released when they are formed, ensuring a very exothermic reaction.

The alkanes—those hydrocarbons containing only single bonds—form a family with a general formula C_nH_{2n+2} and, with naphthenes (cycloalkanes), are the main component of natural gas and crude oil; aromatics and asphaltenes are also present. Some scientists have suggested that fossil fuels originate deep under the earth, moving upward from the mantle and that they are possibly formed by the action of thermophilic bacteria. However, most scientists accept that preserved remains of zooplankton and algae on the bottom of a sea or lake, covered with mud and sediment and exposed to high pressure and temperature over a period of millions of years, first converted to waxy kerogen (a mixture of very large organic molecules; oil shale contains a lot of kerogen). Over a further period of heating, kerogen breaks down to much smaller hydrocarbon molecules.

Methane (3.1) makes up 70–95% of natural gas; the remainder is small alkanes $(C_2–C_4)$.

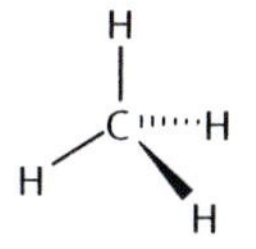

$$(3.1)$$

Beyond natural gas, cows, sheep, and other ruminants have methanogenic bacteria in their rumen, so methane is formed in microbial fermentation of ingested feed. A cow releases up to 300 litres of flatus a day—mainly nitrogen, but with some methane content. Another source is the biogas formed by fermenting sludge and manure; methane is also generated in landfills. It is a very clean fuel, though incomplete combustion yields carbon (soot) or toxic carbon monoxide.

Because methane is a small enough molecule to fit into the spaces within the lattice structure of ice, a good deal of methane is locked up in methane clathrates in cold sediments at bottoms of the oceans; this has the potential to be a major energy reserve. As a demonstration, the methane can be burned in "flaming snowballs". More seriously, because the global warming potential of methane is about 20 times that of CO_2 (over a 100-year time span), there is concern about the effect that release of methane could have on climate change, especially in permafrost regions like Siberia. Levels of CH_4 in the atmosphere have risen from 700 ppb in 1750 to ~1800 ppb at present.

Apart from its use as a fuel, methane is an important chemical resource. Its reaction with steam in the presence of a hot nickel catalyst affords the mixture of CO and H_2 known as "synthesis gas"—a widely used "feedstock" to make chemicals such as hydrogen, methanol, and ethanoic acid, for example.

$$CH_4(g) + H_2O(g) \rightarrow CO(g) + 3H_2(g)$$

Petroleum does not mean petrol; rather, it strictly means what we call **crude oil,** though it is often convenient to think of crude oil and natural gas together. The alkanes CH_4 to C_4H_{10} are gases at ~20°C, though butane liquefies under slight pressure, as can be seen in "disposable" cigarette lighters. The alkanes from C_5H_{12} upward are liquid at ~20°C. Crude oil is usually a black or dark-brown liquid (though occasionally it is yellow, red, or green) and is separated in refining by fractional distillation (Table 3.1).

Around 84% of petroleum is used as an "energy fuel". As the best grades are exhausted, petrol increasingly has to be obtained from heavy oil and bitumen, using processes like cracking, isomerisation, and reforming to generate short-chain alkanes and cycloalkanes. **Octane** (3.2) is familiar in the context of **octane ratings** of fuels. Crude oil would make a dreadful fuel for a car; even the gasoline fraction from refining would not be good enough for high-compression petrol engines in modern cars because it would ignite too soon and not burn smoothly.

TABLE 3.1
Fractions from Crude Oils and Their Uses

Crude Oil	Boiling Point (b.p.) (°C)	C_n	Use
Petroleum gas	<25	1–4	Bottled gas fuel
Gasoline	40–100	6–12	Petrol
Naphtha	100–150	7–14	Source of petrochemicals
Kerosene	150–200	11–15	Aircraft fuel
Diesel oil	200–300	16–20	Fuel for cars, lorries
Lubricating oil	>300	20–30	Lubricating machinery
Heavy fuel oil	>400	30–40	Power stations, etc.
Wax and grease	>400	40–50	Candles and machinery
Bitumen	>400	40–50	Surfaces of roads, roofs

The **octane rating** of a fuel is a measure of the resistance of fuels to autoignition in internal combustion engines. Octane rating is defined by comparing the fuel with the mixture of 2,2,4-trimethylpentane (**iso-octane,** 3.3) and heptane, which would have the same antiknocking capacity as the fuel being tested: The octane number of the fuel is the percentage (by volume) of 2,2,4-trimethylpentane in that mixture. "Straight-chain" alkanes have relatively low octane ratings (heptane = 0), so additives are used—branched alkanes (isooctane = 100), cycloalkanes (cyclohexane = 110), and aromatic hydrocarbons (benzene = 101; toluene = 110)—so the petrol put into the car might only contain 10%–15% of straight-chain alkanes. At one time, tetraethyllead was used to improve octane ratings (see p. 105)

(3.2)

(3.3)

Diesel fuel does not have "octane ratings"; instead, cetane numbers are assigned, based on how easily a fuel ignites under compression. Cetane itself (hexadecane, $C_{16}H_{34}$) is a typical ingredient of diesel fuel and is a liquid at room temperature; octadecane ($C_{18}H_{38}$) is the first member of the alkane series to be a solid at room temperature. Longer alkanes are waxy solids; candle wax is over 90% alkanes in the range $C_{18}H_{38}$ to $C_{40}H_{82}$, with $C_{29}H_{60}$ most abundant. Beeswax is a mixture of compounds—mainly esters but also carboxylic acids and alkanes, principally odd-carbon unbranched alkanes with 23–31 carbons; the most abundant is $C_{27}H_{56}$.

Large alkanes are important to plants; mixtures of alkanes coat areas of the plants exposed to the outside world. Because of their nonpolar nature, they form a hydrophobic barrier that polar water molecules cannot cross; they also protect the plant against insect pests, fungi, and bacterial infection. The main alkane in apple skin is $C_{27}H_{56}$; the presence of waxy alkanes in the skin is why apples can be polished. $C_{29}H_{60}$ predominates in cabbage leaves, whilst $C_{31}H_{64}$ is the principal alkane on tobacco and spinach leaves.

Hydrocarbons also form a cuticular coating in insects that is largely a mixture of linear and branched alkanes, sometimes with some alkenes. Different species have their own blend; they are used for recognition in the nest, and some insects have an amazing ability to control their emissions, as in the case of the paper wasp *Polistes atrimandibularis* (Chapter 6, p. 67).

Larger molecules are components of bitumen, for covering roads or waterproofing roofs. The ultimate large alkane is polyethylene, which involves chains thousands of carbon atoms long.

Naphthenes (cycloalkanes) are saturated, with only single bonds between carbons; unlike alkanes, their general formula is C_nH_{2n}. Cyclohexane, C_6H_{12}, is the most important naphthene (3.4); unlike benzene, the ring is not flat.

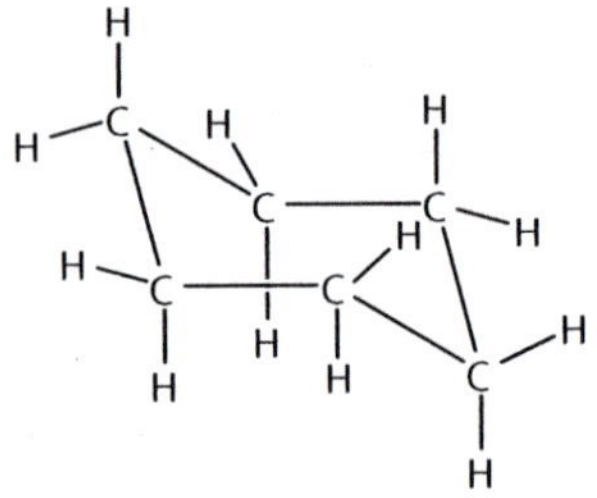

$$(3.4)$$

Cyclohexane has a high octane rating (110), so it is a good petrol additive. It is widely used as a solvent, but most cyclohexane is converted into cyclohexanone or cyclohexanone–cyclohexanol mixtures, raw material to make adipic acid, and caprolactam for making nylon. The 1974 disaster at Flixborough (Lincolnshire), which killed 28 people, was at a cyclohexane plant making caprolactam.

Ethene (3.5), also known as ethylene, is the smallest alkene.

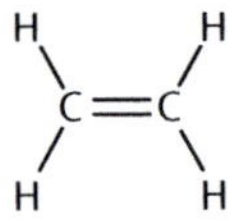

$$(3.5)$$

The double bond in alkenes is stronger than the ordinary C–C single bonds in alkanes, but because it is so electron rich and thus attracts electrophiles, this makes alkenes much more reactive than alkanes. This extra reactivity makes alkenes rare in nature; ethene is an exception. At room temperature, ethene is a colourless gas with a faintly sweet smell and taste that has anaesthetic properties. Geologists have suggested that it was formed in cracking reactions in the bitumen-containing rocks under the ancient Greek shrine at Delphi and that the famous Oracle of Delphi may have been in an ethene-induced trance when she made her predictions. Industry makes ethene, which it calls ethylene, in very large amounts by cracking alkanes; it is used to make many other chemicals (Figure 3.1), such as polythene, ethanol (used as a solvent and fuel), ethylene glycol (ethane-1,2-diol) for antifreeze and some polymers,

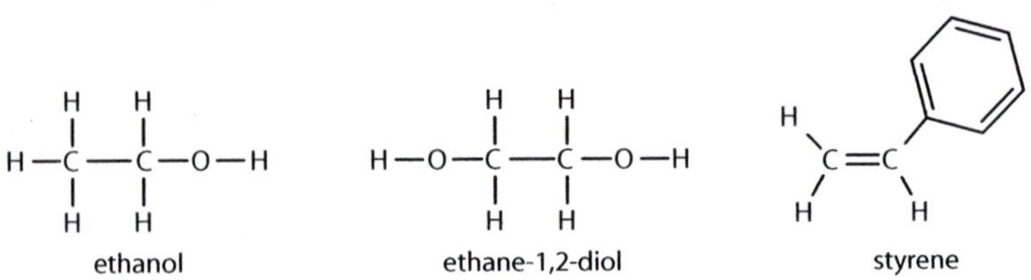

FIGURE 3.1   Many chemicals are made from ethene.

phenylethene (styrene monomer for polystyrene), and chloroethene (monomer for PVC), as well as ethanal, via the Wacker process.

In nature, ethene can be produced in cells from the amino acid methionine. It has an important role as a plant growth hormone. It is produced by ripening fruit and it also brings fruit to ripeness. Nowadays fruits like bananas are picked unripe, stored until they are needed for the market, and then exposed to ethene to bring them to ripeness. Similarly, if green tomatoes are placed in a plastic bag with a ripening banana, they go red. This is why a ripe fruit (which produces ethene) should not be stacked with unripe fruit—because "one rotten apple spoils the barrel".

Isoprene (3.6) is the parent of the terpenes, but has an important life in its own right.

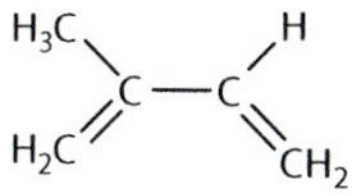

$$\text{(3.6)}$$

Because of its two C=C bonds, isoprene (2-methyl-1,3-butadiene; C_5H_8) is a diene, a colourless, easily vapourised (b.p. 34°C) liquid. It is the basis of the natural hydrocarbons called terpenes, which involve isoprene units joined together, head to tail, so that they contain multiples of five carbon atoms (10, 15, 20...). Isoprene itself is not directly involved in their biosynthesis, which starts with acetyl coenzyme A. Many plants make terpenes (and terpenoid derivatives) as repellents to insects.

Trees like oak emit large amounts of terpenes and even more isoprene, causing "natural" air pollution; their oxidation by atmospheric ozone, OH, and NO_x radicals affords species like aldehydes, hydroperoxides, epoxides, and organic nitrates, which generate aerosols and haze, giving enhanced scattering of shorter wavelength light and making it look blue, as with the hazes associated with the Blue Ridge Mountains of Virginia, in the United States, or the Blue Mountains of New South Wales, Australia.

Limonene, $C_{10}H_{16}$, is one of the most familiar terpenes. It has a chiral carbon atom, and the (R)-enantiomer, D-(+)-limonene (3.7), is found in and extracted from the rinds of citrus fruits (especially oranges) and has a citrus smell. The (S)-enantiomer, L-(−)-limonene (3.8), has a "pine" smell. It has been suggested that these smells are due to impurities in the source from which they are extracted. Chemically, the two isomers are identical.

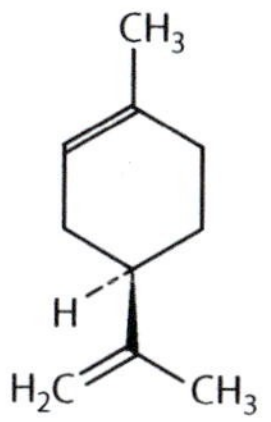

$$\text{(3.7)}$$

R-(+)-limonene

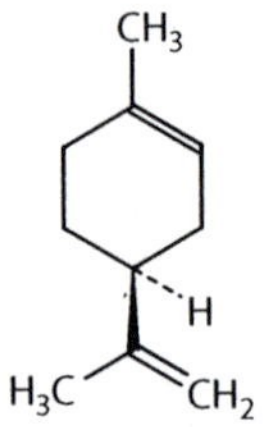

(3.8)

S-(–)-limonene

Pinene also has the molecular formula $C_{10}H_{16}$, and its isomers are an important ingredient of pine resin. The presence of flammable hydrocarbons like pinene
(α-pinene (3.9), β-pinene (3.10)) is the reason for pine cones and Christmas trees
burning well, given the chance; it is also why forest fires involving conifers can
spread fast.

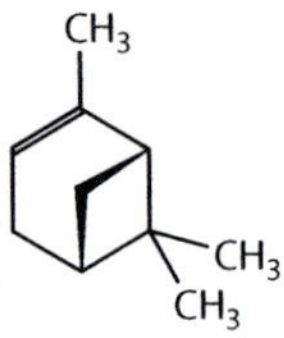

(3.9)

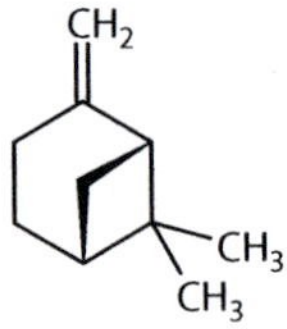

(3.10)

Beta-carotene ($C_{40}H_{56}$; 3.11) is found in many plants and fruits, and it contributes to the colour of carrots. Excess consumption of carrots and other foods and
drinks containing β-carotene can lead to carotenodermia, a harmless and reversible
condition manifest in orange skin, especially in the palms of the hands and soles
of the feet, due to carotene deposits. During World War II, the rumour was spread
that the top night-fighter crews like John "Cat's Eyes" Cunningham and Jimmie
Rawnsley owed their success in 1940 and 1941 to a diet especially rich in carrots, as
a way of concealing the existence of the Airborne Interception Radar (AI) sets in the
Beaufighter night-fighter aircraft (see p. 55).

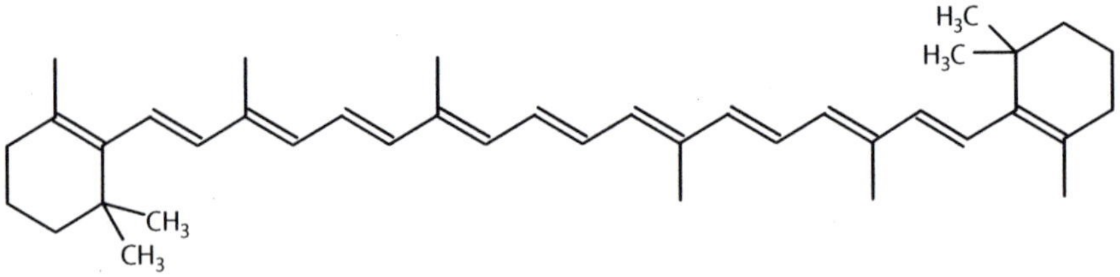

(3.11)

Rubber comes from latex, which is produced by rubber trees (e.g., *Hevea brasiliensis*) as a defence against wounding by insect predators. It is designed to gum up the mouths of predators. Rubber got its name from Joseph Priestley, who called it rubber because it rubbed out pencil marks. It is a polymer (3.12) in which isoprene molecules have been joined in a *cis* configuration, which makes it stretchy.

$$n \quad \begin{array}{c} H_3C \\ H_2C \end{array}\!\!\diagdown\!\!\begin{array}{c} \\ C-C \\ \end{array}\!\!\diagup\!\!\begin{array}{c} H \\ CH_2 \end{array} \longrightarrow \left[\begin{array}{c} H_3C \\ H_2C \end{array}\!\!\diagdown\!\!\begin{array}{c} \\ C=C \\ \end{array}\!\!\diagup\!\!\begin{array}{c} H \\ CH_2 \end{array} \right]_n \qquad (3.12)$$

3.6   3.12

Gutta-percha, obtained in latex from other trees (such as *Palaquium gutta* and other East Asian evergreens), is the *trans*-isomer and is not elastic (gutta-percha is often used as an insulating material for cables, not to mention golf-ball coverings). Chicle, used to make chewing gum, is a polymer with both *cis* and *trans* linkages.

In 1825, Michael Faraday was the first person to identify **benzene** (3.13) as a by-product of the manufacture of (illuminating) gas. Although he determined its empirical formula, it was not until the 1860s that a cyclic C_6H_6 structure was proposed by Josef Loschmidt and August Kekulé. In 1865, Kekulé suggested a cyclic structure with alternating single and double bonds; much later (1890), he said that the cyclic structure was due to a dream of a snake seizing its own tail. It is now recognised that a cyclic structure had been suggested by Loschmidt in 1861. Not until 1929 did Kathleen Lonsdale show the benzene structure was a regular hexagon with all carbon–carbon bonds the same length. It was some time before its low reactivity was reconciled with its unsaturated nature, and the concept of six aromatic π-electrons was only understood in the 1920s.

$$\hexagon \longleftrightarrow \hexagon \qquad \hexagon \qquad (3.13)$$

Benzene is very toxic and can cause leukaemia; it was once used as an aftershave on account of its pleasant smell. Its simple, regular structure means that benzene molecules stack together well. Thus, in contrast to benzene, which freezes at 6°C, the larger and heavier toluene has a much lower freezing point of –93°C; the methyl substituent stops the toluene molecules from packing so well in the solid state. Although it is found in small amounts in crude oil, benzene is such an important chemical that it is manufactured in large quantities by cracking and reforming reactions. It is used as an industrial solvent, as an ingredient of gasoline, and as a source of many aromatic compounds.

The readiness of benzene to undergo controlled replacement of ring hydrogen atoms in electrophilic substitution reactions leads to ethylbenzene, source of the phenylethene monomer used to make polystyrene; cumene ((1-methylethyl)benzene), used to make phenol and acetone; alkylbenzene sulfonates, used in detergents; and nitrobenzene, itself the synthon for phenylamine and amine dyes.

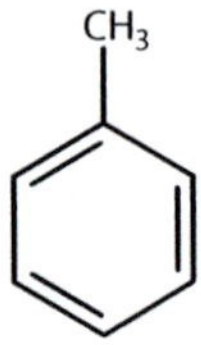

$$CH_3$$

(3.14)

As with benzene, **toluene** (3.14) is found in small amounts in crude oil, but most is made by reforming reactions from molecules like methylcyclohexane. Its electron donating $-CH_3$ group makes it more reactive to electrophilic substitution than benzene as a source of other molecules, notably TNT. There is also the possibility of reactions involving the side chain (e.g., oxidation to benzaldehyde or benzoic acid). A high-octane (111) fuel additive, toluene is a widely employed solvent that is also used in adhesives; unfortunately, it can be abused by inhalation.

New cars have an unmistakeable smell, caused by volatile organic compounds (VOCs) in glues, paints, vinyls, and plastics. Research has indicated worryingly high levels of hexane and of the aromatics benzene, toluene, and xylene, so it is worth keeping the windows open for the first few weeks that you drive a new car.

4 Acids and Alkalis

ACIDS

Everyone knows that acids are, well, acidic, and have a sharp taste. This is due to hydrated **hydrogen ions,** often written $H^+(aq)$ or $H_3O^+(aq)$, though in fact a number of water molecules are actually associated with it in the hydronium or hydroxonium ion. Two representations that have been described are the Eigen structure ($H_9O_4^+$; 4.1) and the Zundel structure ($H_5O_2^+$; 4.2), which are present in crystals of $[H_9O_4]$ $[CHB_{11}H_5Br_6]$ and $[H_5O_2][CHB_{11}Cl_{11}]$, respectively.

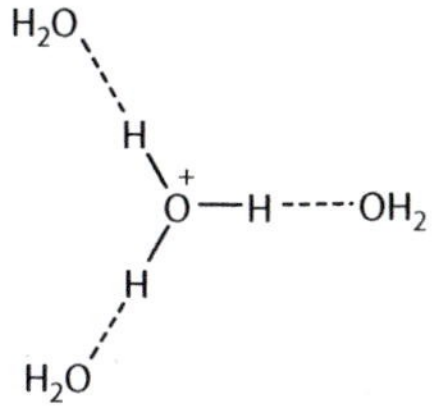

$$(4.1)$$

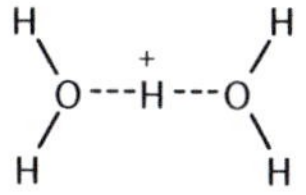

$$(4.2)$$

What is found in a crystal is, of course, no guide to what is present in solution. Vibrational spectroscopy suggests that, on the vibrational timescale, a $H_{13}O_6^+$ ion (4.3) is involved—a "Zundel-type" $H_5O_2^+$ ion with four associated waters involved in spreading out the proton's charge.

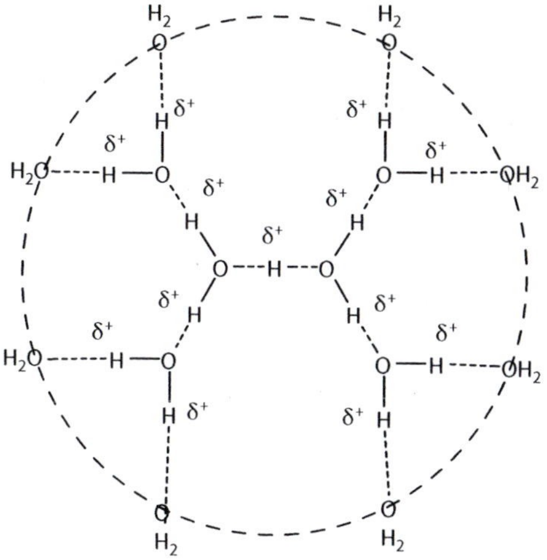

$$(4.3)$$

On the other hand, using terahertz time-domain spectroscopy, studies of a proton moving in solution suggest that there are 20 water molecules associated with it. According to this model, a hydronium ion, H_3O^+, is surrounded by three strongly bound water molecules; whilst these four waters are strongly associated with the proton, the others need to rearrange to facilitate the transfer of the proton.

It is that same $H^+(aq)$ that gives acids their reactivity, attacks many metals and compounds of the metals, and neutralises alkalis. Strong acids (H_2SO_4, HNO_3, HCl, H_3PO_4) are so called because all or virtually all of the molecules are split into ions, whereas weak acids in solution are mainly present as molecules (most organic acids), with very few (maybe 1%) ionised, so the concentration of free hydrogen ions is much lower than the acid concentration.

Sulfuric acid (4.4) is the cheapest acid on the market, and a nation's consumption of H_2SO_4 has been taken as a measure of the success of its economy. Most sulfuric acid is consumed in the manufacture of fertilisers, such as superphosphate and ammonium sulfate. After that come other major uses, including a whole range of chemicals like paints, dyes and other pigments, polymers, soaps and detergents, and, traditionally, lead–acid car batteries.

$$\begin{array}{c} H{-}O \diagdown \quad \diagup O \\ S \\ H{-}O \diagup \quad \diagdown O \end{array} \qquad (4.4)$$

Commercial production of sulfuric acid usually involves burning sulfur or roasting a suitable mineral (like iron pyrites) forming SO_2; this is reacted with oxygen (in air) over a V_2O_5 catalyst at about 450°C. The resulting SO_3 is absorbed into concentrated sulfuric acid to make $H_2S_2O_7$ (oleum) and then water is added to the oleum, forming concentrated H_2SO_4. This is known as the Contact process:

$$2\ SO_2(g) + O_2(g) \rightarrow 2\ SO_3(g)$$

$$H_2SO_4(l) + SO_3(g) \rightarrow H_2S_2O_7(l)$$

$$H_2S_2O_7(l) + H_2O(l) \rightarrow 2\ H_2SO_4(l)$$

Pure concentrated sulfuric acid (traditionally known as oil of vitriol) is a viscous liquid made almost entirely of H_2SO_4 molecules, with very few ions; there is extensive hydrogen bonding between the polar molecules, leading to the high viscosity of the concentrated acid. Upon addition of the sulfuric acid to water (never water to acid), a very exothermic reaction takes place as the molecule is broken into ions, and a very acidic solution is formed:

$$H_2SO_4(l) + 2\ H_2O(l) \rightarrow 2\ H_3O^+(aq) + SO_4^{2-}(aq)$$

Sulfuric acid is a very versatile chemical; it is capable not just of acting as a strong acid (important in industry for cleaning metal surfaces), but also of fulfilling several other roles:

- sulfonating aromatics, with the introduction of a $-SO_3H$ group, which is important in making detergents
- as a key ingredient in nitrating mixture for making aromatic nitrocompounds
- catalysing certain reactions such as esterification and the commercial synthesis of isooctane and caprolactam (precursor to nylon)
- as a dehydrating agent, classically demonstrated by converting sugar to a spongy black mass of charcoal

Concentrated sulfuric acid causes severe damage to skin and is the basis of vitriolic attacks (vitriolage), where acid is thrown into the face of the (usually) female victim. These attacks still take place, notably in parts of Asia where the Acid Survivors Foundation supports victims. Graham Greene uses a vitriolic attack in the plot of *Brighton Rock*, but the most famous incident of this kind occurs in Sir Arthur Conan Doyle's Sherlock Holmes novel *The Adventure of the Illustrious Client*, in which the evil Baron Adelbert Gruner gets vitriol thrown in his face by Miss Kitty Winter (who had been the baron's last mistress), thus destroying the handsome features on which he had relied to seduce innocent women. Sulfuric acid is a major contributor to the acid rain generated by oxidation in moist air of SO_2, which is itself formed by combustion of sulfur-containing carbon-based fuels such as coal.

Sulfamic acid (4.5) is closely related to sulfuric acid, but has the advantage of being a white crystalline solid that can be obtained in a very pure form and which does not absorb water.

$$H_2N \diagdown \diagup O \atop H-O \diagup S \diagdown O \tag{4.5}$$

This means that, unlike sulfuric acid (and all the other strong acids), solutions of known concentration can be made simply by weighing out the solid acid and dissolving it in water—important to analytical chemists. It can be used to make sweeteners like sodium cyclamate and in a range of organic syntheses. Its acidic properties mean that it is used in commercial cleaners to remove rust from metals and limescale from kettles and boilers. In solution, it exists largely as the zwitterion $^-OS(=O)_2NH_3^+$.

Hydrochloric acid (traditionally, muriatic acid or spirits of salt) cannot be obtained as pure "hydrochloric acid". When the small covalent molecule hydrogen chloride (a gas at room temperature; 4.6) dissolves in water, it forms an acidic solution known as hydrochloric acid containing hydrated hydrogen ions and chloride ions:

$$H-Cl$$

$$H-Cl(g) + aq \rightarrow H_3O^+(aq) + Cl^-(aq) \tag{4.6}$$

Hydrogen chloride itself is made by direct synthesis from the elements. The largest industrial use of hydrochloric acid is in removing oxide layers from metals, especially iron; however, other uses are legion, such as the synthesis of chloroethene, the monomer for making the polymer PVC. As a strong acid that is not an oxidising agent, aqueous hydrochloric acid is a good choice for hydrolysing large carbohydrates, like turning starch into glucose, and in obtaining gelatine from bones. Hydrochloric acid is widely used to adjust the acidity of swimming pools to a pH ca. 7.4, which is the best pH for eye comfort. Additionally, if the pH is too low, the water circulation system will corrode; if the pH is too high, scale may form and the water become discoloured. Hydrochloric acid is sometimes used as a descaler for toilet bowls.

Perhaps the most unexpected place to encounter hydrochloric acid is in the human stomach, where gastric acid has hydrochloric acid as its most important ingredient. The pH of the stomach contents is usually between 1.5 and 2, corresponding to a hydrogen ion concentration slightly in excess of 0.01 mol dm^{-3}. The acid denatures proteins, so the protease enzyme pepsin can then hydrolyse peptide links between individual amino acids. Pepsin is a singular enzyme that works well at a pH where most enzymes would be degraded, and the stomach has a protective layer of mucus to protect it from the acidic contents (peptic ulcers can develop when this layer is breached). Once it has left the stomach, the acid is neutralized in the duodenum by sodium hydrogen carbonate.

Nitric acid (4.7) is a strong acid, but its properties are complicated by the nitrate ion, which can be reduced to brown fumes of NO_2. Thus, most metals (Ca and Mg are exceptions) do not form hydrogen gas with even very dilute nitric acid. Some metals, like aluminium and iron, are "passivated" by the concentrated acid, forming a tough oxide layer on their surface.

Traditionally known as aqua fortis, concentrated nitric acid—especially "fuming" nitric acid (above 85% HNO_3)—tends to decompose slowly in light, even at room temperature. The NO_2 produced dissolves to give the acid a yellow colour:

$$4 \ HNO_3 \rightarrow 2 \ H_2O + 4 \ NO_2 + O_2$$

The industrial synthesis of nitric acid by reacting nitrogen dioxide (obtained from ammonia) with air and water is a reversal of this.

$$(4.7)$$

By far, the largest use of nitric acid lies in the manufacture of fertilisers, notably ammonium nitrate, NH_4NO_3; however, nitric acid is also important in organic synthesis. A significant amount is used in converting cyclohexanone into hexane-1,6-dioic acid (adipic acid), an important material in the manufacture of nylon.

Combined with conc. H_2SO_4, conc. HNO_3 nitrates organic molecules to form nitrocompounds, which are often important explosives in their own right (e.g., TNT, nitroglycerine; see pp. 120–122) or perfume ingredients (nitromusks, pp. 82–83); the nitrocompounds can also be reduced to aromatic amines, which are starting materials for making azo-dyes. Concentrated nitric acid will violently oxidise many compounds of nonmetals such as carbon, often with explosive violence and immediate ignition. Thus, it has been used in experimental rocket fuels as the oxidant for fuels including kerosene and hydrazine.

Pure **phosphoric acid,** H_3PO_4 (aka orthophosphoric acid or phosphoric(V) acid; (4.8), is a low-melting (42.35°C) white solid that is readily soluble in water and forms a colourless solution that is viscous when concentrated. It is quite nontoxic in very dilute solution; it is used as a food additive (E338) and its pleasant, sharp taste can be experienced in "cola" drinks, where it is a cheaper alternative to citric acid. The greatest amount of phosphoric acid is employed in the manufacture of fertilisers, but it is also converted into phosphate salts for use in the food, detergent, and toothpaste industries. H_3PO_4 is used for removing oxide layers from metals and, in particular, rust from iron, which it turns into a black coating of $FePO_4$. This can be removed or left on the iron to provide a corrosion-resistant coating. Phosphoric acid also makes a good soldering flux.

$$(4.8)$$

Most acids are weak acids, which has nothing to do with their concentration, but rather is a measure of the proportion of the molecules ionised in solution (p. 30). They are not as corrosive as strong acids, but can be just as—or more—toxic (hydrocyanic acid, solutions of HCN, are an example of this).

Methanoic acid (4.9) is the smallest and simplest of the carboxylic acids, which contain the –COOH group. It was formerly known as formic acid because it was first isolated in the seventeenth century by the distillation of red ants (Latin: *formica*), which use it in their venom. The presence of a hydrogen attached to the –COOH group, rather than an alkyl group, gives methanoic acid some reactions that are not shared by the other carboxylic acids; for one thing, it has reducing properties like aldehydes, giving a positive "silver mirror" test. It can be dehydrated by conc. H_2SO_4 to form water and CO (a convenient synthesis of CO), but catalysts like platinum split it into H_2 and CO_2. The human body forms methanoic acid when methanol is metabolised; it destroys the optic nerve and is therefore associated with the blindness that can result from consumption of methanol. Like propanoic acid, methanoic acid is used as an antibacterial agent in agriculture (e.g., preserving feedstock used for winter feed) and finds minor uses, as in some limescale removers.

$$(4.9)$$

Ethanoic acid (4.10), CH_3COOH, is formed when ethanol is oxidised by *Acetobacter* bacteria under aerobic conditions, generating the familiar vinegary smell:

$$CH_3CH_2OH(l) + O_2(g) \rightarrow CH_3COOH(l) + H_2O(l)$$

$$\underset{H_3C}{\overset{\overset{\textstyle O}{\|}}{\diagdown}}\underset{}{\overset{C}{\diagup}}\underset{O}{\diagdown}H \qquad\qquad (4.10)$$

Its traditional (and more widely used) name, acetic acid, derives from *acetum,* which is Latin for vinegar. Commercially, though, it is mainly manufactured by carbonylation of CH_3OH, using rhodium- or iridium-based catalytic processes developed by Monsanto and B.P. Other routes include catalytic oxidation of acetaldehyde (ethanal) and of short-chain alkanes, though the latter give a mixture of products. Apart from its use in vinegar and foodstuffs (E260), it has a wide range of applications commercially, such as in manufacturing polymers like cellulose acetate and polyvinyl acetate. It is also used to synthesise esters CH_3COOR (R = alkyl, e.g., C_2H_5), which find significant uses as solvents (e.g., $CH_3COOC_2H_5$) in some marker pens and flavourings.

Because pure ethanoic acid freezes at 16.7°C, bottles of it could be seen frozen in laboratories during the winter (in the days before central heating); it was known as "glacial" acetic acid. Like other carboxylic acids, solid ethanoic acid contains dimeric hydrogen-bonded molecules (4.11), also detectable by mass spectroscopy in the gas phase.

$$H_3C-C\overset{\displaystyle O---H-O}{\underset{\displaystyle O-H---O}{\diagup\diagdown}}C-CH_3 \qquad\qquad (4.11)$$

Butanoic acid (4.12), $CH_3CH_2CH_2COOH$, is formed in rancid butter by the hydrolysis of the glyceride of butanoic acid present in small amounts in the butter; its traditional name of butyric acid derives from *butyrum,* the Latin word for butter. It has an unpleasant sweaty smell ("armpit odour") and is indeed present in human sweat, detectable by tracker dogs. Along with isovaleric acid (3-methylbutanoic acid), it is responsible for the characteristic smell of Limburger and other rind-washed (smear) cheeses, due to the activity of *Brevibacterium linens,* which is also found on human skin. In 2006, Bart Knols and Ruurd de Jong shared the "Ig Nobel" prize for biology, having shown that the female malaria mosquito (*Anopheles gambiae*) is attracted to the smell of Limburger cheese and to the smell of human feet, leading to the suggestion of using Limburger cheese as a bait for mosquito traps.

$$\overset{\overset{\textstyle O}{\|}}{\diagup\diagdown\diagup}\underset{O}{\diagdown}H \qquad\qquad (4.12)$$

Citric acid (4.13)—particularly, the hydrogen ions formed from each molecule—is responsible for the acid taste of citrus fruits, especially lemon and lime. It is also used as a food additive and preservative (E330), particularly in soft drinks. It is a useful chelating agent for metals and was at one time used in ion-exchange separations of the fission products in the Manhattan project, before it was supplanted by the more efficient EDTA. Citric acid is commercially made by microbial fermentation of sugars, using the fungus *Aspergillus niger.*

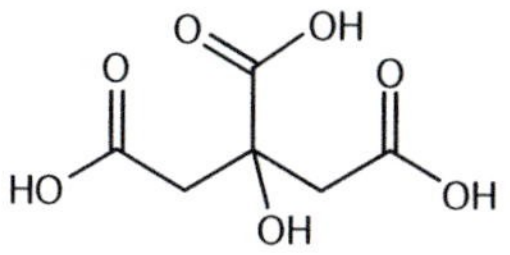

(4.13)

Alka-Seltzer tablets contain citric acid as well as sodium bicarbonate (the blue variety contains aspirin, too). The ingredients do not react in the solid state, but as they dissolve in water, the hydrogen ions from the acid react with the hydrogen carbonate ions from the $NaHCO_3$, forming water and carbon dioxide gas—hence the fizz:

$$H^+(aq) + HCO_3^-(aq) \rightarrow H_2O(l) + CO_2(g)$$

Most importantly, citric acid is a key substance in the Krebs (citric acid, TCA) cycle, the metabolic process by which carbon-based materials (carbohydrates, proteins, fats) are oxidised to carbon dioxide and water in living systems. A CH_3CO (acetyl group) fragment from the breakdown of glucose is joined up with an oxaloacetate group to form citrate, which in the course of the cycle is degraded back to oxaloacetate.

Lactic acid (2-hydroxypropanoic acid) has a chiral carbon; of the two optical isomers, L-(+)-lactic acid (also known as (S)-lactic acid; 4.14), is the important one in living systems.

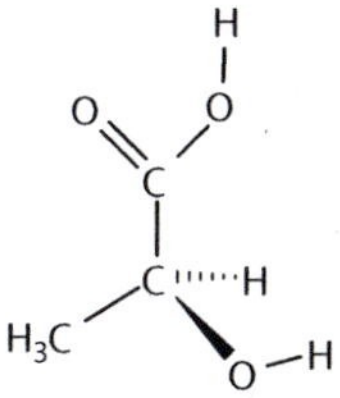

(4.14)

Lactic acid is produced in raw milk by bacterial fermentation of lactose. This process leads to a drop in pH and is responsible for the coagulation of casein and the formation of the curds in cheese-making, a process accelerated by addition of starter bacteria. The sourdough breads popular in the United States are made from dough containing lactobacillus, which produces sour-tasting lactic acid. Lactic acid is generated in the human body by fermentation of pyruvate. In hard exercise (e.g., a 100-metre sprint), it is produced—as lactate—faster than the tissues can oxidise it, though it is not responsible for acidosis.

Benzoic acid (4.15) occurs quite widely in nature (e.g., in fruits), as esters and also as the free acid; it is present as both in gum benzoin (resins like benzoin resin or styrax resin; not to be confused with the molecule benzoin). It was first isolated by heating gum benzoin until it distilled off (dry distillation). Commercially, it is made by catalytic oxidation of methylbenzene (toluene):

$$C_6H_5CH_3 + 3/2\ O_2 \rightarrow C_6H_5COOH + H_2O$$

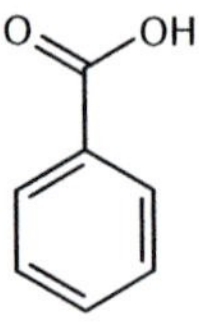

$$(4.15)$$

Both benzoic acid (E210) and salts, particularly sodium benzoate (E211), are widely used as preservatives in food and soft drinks because benzoic acid inhibits growth of moulds and some bacteria. It was reported in 1993 that ascorbic acid (vitamin C, an antioxidant additive, E 300) can reduce benzoic acid to the highly toxic benzene, leading to the presence of benzene in soft drinks. The American FDA follows the US Environmental Protection Agency (EPA) in recommending a maximum benzene level of 5 ppb in water and has tested many soft drinks; the vast majority had benzene levels well below 5 ppb, and the exceptions have been reformulated. Benzoic acid is used to make a wide range of other substances, including plasticisers and other materials in the polymer industry, whilst esters are used as flavourings and in perfumes.

Ascorbic acid (E 300; 4.16) has an important dietary role: It prevents degeneration of connective tissue and is also an antioxidant. Primates cannot synthesise it and thus require it in their diet so that they can make collagen. In the days of long voyages by sailing ship with a restricted diet, sailors were prone to develop scurvy; in 1747, James Lind published a treatise advocating citrus fruit in the diet of sailors. It is now known that this key ingredient is vitamin C, the L-isomer of ascorbic acid, which has a recommended daily allowance of 60 mg, which is substantially exceeded in normal Western diets. Albert Szent-Györgyi and Norman Haworth received their Nobel Prizes in 1937 (for medicine and chemistry, respectively) largely for their researches on vitamin C and ascorbic acid. Despite his best efforts to win an unequalled third Nobel Prize for these researches, the late Linus Pauling was unsuccessful in showing that large doses of vitamin C could prevent the common cold. Ascorbic acid does not contain a –COOH group, but rather owes its acidity to the conjugation of the C=O, C=C, and the lone pair on the OH group.

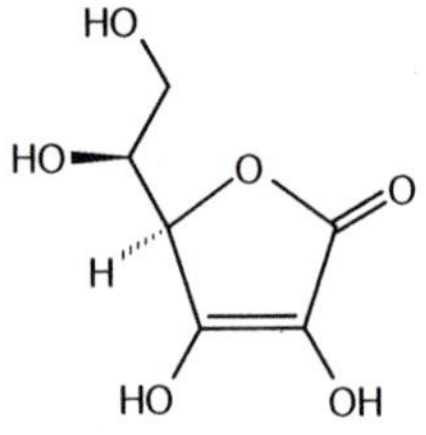

$$(4.16)$$

ALKALIS

Alkalis are alkaline because they contain **hydroxide ions;** when dissolved in water, alkalis create an excess of $OH^-(aq)$ over $H^+(aq)$ ions. It is thought that, in solution, the OH^- ion is surrounded by four water molecules (4.17).

$$\begin{array}{c}
H_{\diagdown O} \\
| \\
H \\
\vdots \\
O-H\text{------}O\text{------}H-O{\diagup}^{H} \\
{\diagup}H \qquad\qquad \diagdown H \\
\vdots \\
H \\
| \\
O_{\diagdown H}
\end{array} \qquad (4.17)$$

Sodium hydroxide, traditionally known as caustic soda and in America as lye (short for alkali), is the most important alkali; it is entirely composed of the ions Na^+ and OH^-, resulting in strongly alkaline solutions. Sodium hydroxide is produced commercially by electrolysis of sodium chloride solution (brine), either using a mercury cell or by a more environmentally desirable route (diaphragm or membrane cell). It is used in applications requiring a strong alkali, including the paper and textile industries. In the purification of aluminium oxide in the electrolytic extraction of aluminium, NaOH dissolves amphoteric Al_2O_3 but insoluble impurities are left behind and separated off; the aluminium-containing solution is then acidified to reprecipitate pure Al_2O_3. In the making of soap (Figure 4.1), sodium hydroxide hydrolyses fats, which are triglyceride esters of long-chain carboxylic acids, forming the sodium salts of the acids (soaps). The soapy feeling on your skin when you get even a trace of dilute NaOH solution on your finger is due to sodium octadecanaote (sodium stearate = soap) formed by hydrolysis of lipids in your skin.

As a strong alkali, NaOH reacts with grease, fats, protein, and some oils, hydrolysing the molecules into smaller and more soluble molecules. It is used in oven cleaners with the addition of surfactants, detergents, and abrasives. Other bases like ethanolamine, $H_2NCH_2CH_2OH$, are sometimes added.

Clogged drains can also be unblocked using solid sodium hydroxide. In commercial drain cleaners, a mixture of sodium hydroxide, sodium nitrate, and aluminium turnings is often used. Heat is evolved when NaOH dissolves in water; this can melt a solid grease, and the higher temperatures will speed up the hydrolysis reactions. Apart from saponifying ester linkages in fat, the hot alkali can hydrolyse peptide links in any waste hair present. After the hot alkali has stripped the oxide layer off the aluminium's surface, it reacts very exothermically with the metal generating a vigorous effervescence of hydrogen gas; a combination of gas pressure and swirling pieces of alumin-

$$\begin{array}{l}
CH_2O.CO.C_{17}H_{35} \\
| \\
CHO.CO.C_{17}H_{35} \quad + \quad 3\,NaOH \quad \longrightarrow \\
| \\
CH_2O.CO.C_{17}H_{35}
\end{array}
\qquad
\begin{array}{l}
CH_2OH \\
| \\
CHOH \quad + \quad 3\,C_{17}H_{35}COONa \\
| \\
CH_2OH
\end{array}$$

FIGURE 4.1   Making soap.

ium will assist in clearing the blockage. Under these conditions, nitrate gets reduced to gaseous ammonia, thus lessening the amount of flammable hydrogen generated:

$$2\ Al(s) + 2\ NaOH(aq) + 6\ H_2O(l) \rightarrow 3\ H_2(g) + 2\ NaAl(OH)_4(aq)$$

$$2\ NaNO_3(aq) + 9H_2(g) \rightarrow 2NH_3(g) + 6H_2O(l)$$

$$3\ NaNO_3(aq) + 8Al(s) + 18\ H_2O(l) \rightarrow 3\ NH_3(g) + 8\ Al(OH)_3(aq) + 3\ NaOH(aq)$$

Chemists who have used old-style reagent bottles with glass stoppers will recall the stoppers sticking when not in regular use, owing to the formation of sodium silicate from an acid–base reaction. Similarly, bottles of sodium hydroxide solution after a while develop a white precipitate of $NaHCO_3$ through reaction with atmospheric carbon dioxide.

Ammonia (4.18) was first isolated in 1774 by the English chemist Joseph Priestley; in 1787, the Frenchman Claude Louis Berthollet showed that it was NH_3. Ammonia is a pyramidal molecule. There are four pairs of electrons attached to the nitrogen atom; three make up the three N–H bonds. The fourth is not involved in bonding and repels the other pairs slightly more strongly, so they are not quite evenly spread in space around the N atom (H–N–H angle 107°).

(4.18)

Ammonia has a quite characteristic, pungent smell and is the active ingredient of smelling salts (ammonium carbonate, $(NH_4)_2CO_3$)); inhaled ammonia triggers an inhalation reflex and stimulates breathing. The same smell is associated with some domestic cleaning products. Although it is a gas, ammonia is exceptionally soluble in water, producing a solution that is a weak alkali:

$$NH_3(g) + H_2O(l) \rightleftharpoons NH_4^+(aq) + OH^-(aq)$$

The position of equilibrium is well to the left, which is why the solution smells strongly of ammonia (it can be driven off by boiling); the low concentration of hydroxide ions means that the solution is a weak alkali. Ammonia solution can be neutralised with acids, forming ammonium salts, which are often used in fertilisers and are the main reason for well over 100 million tons of ammonia being manufactured each year.

Throughout time there have been famines, yet the world is capable at present of feeding 7 billion people. For that, chemists and ammonia are largely responsible. Human life is not possible without nitrogen compounds. They are present in all cells, in RNA and DNA, and in amino acids, proteins, and enzymes—not to mention hemoglobin and chlorophyll. Shortage of nitrogen limits crop production, and any nitrogen deficiency shows up in slow growth and stunted crops. Most plants cannot make use of atmospheric nitrogen, apart from some leguminous plants including bean, pea, vetch, and clover; they need to absorb nitrate through their roots.

In the nineteenth century, increasing demand for nitrogen compounds was matched first by Peruvian "guano" (bird droppings) and then by Chilean supplies of sodium nitrate, used mainly for fertilisers but also for making nitric acid, a key chemical in the manufacture of explosives. Germany needed to become independent of nitrate imports. Yet, although nitrogen gas makes up nearly four-fifths of the atmosphere, it is as extremely unreactive N_2 molecules, and Fritz Haber (1868–1934) was faced with problems in making it combine with hydrogen:

$$N_2 + 3\,H_2 \rightleftharpoons 2\,NH_3$$

At high temperatures, the reaction was fast, but hardly any ammonia was present in the equilibrium mixture; at low temperatures, the yield of ammonia was better, but reaction was incredibly slow. Haber found that, by using very high pressure, a better yield was obtained, helped by the use of iron catalysts. Carl Bosch (1874–1940) made the process work on an industrial scale and a large-scale plant was producing 10 tons of ammonia a day in October 1913. A year later, Germany was at war and moved to build plants to convert ammonia into nitric acid, as well as to expand ammonia production. It was the key to their production of explosives that kept Germany in the war. Both Bosch and Haber were later to win Nobel prizes; although this should be known as the Haber–Bosch process, Bosch's name is largely forgotten today. Thus, ammonia can be used to make fertiliser to help people live and, without ammonia, it would be much harder to have explosives.

Calcium oxide, CaO, and **calcium hydroxide, Ca(OH)$_2$,** are two closely related compounds; calcium oxide is one of the chemicals manufactured in the greatest amounts worldwide. Neither occurs naturally; they are made from the abundant calcium carbonate (chalk, limestone, marble). Confusingly, several substances, including calcium oxide, calcium hydroxide, and calcium carbonate, can all be referred to as "lime". Calcium oxide is manufactured by heating calcium carbonate at above 1000°C ("lime burning"):

$$CaCO_3 \rightarrow CaO + CO_2$$

If calcium oxide is heated above 2000°C, it emits a bright white light that was once used in stage lighting (the "limelight"). Calcium oxide reacts slowly with CO_2 to reform $CaCO_3$, but very vigorously with water to form $Ca(OH)_2$. This has been used commercially in self-heating food cans; these contain a chamber of the food (or drink) and separate chambers of two chemicals that are mixed by pressing a button or pulling a ring:

$$CaO + H_2O \rightarrow Ca(OH)_2$$

Calcium oxide is traditionally known as quicklime, probably because its reaction with water gives the impression that it is "alive" ("quick" in mediaeval parlance); calcium hydroxide is "slaked lime" because the reaction has slaked the "thirst" of the calcium oxide. They are both slightly soluble in water to give an alkaline solution of $Ca(OH)_2$ and are used in applications where a stronger alkali, like NaOH, would be unsuitable.

CaO is used to make mortar ("lime mortar") by mixing it with sand and water; its slow hardening is due to the formation of $CaCO_3$:

$$Ca(OH)_2 + CO_2 \rightarrow CaCO_3 + H_2O$$

Cement is made by heating $CaCO_3$ with clay and sand to 3000K. Traditionally, glass is made by heating CaO with sodium carbonate and silica, a process first employed some 4,000 years ago. Other applications put the basic properties of calcium oxide to use; CaO is widely used as a slagging agent in the extraction of metals from their ores because the basic calcium oxide combines with acidic nonmetal oxides like SiO_2. CaO and $Ca(OH)_2$ are also used to reduce SO_2 emissions from power stations (e.g., flue-gas desulfurisation). Scientists have suggested that adding lime to seawater would increase its pH and thus remove CO_2 from the atmosphere. Liming agents are employed in agriculture to raise the pH of soil, but powdered $CaCO_3$ is most often used rather than CaO or $Ca(OH)_2$. Other uses include calcium hydroxide as a flocculant that helps clarify the water in water treatment.

5 Steroids and Sex

INTRODUCTION

Cholesterol is one of the most important molecules in the body, as well as the most important steroid. As Brown and Goldstein put it in their 1985 Nobel Lecture, "[Cholesterol is] the most highly decorated small molecule in biology. Thirteen Nobel Prizes have been awarded to scientists who devoted major parts of their careers to cholesterol". All steroids have four carbon rings fused together in a similar geometry (5.1); three of these are six membered and one is five membered.

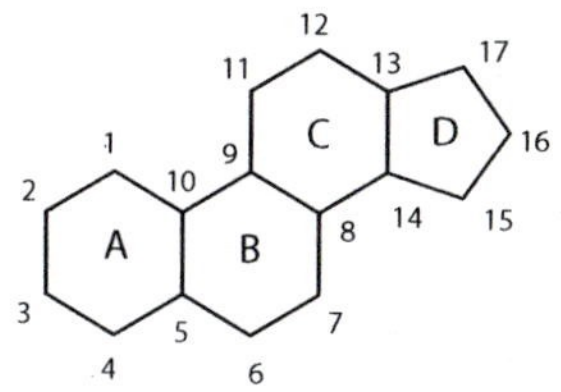

(5.1)

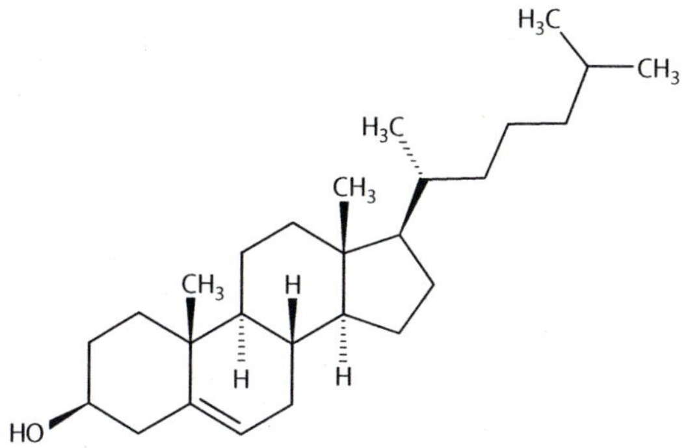

(5.2)

The human body contains around 100–150 g of cholesterol (5.2). Within the body, it is found in all cells (in the membranes); it is the precursor of testosterone and the other sex hormones, as well as the source of the bile acids used to digest lipids. It is also needed to make vitamin D, absorb calcium, and regulate calcium levels. The body must make about 0.8–1.0 g of cholesterol a day—much more than we take in from the diet (around 100 mg.).

The body makes cholesterol, mainly in the liver, from acetyl coenzyme A. A series of enzyme-controlled reactions yields a key intermediate, 3-isopentenyl pyrophosphate, which in turn is converted into the C_{15} species farnesyl pyrophosphate. Another enzyme, squalene synthase, catalyses the reductive tail-to-tail joining of two farnesyl pyrophosphates forming the isoprenoid hydrocarbon squalene, $C_{30}H_{50}$

(a molecule obtained commercially largely from sources like shark liver oil). The squalene is then converted to the epoxide 2,3-oxidosqualene, which is held in the active site of the enzyme oxidosqualene cyclase in just the right conformation to facilitate conversion into lanosterol; this is then turned into cholesterol in a 19-step process involving loss of three methyl groups, reduction of a double bond, and migration of another double bond. The human body is a remarkable organic chemist.

Important steps in this transformation are summarised in Figure 5.1.

Cholesterol is insoluble in water (which aids its suitability for making membranes) and therefore requires low-density lipoprotein (LDL) to transport it through the blood to tissues. LDL carries an apolipoprotein B in its outer shell; this is recognised by LDL receptors in various cells throughout the body so that the LDL can deliver the cholesterol that the cell needs for membrane building. High-density lipoprotein (HDL) scavenges cholesterol from the tissue surfaces of dying cells to return it to the liver or to pass it on to tissue that will use it to make steroid hormones. High levels of

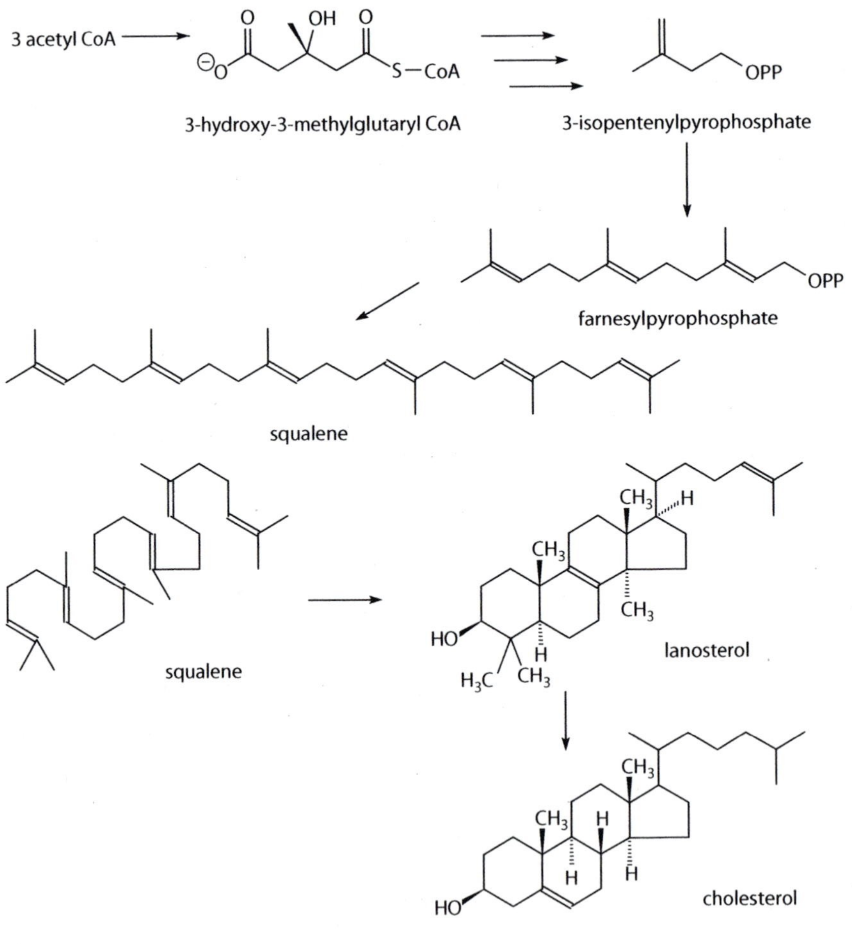

FIGURE 5.1 Production of cholesterol in the body.

cholesterol in blood serum are unhealthy because they create atherosclerotic plaques in the arteries, leading to arterial narrowing and an increased risk of heart attacks.

LDL has been classed as "bad" because of its high cholesterol content and HDL as "good" for its high protein content.

TESTOSTERONE

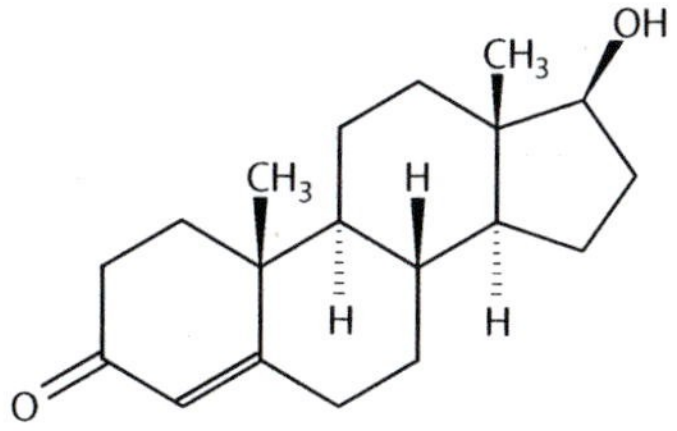

(5.3)

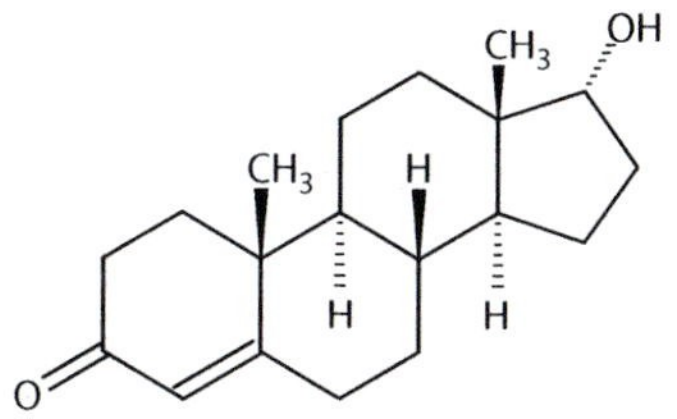

(5.4)

In the early twentieth century, treatments with hormonal products, "monkey-gland" transplants, and insertion of testicular material were fashionable, but it was not until 1935 that the male hormone was isolated and synthesised. Adolf Butenandt (Germany) and Leopold Ružička (Switzerland) shared the 1939 Nobel Prize for Chemistry for their work on steroids. Testosterone is the principal male sex hormone. Both sexes carry some of each of the principal sex hormones, **testosterone** (5.3) and estradiol, but there are generally significantly higher levels of testosterone in men than in women and, conversely, more of the estrogens such as estradiol in women. Testosterone has an inactive epimer (5.4), which has a different configuration at carbon 17 (see p. 44).

Testosterone is synthesised from cholesterol, particularly in the male testes, from puberty onward; it is anabolic, stimulating the synthesis of muscle protein, leading to greater muscle mass and strength. It is linked with improved athletic performance. It is metabolised by oxidising it to androstenone, which is excreted in the urine. Major roles for testosterone include responsibility for the development of growth and strength, building bones and muscle (anabolic effects), and development of the male sexual organs, deepening voice, and growth of hair in the facial, chest, and pubic regions. To convert cholesterol into testoterone, the long side chain has to be removed. Removal of six carbons of the side chain by a P450 enzyme called 20,22-desmolase (CYP11A1) results in **progesterone** (5.5), whilst 17,20-desmolase (CYP17A1) removes the other two carbons, ultimately forming **androstenedione** (5.6) and then testosterone.

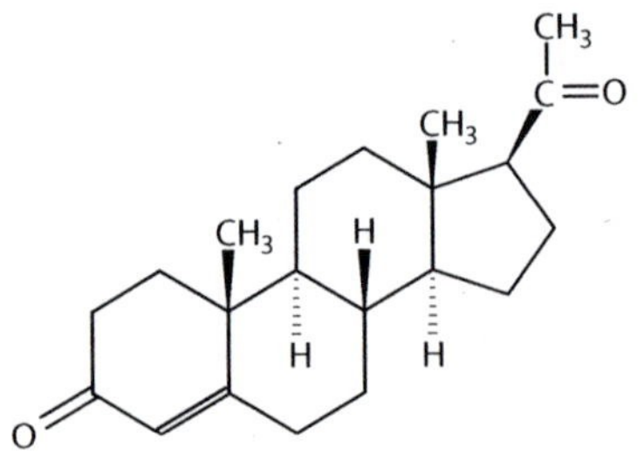

(5.5)

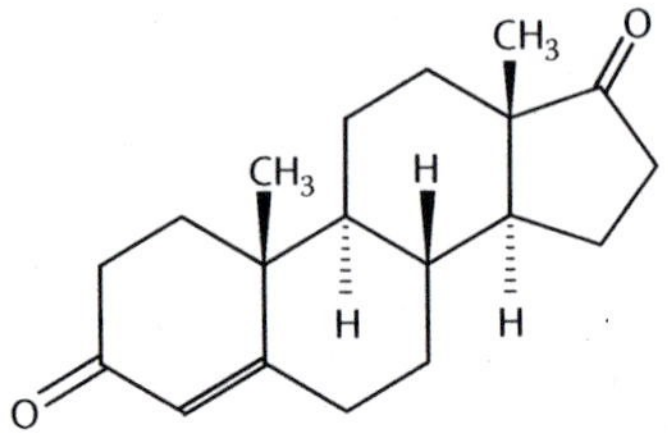

(5.6)

Testosterone "works" by binding to the androgen receptor in the cytoplasm of cells; this produces structural changes that enable this testosterone–androgen receptor complex to move into the cell nucleus. Once there, it binds to chromosomal DNA and promotes genes carrying out transcription and translation, increasing protein synthesis in cells and leading to the buildup of muscular tissue characteristic of anabolic effects.

Once the structure of testosterone was known, chemists started to modify its structure by esterification and alkylation. By the late 1930s, esters like the propionate (propanoate) were available for experiment. Testosterone enhances athletic performance, but there are risks associated with abuse, shared with other anabolic steroids. When the testes are no longer needed to synthesise testosterone, they shrink. At increasing levels of testosterone, the risks are acne and hair loss; mood swings, aggression, and possibly psychiatric disorder; liver and kidney tumours; and heart problems. Paul de Kruif's 1945 book, *The Male Hormone,* brought testosterone to a wider audience. Medal-winning performances by Soviet weightlifters at the 1952 Helsinki Olympic Games aroused suspicions and, within a couple of years, American sportsmen were exploring testosterone use too; it was estimated that steroids were used by one-third of the US track and field team for the 1968 Mexico City Olympics at their pregames training camp. Steroids were not tested for until the 1976 Montreal Olympic Games; previous testing concentrated on drugs like amphetamines and on chromosomal testing of female athletes.

Testing for testosterone abuse involved measuring the relative amounts in urine samples of testosterone and its epimer, **epitestosterone** (5.4), which has no anabolic activity. They differ only in the configuration at carbon 17; testosterone is the *S*-isomer and epitestosterone is the *R*-isomer. The natural T:E ratio is around 1:1 and rarely above 4:1. Thus, the International Olympic Commission (IOC) decided that, from 1982, a 6:1 ratio would result in immediate disqualification.

One way round that was to give athletes extra synthetic epitestosterone as well as testosterone, but this trick may be detected using a carbon isotope ratio (CIR) test that measures the $^{13}C{:}^{12}C$ ratio in the sample. Because body-made testosterone and synthetic testosterone are made by different biochemical pathways, they have different $^{13}C{:}^{12}C$ ratios. The highest profile victim of epitestosterone testing was the American cyclist Floyd Landis, who finished first in the 2006 Tour de France, but lost the title because of an unusually high T:E ratio (reportedly 11 to 1); in 2010, Landis admitted to having taken drugs, including testosterone patches.

MAN-MADE ANABOLIC STEROIDS

The bodies of people with severe burns stop producing testosterone, so the analogue methandrostenolone (5.7; **dianabol***)* was released onto the market in 1958 as an aid to their recovery. Dianabol is a 17α-alkylated steroid—a tertiary alcohol that is harder to oxidise than testosterone—and it is also harder for enzymes to hydroxylate it so that it can pass through the liver more effectively and be a more active drug. The corollary is that drugs of this type are more likely to cause liver damage. Dianabol was succeeded in 1965 by **turinabol** (5.8), an East German product.

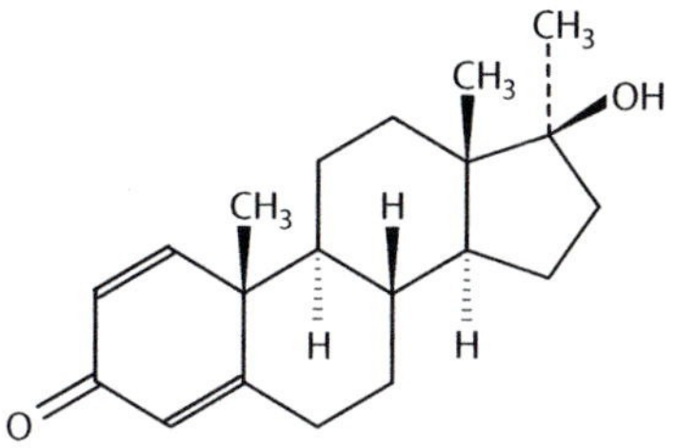

(5.7)

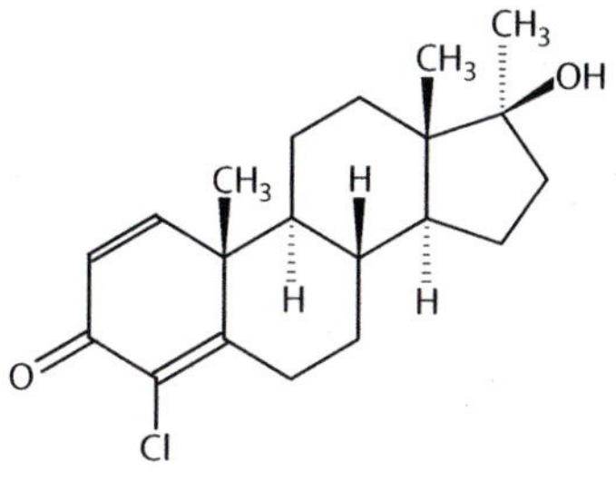

(5.8)

During the 1970s and 1980s, turinabol was given to many young East German swimmers and athletes, who were told that the blue pills were vitamins. The East German women won 11 out of the 14 swimming gold medals at the 1976 Montreal Olympics, but the cost was only apparent later. Aged 15, Rica Reinisch won three backstroke gold medals at the 1980 Moscow Olympics. Inflamed ovaries forced her to retire at the age of 16, after just 2 years of competition. Her case is typical of dozens. After the fall of the Berlin Wall, the two leaders of the East German doping programme, Manfred Ewald and Manfred Hoeppner, were tried in 2000 and convicted on charges of inflicting bodily harm upon more than 100 young women athletes.

Other abused steroids include **stanozolol** (5.9) and **nandrolone** (5.10). The Canadian sprinter Ben Johnson lost the Olympic gold medal that he received for finishing first in the 100-metre race at the 1988 Seoul Olympics after testing positive for stanozolol in urine samples. Numerous sportsmen have tested positive for nandrolone, such as the tennis player Petr Korda, Pakistani Test cricketers Shoaib Akhtar and Mohammed Asif, the Dutch football players Edgar Davids and Jaap Stam, and the British athlete Linford Christie.

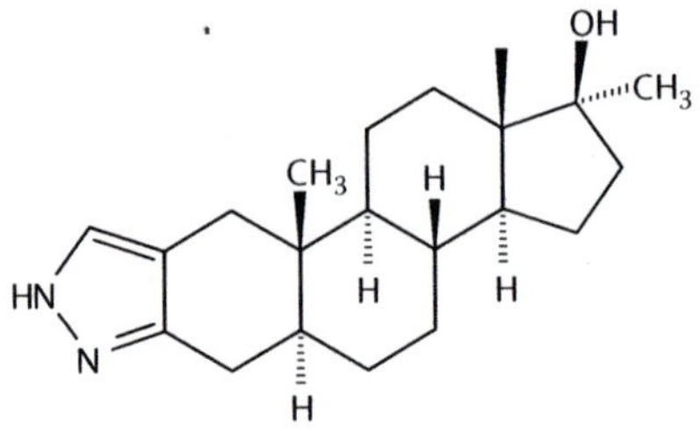

(5.9)

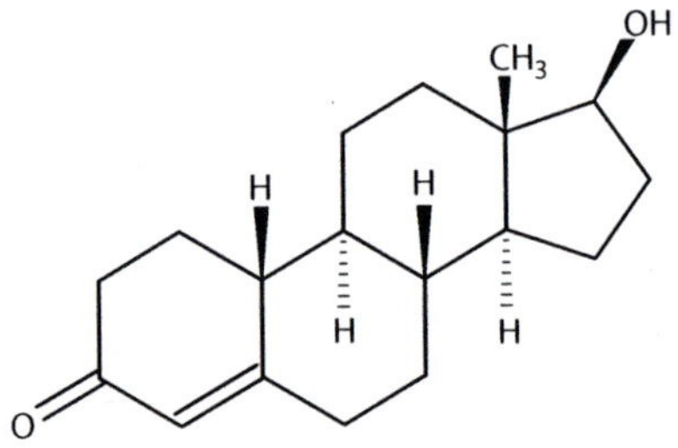

(5.10)

Steroid abuse became front page news thanks to the BALCO scandal. The Bay Area Laboratory Co-Operative (BALCO) was a sport supplement company founded by Victor Conte. From around 2000, BALCO supplied sportsmen with "undetectable" performance-enhancing steroids developed by the chemist Patrick Arnold, who synthesised anabolic steroids that had previously been reported but never distributed. **Madol** (5.11) and **norbolethone** (5.12) were detected by Don Catlin's team at the UCLA Olympic Analytical Laboratory in Los Angeles, in 2005 and 2002, respectively.

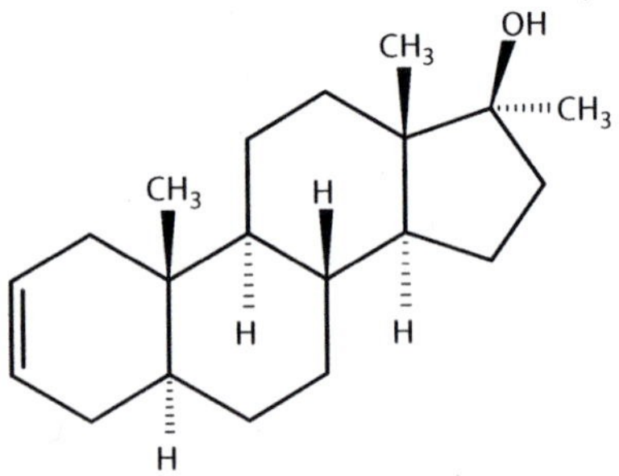

(5.11)

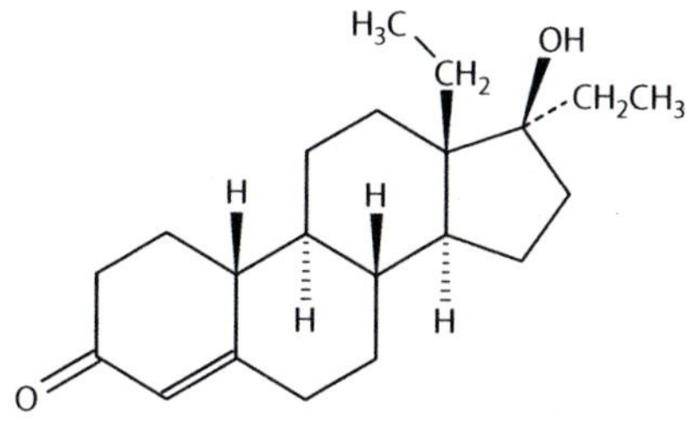

(5.12)

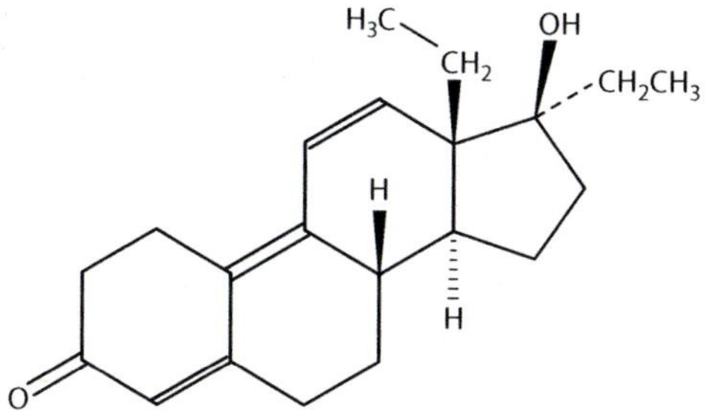

(5.13)

The third of these steroids, **THG** (5.13), had never been made before. In June 2003, a "used" syringe containing an allegedly undetectable anabolic steroid ("The Clear") was anonymously handed to the authorities. Catlin's team found that "The Clear" defied the usual way of analysing for a steroid. First, they converted it into its trimethylsilylated derivative, in which Me_3Si replaced the hydrogens in –OH groups, to make a more volatile derivative that could be more readily studied by mass spectroscopy. The spectrum showed many unidentifiable peaks, but they did spot a peak due to the trimethylsilyl derivative of norbolethone. Reasoning that "The Clear" might have a similar carbon skeleton to this, they looked at the mass spectrum of the pure compound and were able to deduce that it was the hitherto unknown tetrahydrogestrinone (THG). Once they had prepared THG by hydrogenation of the known molecule gestrinone (Figure 5.2), they found that its properties were identical to those of "The Clear".

Scientists have explained the androgenic effects of THG by experiments measuring the interaction of steroids with the human androgen receptor. THG had the strongest affinity for the receptor, about twice as strong as dihydrotestosterone. An explanation for this is found in the structures of the adducts formed between the

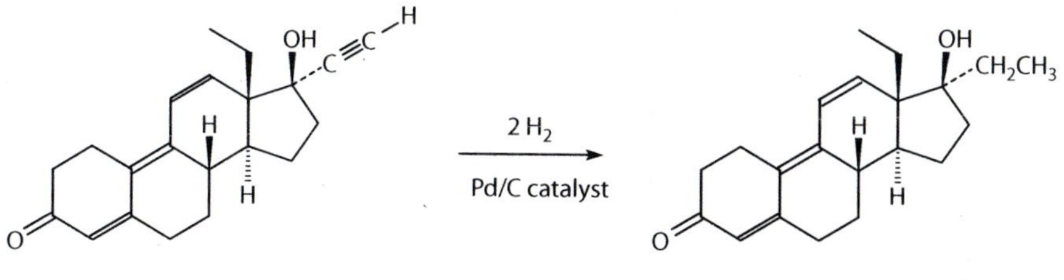

FIGURE 5.2 Preparation of THG by hydrogenation of the known molecule gestrinone.

human androgen receptor and various steroids, which show that THG makes the greatest number of van der Waals contacts with the receptor, owing to the presence of the 17α-ethyl and 18-methyl groups.

Estradiol (5.14); its oxidation product, **estrone** (5.15); and **estriol** (5.16) are the significant female sex hormones responsible for the development of the secondary characteristics, such as breasts. There is an important structural difference between them and other steroids such as testosterone, since in all three female hormones, ring A is a flat aromatic ring. Biosynthesis of the female steroids estradiol, estrone, and estriol proceeds via androstenedione; some of the androstenedione is turned into testosterone, which is then changed into estradiol by the aromatase enzyme. The remaining androstenedione is converted into estrone, which is then reduced to estradiol. One essential step in the biosynthesis is the removal of a methyl group from carbon 10, without which the aromatic ring cannot be created (Figure 5.3).

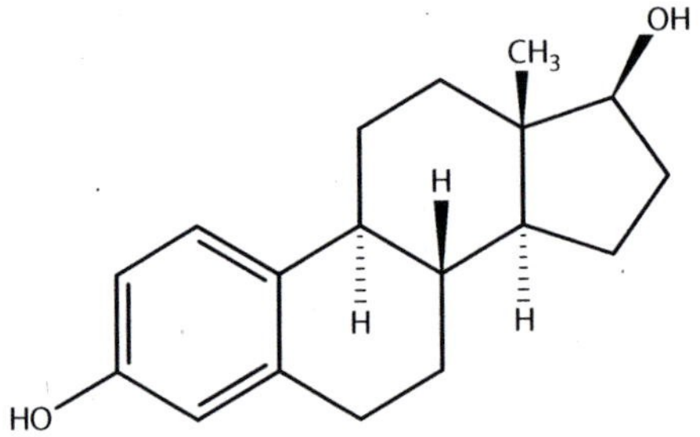

(5.14)

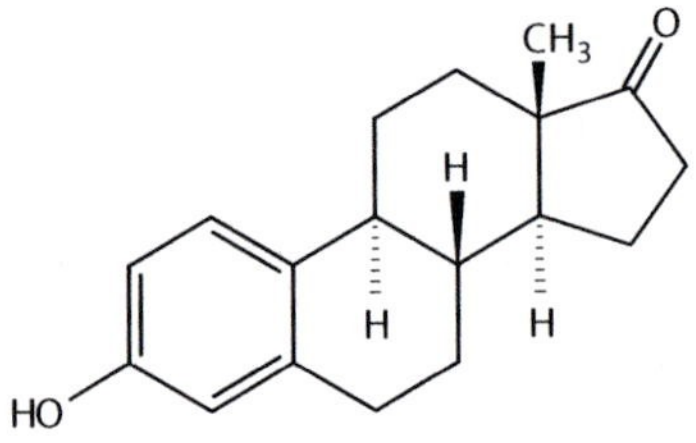

(5.15)

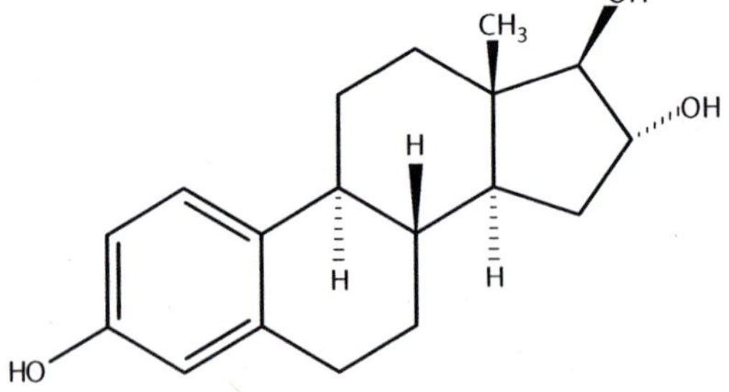

(5.16)

Estriol, formed from both estrone and estradiol in the placenta, is only made in significant amounts during pregnancy. A fourth steroid, progesterone (5.5), is also important in the female menstrual cycle. Progesterone concentrations go up sharply after ovulation so that the uterine wall is ready to have the fertilised egg

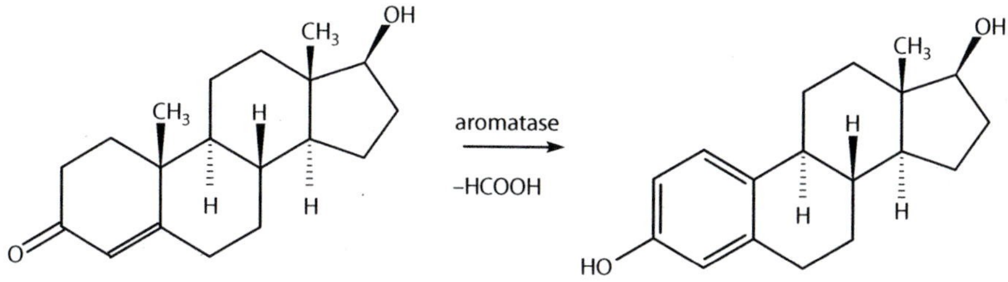

FIGURE 5.3 One essential step in the biosynthesis of female sex hormones is the removal of a methyl group from carbon 10.

implanted; if an egg is not implanted, then the extra blood cells ready for implantation are lost in menstruation. High levels of progesterone throughout pregnancy prevent further ovulation.

Steroids were first isolated and identified in the decade between 1925 and 1935 and at first were very expensive to produce. A breakthrough occurred in 1941, when Russell Marker developed a synthesis of progesterone from diosgenin, which is found in the Mexican yam (*Dioscorea macrostachya*). In 1951, Carl Djerassi discovered a synthesis of **cortisone** (5.17), also from diosgenin; cortisone is medicinally important as an immune system suppressant that reduces inflammation, pain, and swelling. Floyd Landis was taking cortisone injections for a hip complaint during the 2006 Tour de France. It can be used for short-term relief by athletes, but risks attend its long-term use. More importantly, Djerassi went on to make norethindrone (aka norethisterone or 19-nor-17α-ethynyltestosterone), the first steroid capable of acting as an oral contraceptive because it mimics the action of progesterone and prevents further ovulation.

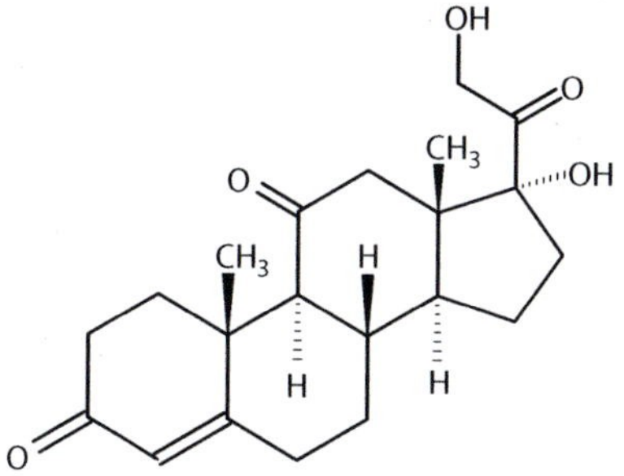

(5.17)

CONTRACEPTION

We think of contraceptive compounds as something new, but people have sought such molecules for thousands of years. A plant named silphium, of the genus *Ferula*, made the North African city of Cyrene very prosperous; its stalk was eaten for food, its flowers were used to make perfume, and a resin from it was used in medicine. The stem was said to be as long as that of giant fennel and its leaf similar to that of celery, but no one is sure because it became extinct in the first century AD owing to demand, not least for its reputed contraceptive properties. Ferujol, a compound made from another *Ferula* plant, is reported to have contraceptive activity when given to female rats.

Modern oral contraceptive pills (OCs) are a combination of a synthetic oestrogen, most usually **ethynylestradiol** (5.18), and a synthetic progestogen, such as **norethindrone** (5.19). In its structure, norethindrone resembles progesterone, and ethynylestradiol resembles estradiol, with its alkynyl group making it much more difficult to be degraded (e.g., by enzymes in the liver). The most controversial contraceptive steroid is **mifepristone** (RU-486; 5.20), an effective postcoital contraceptive; moreover, if it is taken by a pregnant woman in conjunction with misoprostol (a prostaglandin) within the first 7 weeks of pregnancy, it causes the embryo to be aborted.

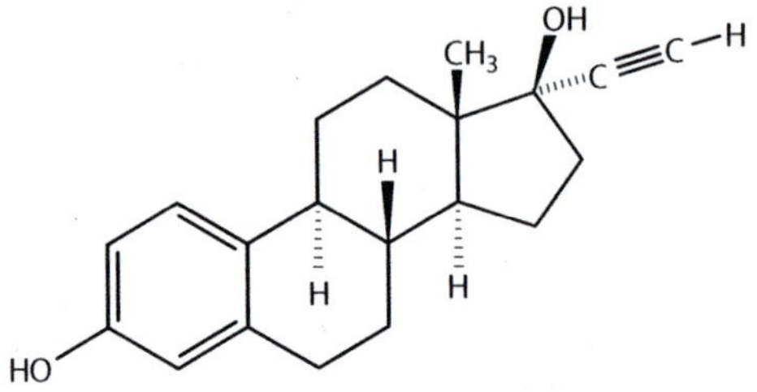

(5.18)

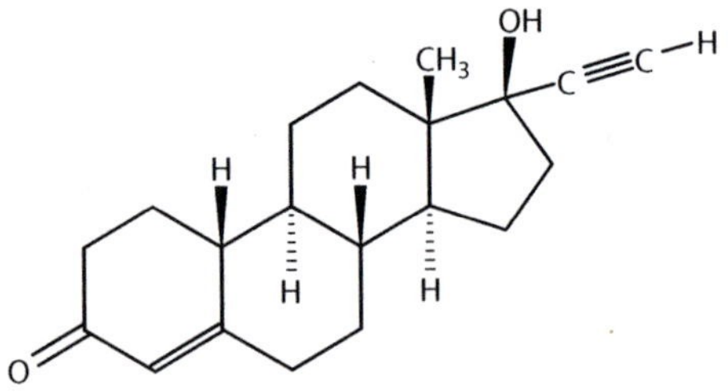

(5.19)

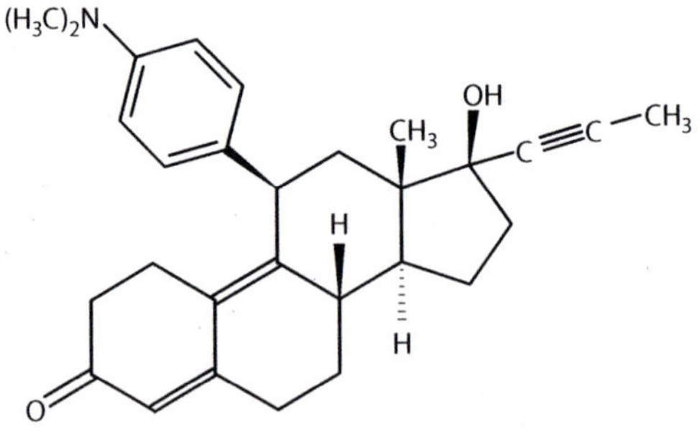

(5.20)

Mifepristone binds to progesterone receptors in the wall of the uterus; it acts as an antagonist, blocking the action of a woman's progesterone and stopping transcription and synthesis of the proteins needed to maintain pregnancy, triggering the shedding of the uterine wall. If a prostaglandin pill is taken 2 days after RU-486, the dead foetus is expelled. Pro-choice groups and most feminists have been in favour of RU-486. They say that it gives women wanting an abortion more control because surgery is not needed; they also point out that it eliminates medical complications caused by surgery and that it is less invasive. Some feminists, however, have been opposed, seeing it as trialling a relatively untested and synthetic chemical on women. The sight of their dead foetus has traumatised some women who have experienced these abortions. Pro-life groups view it as another kind of abortion; the radical feminist Camille Paglia, while applauding FDA approval of the use of RU-486 in the United States, has admitted that "abortion is murder". Others have called it "a chemical to destroy a child in the womb"; the Vatican newspaper has referred to it as "the Pill of Cain; the monster that cynically kills its brothers".

SEX MEANS MORE THAN STEROIDS

Oxytocin (5.21) is a peptide that acts as a hormone. It is made in the hypothalmus of the brain and released from the posterior pituitary gland. The American chemist Vincent de Vigneaud won the 1955 Nobel Prize for Chemistry for its synthesis; oxytocin is rather similar in structure to vasopressin, an antidiuretic hormone, differing in two amino acids. Where oxytocin is cys-tyr-ile-gln-asn-cys-pro-leu-glu, vasopressin is cys-tyr-phe-gln-asn-cys-pro-arg-glu. This affects the stabilising forces; for oxytocin, hydrogen bonding is more important, whilst in vasopressin hydrophobic bonding is more important. This in turn alters the three-dimensional structure so that two rather similar polypeptides bind to different receptor proteins.

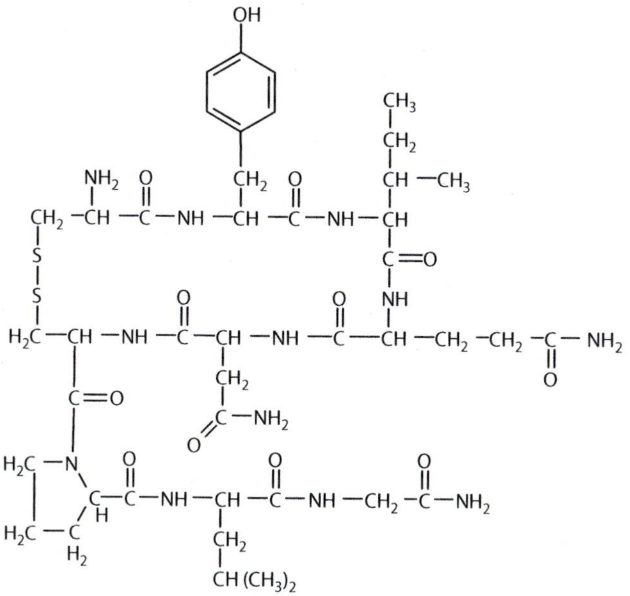

(5.21)

Oxytocin is produced when breast milk is released by women and is also responsible for the contraction of the uterus during birth. Under the name Pitocin, it is used in maternity wards to cause uterine contractions, which in turn induce labour. Because it stimulates milk production, oxytocin has also been given to young mothers who are finding breast feeding of their babies difficult. Known as the "cuddle hormone", oxytocin is thought to be responsible for maternal feelings toward babies, pair-bonding mothers to their offspring, and seems to act by reducing stress. A study reported in 2010 indicates that oxytocin affects men and women differently, making mothers "more loving" and fathers "more encouraging of exploration".

Recent research suggests that oxytocin is important in several areas of social behaviour. It is linked with recognition memory for faces, and several studies show that it improves the ability to deduce people's mental state from "social cues" in the eye region—to read people's minds, if you like. It is now being suggested that oxytocin may be useful in the treatment of shy people and of sufferers from autism. Research using investment games indicates that it is responsible for trusting behaviour between humans.

Many of the studies of the effect of oxytocin on behaviour have involved the prairie vole (*Microtus ochrogaster*); in contrast to the similar montane vole (*Microtus montanus*) and the meadow vole (*Microtus pennsylvanicus*), the prairie vole forms a lasting monogamous attachment to just one opposite-sex partner. Unlike the two other types of vole, female prairie voles have many oxytocin receptors in their brains, whilst males have many receptors for both oxytocin and vasopressin. There are a lot fewer in the montane vole. Blocking these receptors in prairie voles prevents them from forming their usual pair bonds.

At one time, it was believed that oxytocin was implicit in sexual fidelity and that prairie voles were a model for human love, but then paternity tests on prairie voles found that up to a quarter of the litters had not been fathered by the life partner. In other words, prairie voles are socially monogamous but not genetically monogamous. As Professor Ophir and his co-researchers at the University of Florida commented, "Somewhat ironically, this distinction between prairie voles and other monogamous rodents, the dissociation of social and sexual fidelity, leads us to suggest that prairie voles are even better models of human attachment than has been appreciated".

Viagra is not the first molecule credited with curing male impotence, but it is certainly the best known. For hundreds of years, people have claimed that many other substances do the same thing as Viagra. **Cantharidin** (Spanish fly; 5.22) is reported to be effective, but also very poisonous. Some people swear by **yohimbine** (5.23), but there is no medical proof. Other substances, including **GABA** (γ-aminobutyric acid; 5.24), are also said to work.

(5.22)

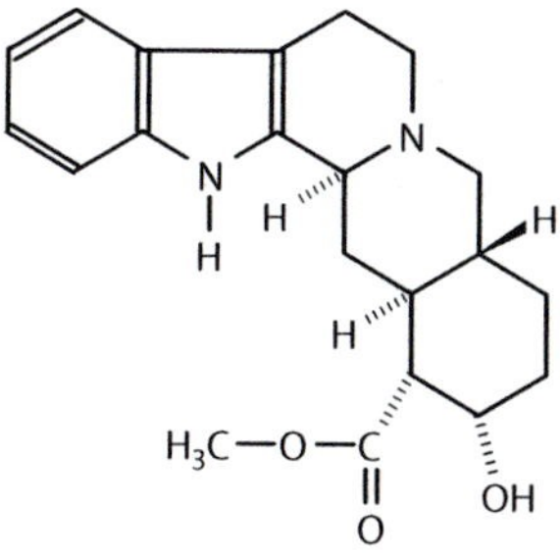

(5.23)

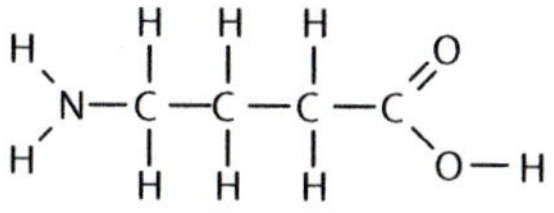

(5.24)

The discovery of Viagra can be traced back to nitroglycerine and its popular-iser, Alfred Nobel (pp. 120–121). Apart from being very explosive, nitroglycerine is also used as a treatment for angina pectoris, the chest pains due to a restricted oxygen supply to the heart muscle; other organic nitrates used as a treatment include amyl nitrate. Shortly before he died, Nobel was recommended nitroglyc-erine as treatment for his angina, but he refused it. It was not until the 1980s that it was found that the active substance responsible for the drug action of nitroglyc-erine and enabling an increased blood flow to the heart was the simple diatomic molecule NO, nitric oxide.

Chemists researching for Pfizer in the UK examined a phosphodiesterase (PDE5) enzyme present in vascular smooth muscle cells and platelets because they had under-stood that inhibiting it would dilate coronary arteries, as well as make the platelets less likely to stick to damaged arteries, thus easing the symptoms of angina. One of the molecules they made was **sildenafil** (later to be generally known as Viagra; 5.25); its trials as an angina treatment were not promising, but a serendipitous side effect was observed: Men taking the drug (as the citrate salt for improved solubility) reported an increased tendency to get erections.

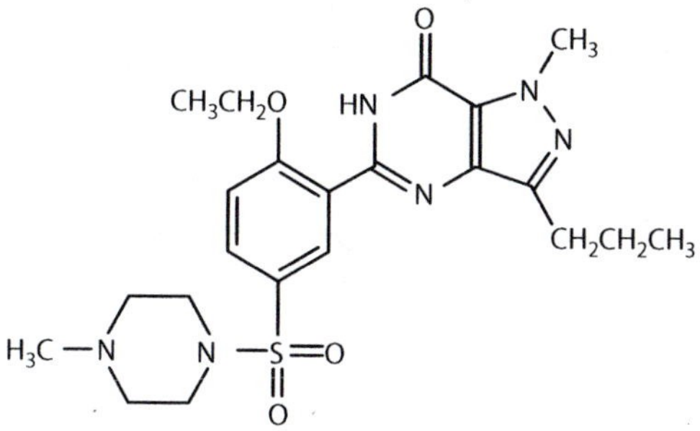

(5.25)

Male sexual excitement leads to increased levels of NO, which stimulates gua-nylate cyclase to convert guanosine triphosphate (GTP) into cyclic guanosine monophosphate (cGMP), leading to smooth muscle relaxation in the erectile tissue of the penis and thus causing an erection. PDE enzymes catalyse the hydrolysis of cGMP back into GTP and thus restrict blood flow. People with erectile dysfunction produce low levels of NO, which in turn leads to low concentrations of cGMP, readily degraded by PDE5. Sildenafil inhibits (blocks) the enzyme PDE5, so higher levels of cGMP are reached, leading to an erection. Sildenafil was approved by the FDA for use in 1998 and was the first effective oral treatment for erectile dysfunction; it thus had a head start over other treatments.

Two other molecules that work in the same way as Sildenafil received approval in 2003 and now provide stiff competition to Viagra. **Vardenafil** (Levitra, 5.26) has a very similar structure to Viagra, in contrast to **tadalafil** (Cialis; 5.27). Sildenafil and vardenafil have half-lives in the body of 4–5 hours, whereas tadalafil has a half-life of about 17 hours, leading it to be known as "the weekend pill".

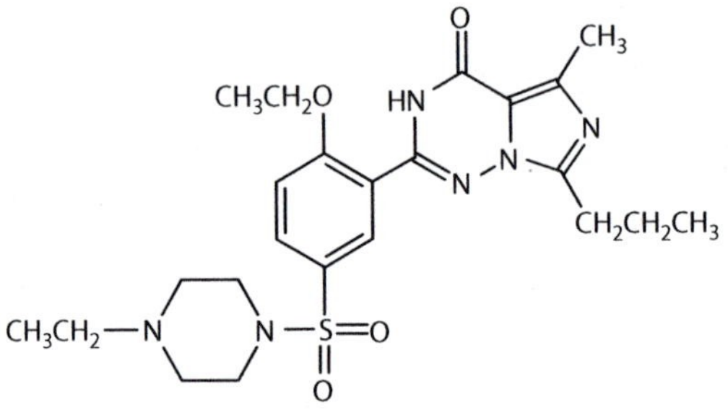

(5.26)

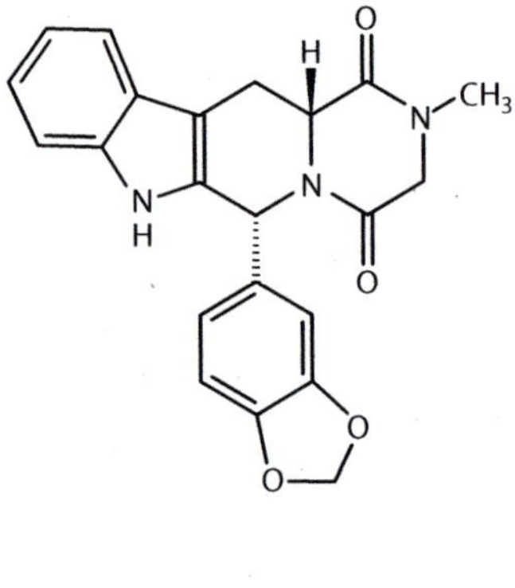

(5.27)

6 The Senses

Tempting though it may be to think no further than "smell" when it comes to associating the senses with chemistry, sight and taste are also deeply involved.

CAROTENE AND VISION

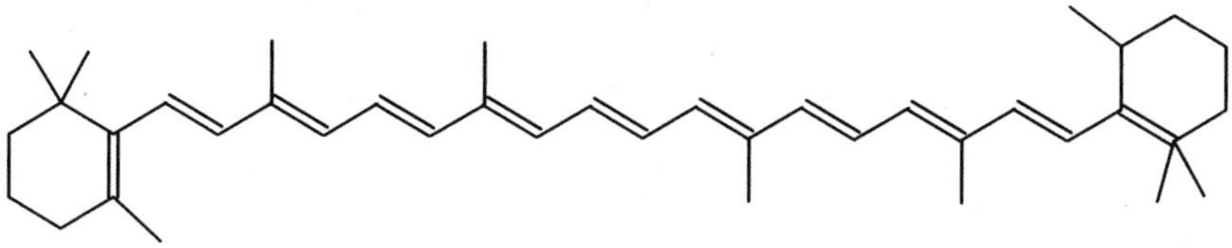

$$(6.1)$$

Beta-carotene (6.1) is a pigment found in many fruits and vegetables, including carrots, broccoli, spinach, apricots, and nectarines. It absorbs light in the blue–indigo region of the visible spectrum, reflecting red and yellow so that it is orange in colour. Its importance lies in its being a precursor to vitamin A; when eaten, carotene undergoes oxidative cleavage by enzymes, forming two molecules of retinol, vitamin A. This in turn can be oxidised to retinal.

When *cis*-retinal (more properly, 11-*cis*-retinal) joins to an opsin protein, it forms rhodopsin, the light-sensitive pigment in the retina (Figure 6.1). Rhodopsin in the rods of the eye absorbs light over much of the visible spectrum (peaking ~ 500 nm); when a photon hits a rhodopsin molecule, it breaks the π-bond between carbons 11 and 12, enabling free rotation to occur and convert the retinal into the more stable *trans*-isomer. This does not fit the binding pocket of the opsin and breaks free, producing nerve impulses transmitted down the optic nerve to the brain, which are visual signals. Free *trans*-retinal molecules are converted back enzymatically to the *cis*-isomer, ready to bind to another opsin again. Cone cells contain photoreceptor proteins that have absorption maxima at 426, 530, and 560 nm, corresponding to the blue, green, and red regions of the spectrum, that are responsible for colour vision.

Beta-carotene can be stored in the liver and body fat, so it is available for retinal synthesis. In 1940 and 1941, the Royal Air Force took delivery of Bristol Beaufighter aircraft equipped with the first airborne AI (airborne interception) radar sets and enjoyed great success in destroying German night bombers—most famously, John "Cat's-Eyes" Cunningham. To conceal the fact that the RAF had radar, the story was promoted that the success of these night-fighter crews was due to their having eaten lots of carrots for several years, giving them enhanced night vision. This story had some plausibility and also helped Lord Woolton, the British food minister, to promote the value of vegetables in the wartime diet.

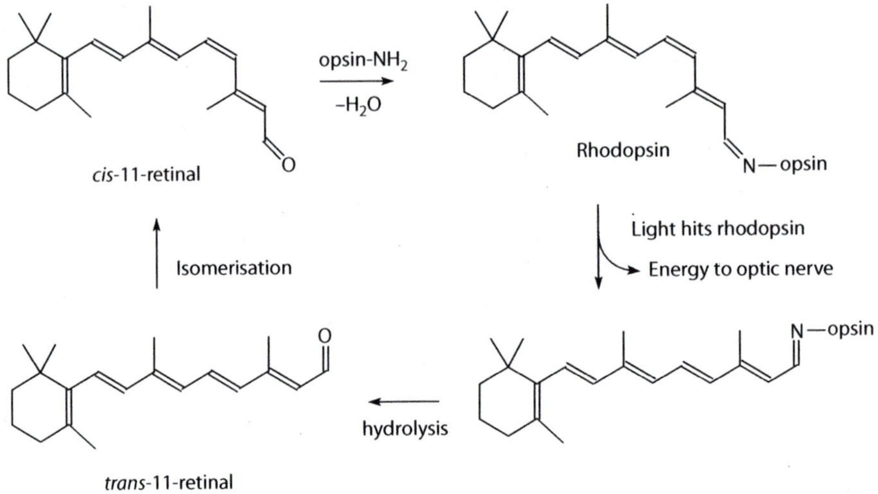

FIGURE 6.1 Rhodopsin.

In fact, excessive intake of carrots or other sources of β-carotene can lead to carotenemia, a condition in which the skin turns yellow or orange (notably the palms and soles) due to excess carotene being deposited in the skin. There are authenticated accounts of a 4-year-old child's hands and face turning bright orange and yellow after regularly drinking large quantities of a soft drink fortified with large amounts of β-carotene as a colourant. In 2007, a student spent a month with an entirely carrot-based diet, finding that at the end of the month her palms and the soles of her feet had turned orange. The livers of the polar bear, husky, walrus, and seal, as well as some other animals, are very rich in vitamin A. If humans eat them, it can lead to hypervitaminosis A, with symptoms including skin peeling and, at high doses, fatality. The Arctic explorer Xavier Mentz died in 1913 after eating the livers of his huskies.

Carotene is found in the leaves of trees, along with chlorophyll; chlorophyll absorbs red–orange and blue–violet light strongly, reflecting green. The absorbed light energy is used to drive photosynthesis, converting carbon dioxide into sugars. The high concentration of chlorophyll in leaves makes them look green, masking other colours, such as that of carotene, which is present with other carotenoids like the xanthophylls that also absorb light and have a protective role. Anthocyanins (which may have a role in protection from UV damage) are also found in leaves; these absorb light in the regions of the visible spectrum from blue to green and thus tend to be red to brown in colour; they are responsible for the colours of apples and grapes. Light exposure is needed to produce anthocyanins; this is why there are apples that are red on one side and green on the other.

The formation of abscission layers in the base of the leaves in autumn seals the leaf from the tree and reduces flow of nutrients to the leaf. Autumnal breakdown of chlorophyll means that the leaves lose their intense green colour and the yellow and orange colours of the carotene and xanthophylls can be seen. Brown colours tend to be associated with cell death and are due to polymerisation of quinones, as well as to other molecules such as tannins. It is now known that anthocyanins are produced in autumnal leaves shortly before they fall; some suggest that they protect leaves against harmful effects of light.

Anthocyanins are generally red, but their colour does depend upon the pH of the sap. Many plant materials have pH-dependent colours, as with cyanidins, which are responsible for the red colours of apples, roses, and many red berries, like cherry and cranberry. They also make violets and blueberries blue and cause blackberries and grapes to be purple. Anthocyanins are regarded as antioxidants; the commonest anthocyanin in leaves is cyanidin-3-glucoside (6.2), which has been shown to inhibit tumour promoter-induced carcinogenesis and tumour metastasis in vivo. Autumn colours have become a major tourist attraction for some US states, notably New England, Michigan, and Wisconsin.

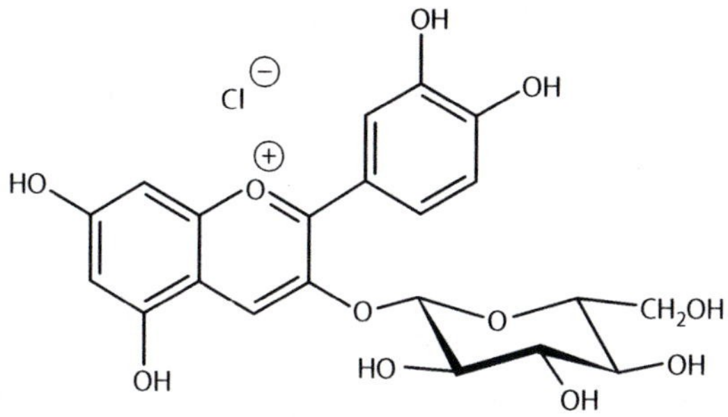

(6.2)

Below pH 3, anthocyanidins are mainly present as the red flavylium cation (2-phenylchromenylium cation). As the pH is increased (Figure 6.2), it first forms the carbinol pseudobase (colourless) and then the purple quinoidal anhydrobase, as the conjugated system is reformed. The anhydrobase becomes a purple anion above pH 7 and a yellow chalcone above pH 8. So far, the changes are reversible, but if the pH becomes too high, the next product, an α-diketone, undergoes an irreversible change to a carboxylic acid and a hydroxyaldehyde.

TASTE

Although smell contributes to flavour, involatile substances that cannot be smelled can still be detected by their taste. Just as bad smells can tell us that food is "off", so bitter tastes warn that there may be toxic molecules in the food. The surface of the tongue contains bumps (papillae), which have taste buds embedded in them that detect all five flavours: sweet, sour, bitter, salty, and umami. It was once widely believed that different zones of the tongue detected different types of taste, but this is now known not to be the case.

Sour taste is due to the (hydrated) hydrogen ion $H^+(aq)$ and salty taste mainly to Na^+ ions (some other metal ions like Li^+ and K^+ have weaker salty tastes). Both are detected by ion channels; in contrast, bitterness and sweetness are

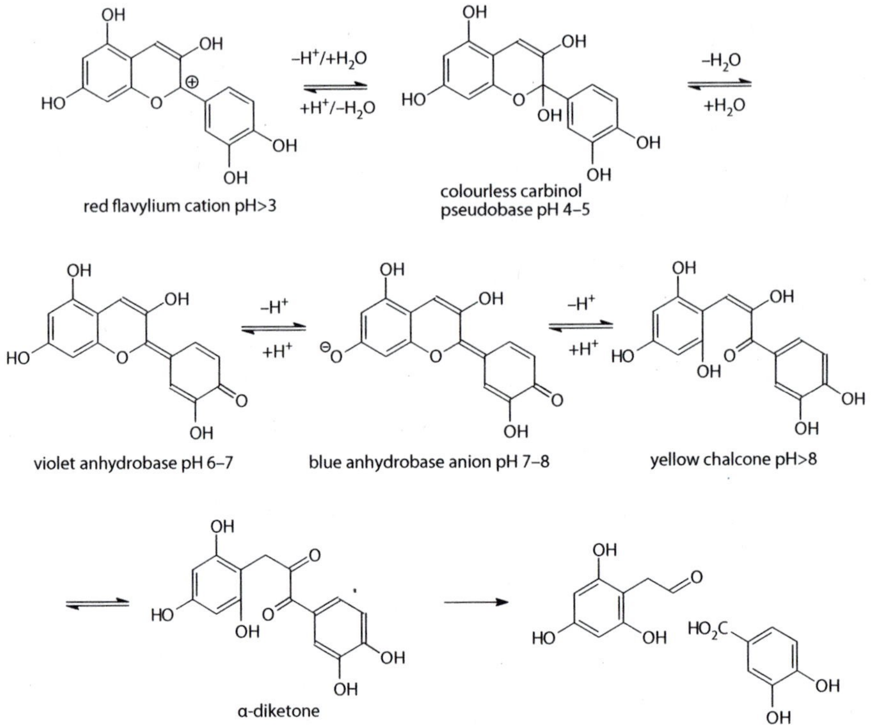

FIGURE 6.2 The colour of anthocyanins depends upon pH.

picked up by 7TM receptors (seven-transmembrane domain receptors, aka G protein-coupled receptors, GPCRs). Umami taste is caused by the glutamate ion ($^-$OOC(CH$_2$)$_2$CH(NH$_3$$^+$)COO$^-$), which is detected by a modified glutamate neurotransmitter receptor (mGluR4).

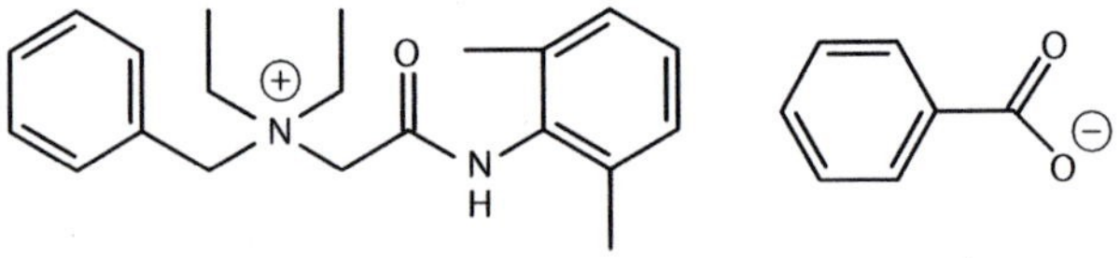

(6.3)

Bitrex (denatonium benzoate; 6.3) is said to be the bitterest compound known. It is used as an additive to products of the automotive and agrochemical industries, cleaning products and denatured alcohol, and many other substances. Its purpose is to deter people—especially young children—from drinking toxic substances like sweet-tasting antifreeze.

It has been increasingly recognised that a large range of compounds can have sweet tastes (Chapter 2): sugars like glucose and sucrose, artificial sweeteners like saccharin, amino acids, peptides, and even a few proteins (e.g., brazzein). Chemists have long attempted to find a unified theory of sweetness related to molecular structure.

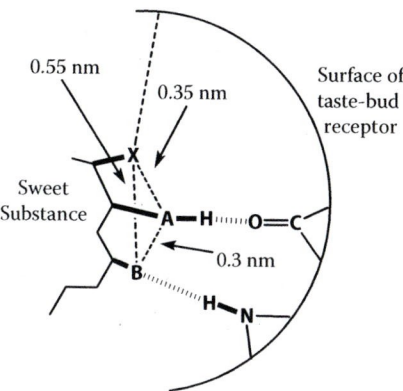

FIGURE 6.3 A "triangle of sweetness" based on the AH–B model.

The AH–B theory (1963) of Shallenberger and Acree involved interaction between a hydrogen-bond donor (AH) and an acceptor base (B): the diagram in Figure 6.3 indicates a "triangle of sweetness" based on the AH–B model. A more sophisticated model of Tinti and Nofre (1991) suggested that there are eight available interaction sites. The sweet receptor is now believed to be a dimeric G-protein-coupled receptor, though its structure has not yet been determined.

A 2005 modelling study of the heterodimeric G-protein-coupled receptor (GPCR) T1R2-T1R3, believed to be the sweet-taste receptor, suggests that there are four independent binding sites, including pockets that accommodate the small molecules as well as a secondary site on the surface that can accept sweet proteins that are too big to fit into the pockets. The receptor can bind different sweeteners independently; this explains how sweet tastes can be additive (and, possibly, in a different context, how "flavour enhancers" work). Some sweeteners, like saccharin, also taste bitter, probably due to their activation of a bitter receptor.

SMELL

Smell is the most chemical of the senses. It is commonplace to mention Proust's aunt and the madeleine. Most people will experience a particular smell evoking an event or place from the past—such is the intimate link between smell and memory.

In order for us to be able to smell a molecule, it has to be relatively small ($M_r <$ 300) *and* to be volatile (some small molecules, like glucose [$M_r = 180$], are involatile). As molecules pass up the nose, they reach the olfactory epithelium, a region of tissue with an area of around 3–4 cm^2; its sensory neurons are covered in cilia (like hairs) immersed in mucus. The odorant molecules are transported across the mucus to dock in the olfactory receptors in the cilia, which are seven transmembrane (7TM) proteins that bind to the receptors weakly and reversibly (via hydrogen bonds and van der Waals-type forces). The act of binding activates a coupled guanine nucleotide-binding protein (G-protein) and opens ion channels in the cell membrane; Na$^+$ and Ca^{2+} ions enter the cell, altering the membrane potential and sending a signal from the olfactory neuron into its glomerulus in the olfactory bulb of the brain, where signals from identical receptors converge in the same glomerulus. The pattern of signals received by the brain is what we call "the smell".

Nobel Prize-winning research by Richard Axel and Linda Buck first identified (1991) around 350 genes coding for active olfactory receptors (ORs). Buck's team then showed that a molecule activated a combination of receptors (1999), leading to a particular pattern of signals arriving at the brain. Each olfactory sensory neuron expresses just one OR gene, one olfactory protein. All neurons expressing the same receptor send their signal to the same one glomerulus in the olfactory bulb of the brain. Buck's researchers examined a range of C_4 to C_9 primary alcohols and carboxylic acids, as well as the corresponding bromocarboxylic acids and dicarboxylic acids. They established that a combinatorial strategy operates:

- A single receptor can recognize many odorant molecules.
- A single odorant is recognized by multiple receptors.
- Different odorants are recognized by distinct combinations of receptors.

It is not a case of one odorant = one receptor; each different molecule activates a different combination of ORs and thus sends a different signal to the brain. Even small changes in odorant, such as using octan-1-ol instead of octanoic acid, produce a change in the combination of ORs detecting it. Because an individual compound usually activates a number of receptors, the number of different combinations is immense; this explains why over 10,000 different smells can be distinguished by using only about 350 different receptors.

Some people cannot smell at all (general anosmia); this has recently (2011) been linked with loss-of-function mutations in the sodium channel $Na_v1.7$, the tetrodotoxin-sensitive channel (pp. 94–95) that is also necessary for pain sensation. A type of smell is sometimes associated with a particular functional group; however, whilst some esters have "fruity" smells, others have very different smells (e.g., methyl salicylate, wintergreen). Similarly, small aldehydes have a sharp, pungent odour with perhaps a hint of fruitiness; ethanal gives the sharp note to fresh-squeezed orange juice and butanal has been called "sweaty" or "cheesy", whilst octanal is "orange" and tridecanal is fatty with a fresh citrus note. An MRI study of the mouse olfactory bulb showed that each one of the C_4 to C_8 aldehydes caused different globular activity patterns corresponding to different "smells".

CHIRAL MOLECULES AND SMELL

Although still little is known about the olfactory receptors and much is based on modelling studies, it is a fact that the receptors are proteins made of amino acids. Since natural amino acids all have the (S)-configuration, the pockets that bind odorants will be chiral. It is therefore no surprise that enantiomers can sometimes have different smells. The best known pair (6.4) are S-(+)-carvone, which smells of spearmint, and R-(–)-carvone, with a caraway smell.

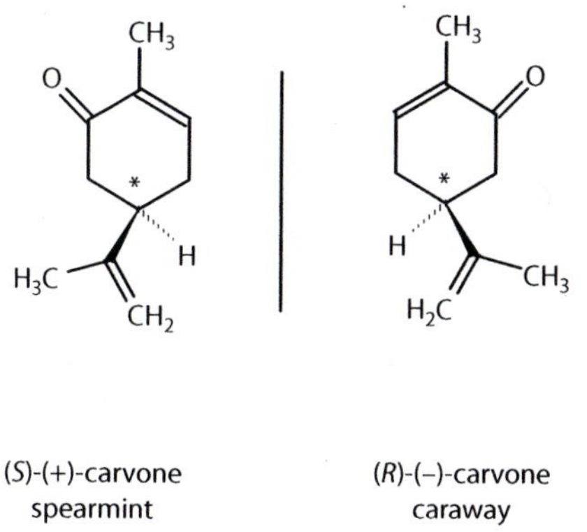

(6.4)

(S)-(+)-carvone
spearmint

(R)-(–)-carvone
caraway

As another example (6.5), (S)-(+)-1-octen-3-ol has a mouldy, grassy smell, whilst (R)-(–)-1-octen-3-ol is the molecule responsible for the smell of mushrooms.

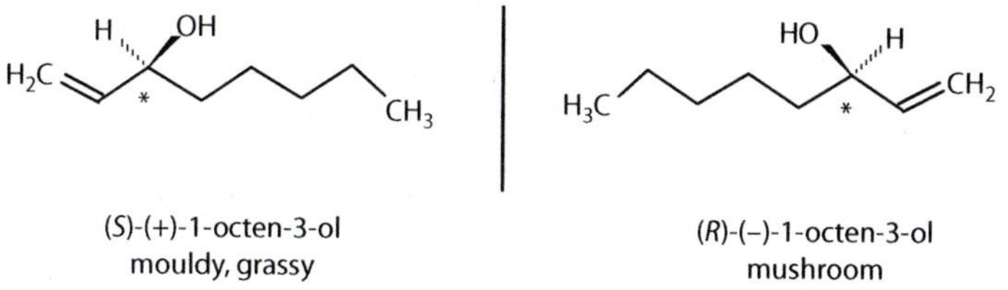

(6.5)

(S)-(+)-1-octen-3-ol
mouldy, grassy

(R)-(–)-1-octen-3-ol
mushroom

The smells of the two enantiomers of limonene are sometimes described as "orange" ((+)-limonene) and "lemon" ((–)-limonene). This is only partly correct. As obtained commercially, (+)-limonene is "orange" and (–)-limonene has a "turpentine" smell, with a citrus note. (+)-Limonene occurs widely in citrus fruits, including both lemon and orange, and much is made from orange sources commercially. Careful purification is reported to reduce its smell, and it has been suggested that the odour of commercial (+)-limonene is mainly due to impurities (probably aldehydes such as octanal). It is also possible that the smell of (–)-limonene owes something to its pine sources.

PYRAZINES AND SMELL

The pyrazines exemplify how small changes in molecules can make an amazing difference to their smell. Simple alkylpyrazines like 2,3-dimethylpyrazine (6.6) and 2,6-dimethylpyrazine (6.7) contribute to the nutty smells of roasted peanuts and baked bread. With just a small tweak to the molecule, 2-ethyl-3,5-dimethylpyrazine is one of the "chocolate" molecules that rounds out the roasted smell of coffee (6.8).

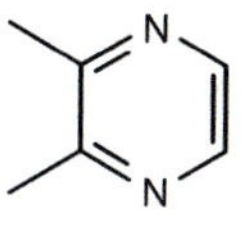

(6.6)

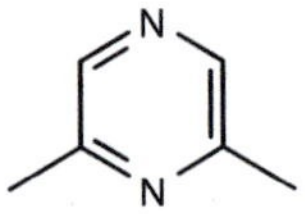

(6.7)

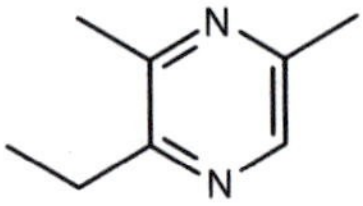

(6.8)

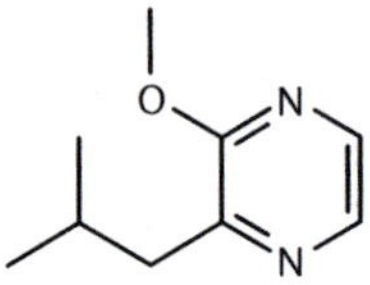

(6.9)

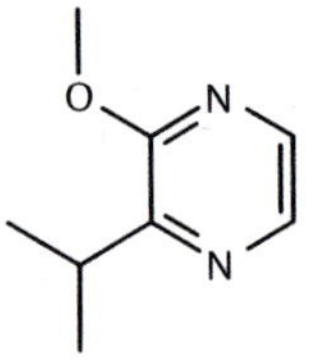

(6.10)

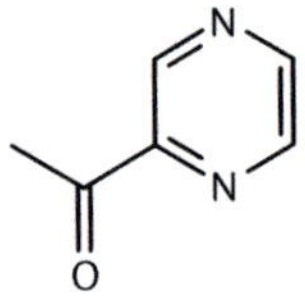

(6.11)

Substituting slightly different groups in the ring results in 2-isobutyl-3-methoxy-pyrazine (IBMP; 6.9). This gives green bell peppers a quite different smell; it is detected by the nose at incredibly low levels—as low as 0.002 ppb. It is also found particularly in sauvignon grapes and cabernet sauvignon, merlot, and sauvignon blanc wines, giving them green pepper aromas; IBMP is present at very high levels in unripe grapes, but its concentration decreases markedly upon ripening.

Another methoxypyrazine, found particularly in green peas and broad beans, is 2-isopropyl-3-methoxypyrazine (IPMP; 6.10); IPMP is a sex attractant in the seven-spot ladybird, *Coccinella septempunctata*. It is also a defence molecule responsible for the characteristic odour of the multicoloured Asian ladybird, *Harmonia axyridis*, which is found in many wine-producing areas; if the ladybird is processed along with the grapes, IPMP (and other 2-alkyl-3-methoxypyrazines) is released and can cause a "ladybug taint" to the wine, even making it undrinkable. 2-Acetylpyrazine (6.11) smells of roasted popcorn, though it is not a major popcorn odorant.

The biosynthesis of pyrazines is still being unravelled. Syntheses of molecules like trimethylpyrazine and tetramethylpyrazine in *Corynebacterium glutamicum* involves transamination and oxidation of acetoin (3-hydroxybutan-2-one) and its analogues to α-aminoketones, followed by condensation and oxidation. In the case of 2-alkyl-3-methoxypyrazines, condensation reactions between amino acids to form

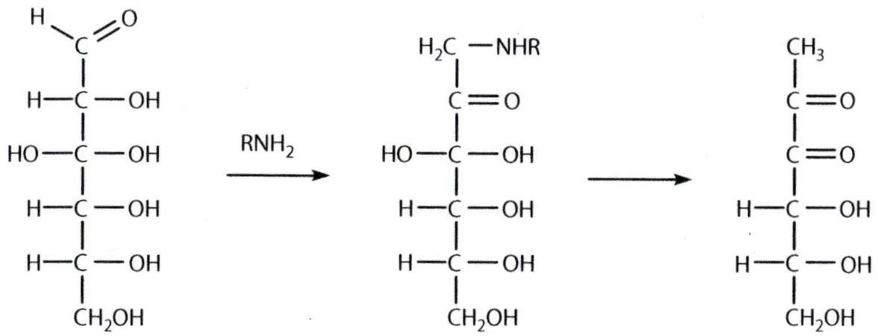

FIGURE 6.4 Formation of dicarbonyl compounds.

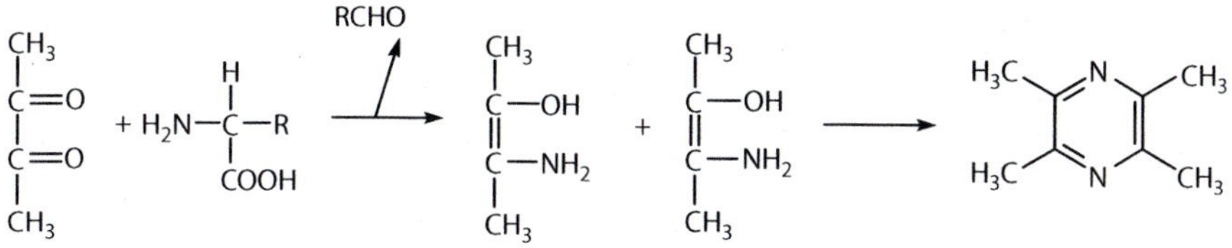

FIGURE 6.5 Alkylpyrazines.

heterocycles are believed to occur. The presence of pyrazines in roast foods like nuts, coffee, and bread results from reactions of aldoses with amino acids to form, initially, Amadori compounds whose decomposition products include dicarbonyl compounds. The latter react with amino acids in condensation reactions, ultimately to form alkylpyrazines; the overall process is known as the Maillard reaction. Many other odorant molecules are formed in these reactions (such as Strecker aldehydes RCHO); a grossly oversimplified view of the type of reaction involved is shown in Figures 6.4 and 6.5.

PHEROMONES

A pheromone is a chemical released by an animal (often but not always an insect) that affects the behaviour of other members of that species. Pheromones are used by social insects—bees and ants particularly, but also by other insects, including moths, wasps, cockroaches, and flies. The main types of pheromone are sex pheromones, alarm pheromones, trail pheromones, aggregation pheromones, recognition phero-mones, and disarming pheromones.

(6.12)

Bombykol (6.12) was the first pheromone to be identified (1959) as the sex attrac-tant emitted by the female silk moth, *Bombyx mori*. As female moths flutter their wings, bombykol molecules are spread out on the air. The male moth's antennae detect bombykol downwind of the female, so the male moves in a zigzag path up the plume of the female's pheromone in the direction of increasing concentration, until it meets its mate-to-be. The research was led by the great German chemist Adolph Butenandt (1903–1995), who had already won the 1939 Nobel Prize for his work on steroids; identifying bombykol took him 20 years' research. He extracted the pheromone from 500,000 silk moths, obtaining 6.4 mg of pure bombykol. There are two double bonds in the molecule, each capable of adopting the E- or Z-configuration. Butenandt made all four possible isomers, finding that the "natural" (10E, 12Z)-hexadeca-10,12-dien-1-ol is at least a thousand million times more effective than any of the others.

A bombykol molecule arriving at the male moth diffuses through open pores in the hairs on the antennae and docks in a pheromone binding protein (PBP); the PBP chaperones bombykol across the aqueous "sensillar lymph" that surrounds the receptor cells until it arrives at a receptor. When the bombykol binds to the receptor, it produces an electrical change in the receptor, causing a nerve impulse to be sent to the male's brain. The instant that happens, the receptor releases the bombykol, which is degraded by enzymes in the lymph fluid; this enables the moth to respond to fresh messages. The female moth makes bombykol from hexadecanoic acid (palmitic acid).

Honey bees employ a remarkable number of pheromones involved in roles such as sexual attractant and mating, alarm, togetherness in the hive, and preventing rearing of queens. The most important of all these molecules is 9-oxo-2-decenoic acid (9-ODA; 6.13), long known to be the sex attractant used by virgin queens.

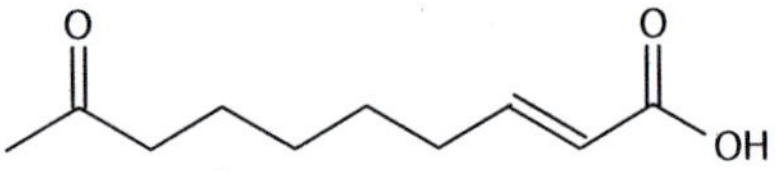

$$(6.13)$$

It is now known that queens make a mixture (Figure 6.6) known as queen mandibular pheromone (QMP), whose major components are 9-oxo-2-decenoic acid (9-ODA), two isomers of 9-hydroxy-2-decenoic acid (9-HDA), methyl-p-hydroxybenzoate (HOB), and 4-hydroxy-3-methoxyphenylethanol (HVA). QMP controls the attraction of drones for mating, swarming, pollen and nectar foraging, brood rearing, inhibiting ovary development in workers (so that the queen is the only reproductive female),

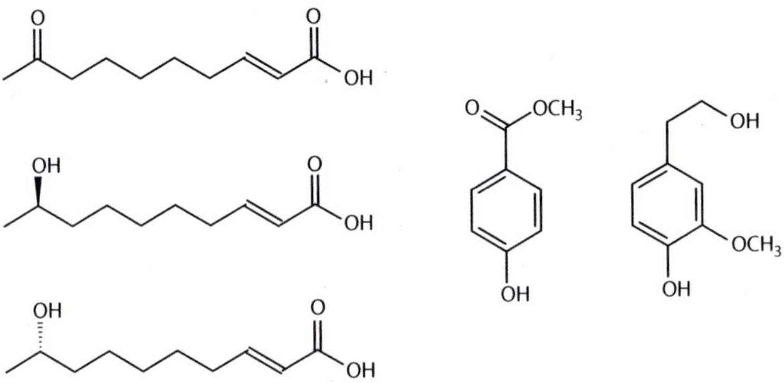

FIGURE 6.6 Queen mandibular pheromone.

and social aspects of the hive, as well as honeycomb construction. It is passed around the hive by food sharing and tells the workers that the queen is there. 9-ODA by itself is not as effective as the mixture. When QMP is mixed with four more chemicals (linolenic acid, hexadecan-1-ol, coniferyl alcohol, and methyl (Z)-octadec-9-enoate), the resulting mixture is known as queen retinue pheromone (QRP). QRP ensures that workers are attracted to form a retinue round the queen.

After they have stung another animal, bees exude an alarm pheromone to attract other bees; released from the Koschenikov gland, near the sting, it is composed of over 40 compounds including several esters and alcohols—notably, isopentyl acetate (6.14), *n*-pentyl acetate, butyl acetate, octyl acetate, hexan-1-ol, butan-1-ol, and octan-1-ol. Isopentyl acetate is particularly important. Heptan-2-one (6.15) has long been believed to be a bee alarm pheromone, though it is now thought that it is a forage marker.

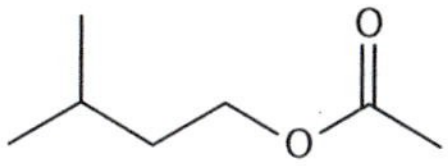

$$(6.14)$$

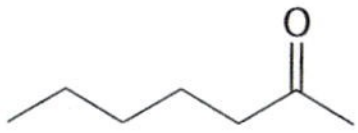

$$(6.15)$$

Workers release Nasonov pheromone, a mixture that includes citral (the *E*-isomer, geranial; 6.16), geraniol (6.17), nerolic acid (6.18), and geranic acid, in order to show other bees the way back to the hive. The workers spread it by raising their abdomen and fanning their wings; Nasonov pheromone is also used to mark flowers that are worth visiting.

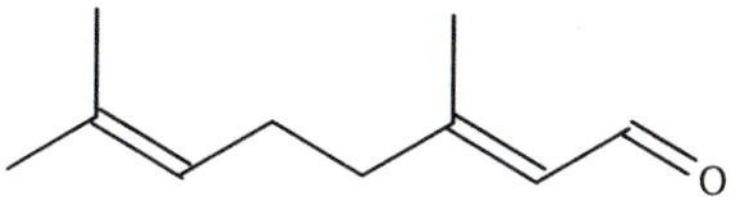

$$(6.16)$$

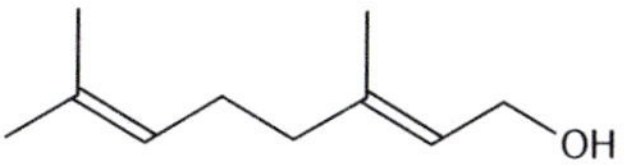

$$(6.17)$$

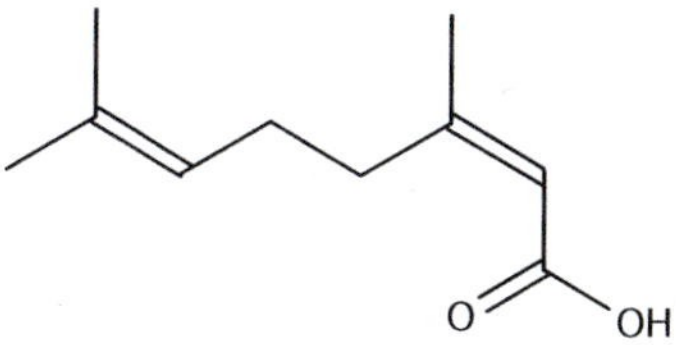

$$(6.18)$$

When a worker bee dies, its body gives off oleic acid ((9Z)-octadec-9-enoic acid; 6.19); other workers recognize the signal and move the body toward the hive entrance for disposal; this has been termed a "posthumous pheromone". This is so effective that workers will try to move a living worker "marked" with oleic acid. Similar behaviour is seen in ants.

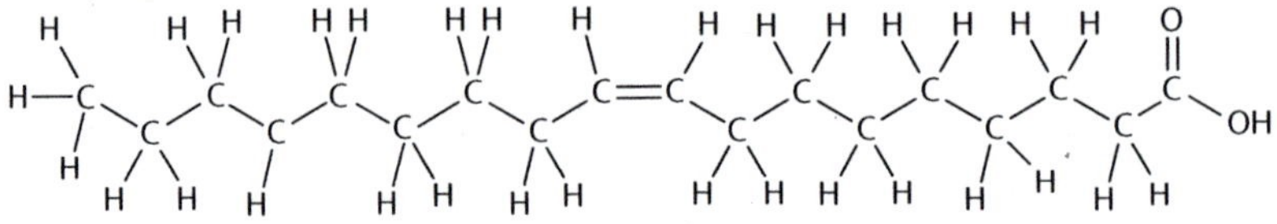

(6.19)

Like bees, ants use a wide range of pheromones to convey messages. As ants return to their nest with food, they lay down a trail pheromone for other ants to follow; though the molecules evaporate, the trail is renewed by later arrivals for as long as the food supply holds out and ants carry on coming to the site. The pheromone evaporates quickly, so ants are not confused by old trails when food is found elsewhere. Thus, an urban pest, the pharaoh's ant (*Monomorium pharaonis*), uses (3*S*,4*R*)-(+)-faranal ((+)-(3*S*,4*R*)-(6*E*,10*Z*)-3,4,7,11-tetramethyl-6,10-tridecadienal) as its trail pheromone (6.20); only this one of the four stereoisomers is active.

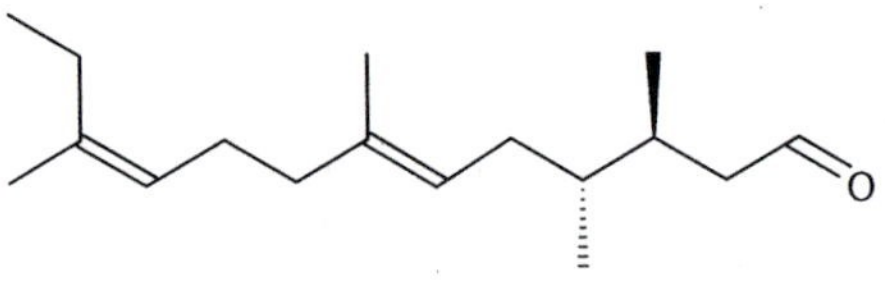

(6.20)

Similar specificity can be found in alarm pheromones. Thus, the leaf cutter ant *Atta texana* only synthesises and uses (4*S*)-4-methylheptan-3-one (6.21) as its alarm pheromone; the optical isomer (6.22) is 400 times less active.

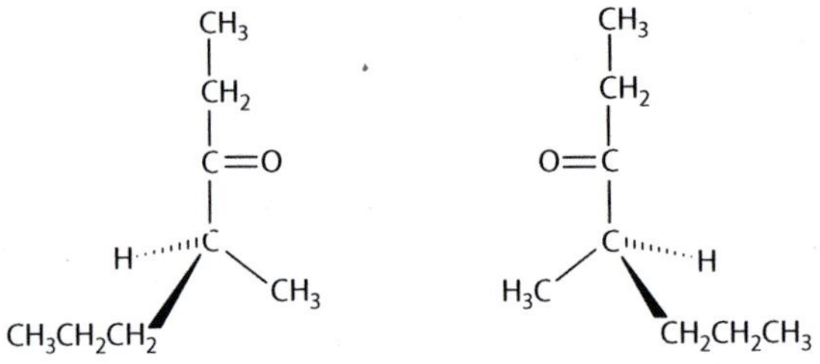

6.21   6.22

Like dead bees, dead ants emit oleic acid as a signal that makes workers remove their bodies. However, living workers of the Argentine ant, *Linepithema humile*, emit two compounds while they are alive: dolichodial and iridomyrmecin. These two molecules disappear within an hour of death as a signal to workers to remove the body.

Sometimes insects just use hydrocarbons as pheromones. The house fly (*Musca domestica*) uses the alkene (Z)-9-tricosene (6.23) as the main component of its sex pheromone.

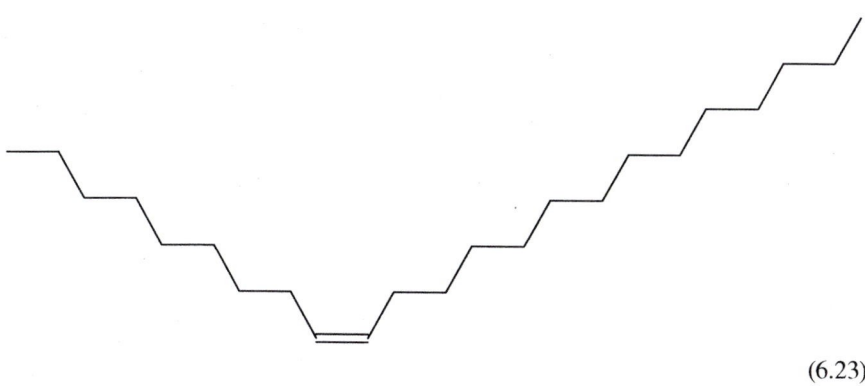

$$(6.23)$$

Insects use recognition pheromones to ensure that they are recognised by their "smell" by others of the same species in their nest; cases are known where other species can mimic this smell in order to become parasites of a nest. The paper wasp, *Polistes atrimandibularis*, is a social parasite of another *Polistes* species, *P. biglumis bimaculatus*. Normally, the *P. atrimandibularis* cuticular signature has a large number of alkenes, whereas the *P. biglumis bimaculatus* emissions are only alkanes. *P. atrimandibularis* queens start searching for a host comb in early June, about a month after the host queen, *P. biglumis bimaculatus*, has founded her nest. The alkene emissions from the parasite soon vanish, and within a month or so, the cuticular signatures of the two species are indistinguishable.

Most insects synthesise within their bodies the chemicals they use; remarkably, solitary euglossine bees collect small organic molecules from decaying wood and sap on wounded plants and faeces, as well as from plants that they pollinate, and move them to storage pouches on their tibia. These euglossine bees appear to use these molecules as short-range attractants, a kind of pheromone.

HUMAN PHEROMONES?

Scientists have argued for years over the existence of pheromones in humans. In 1971, Martha McClintock, then at the University of Chicago, claimed that women sharing a dormitory tended to synchronise their menstrual cycles; more recently, it was claimed that compounds collected from women's axillary secretions in their armpits affected the cycles. However, other groups of researchers have questioned the statistical analysis and validity of the results.

Researchers have argued over whether humans have a functioning VNO (vomer-nasal organ) that could detect pheromones, and the balance of opinion is that they do not. Despite the claims in adverts in men's magazines, the current view is that no chemical has yet been shown to act as a pheromone influencing human behaviour in rigorous peer-reviewed research.

7 Cosmetics and Perfumes

SHAMPOO

Hair looks much better when it is washed regularly and, until recently, this was done using soap. In much of the world, shampoos have now taken over, and they are much more complicated compositions. Apart from the surfactant detergent, they contain a chelating agent like Na_4EDTA to remove Ca^{2+} and Mg^{2+} (which would otherwise remove some surfactants), a pH adjuster (e.g., sodium citrate) to keep the pH slightly on the acid side of neutral to give a smooth cuticle, foam stabilisers, bactericides, and preservatives—not to mention fragrances, dyes, and other additives to improve the look for marketing purposes. It is the detergents that are the most important; they must contain a hydrophilic head and at least one hydrophobic tail. These can be anionic, cationic, or zwitterionic. Anionic detergents tend to be the main foaming and cleansing component; the most widely used types are exemplified by sodium lauryl sulfate (sodium dodecyl sulfate; 7.1), sodium stearate (sodium octadecanoate; 7.2), sodium dodecylbenzenesulfonate (7.3). Cocamidopropyl betaine (CAPB; 7.4) is an important zwitterionic surfactant, with a mild detergent action that is also more viscous than the other types of surfactant.

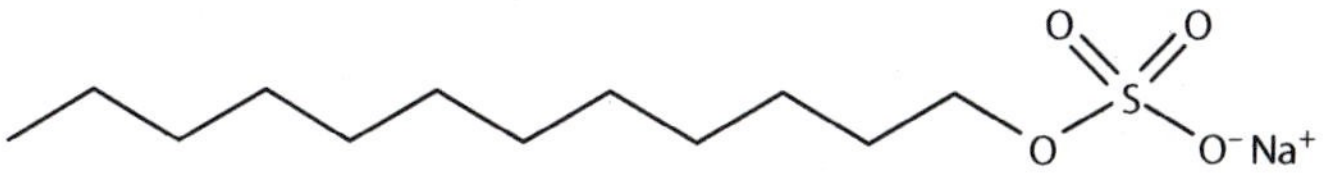

(7.1)

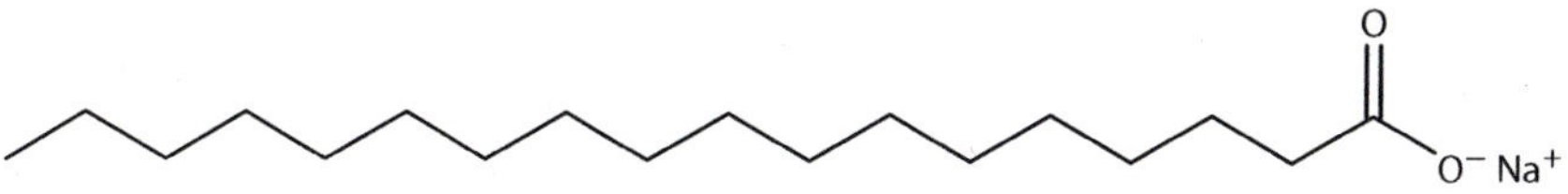

(7.2)

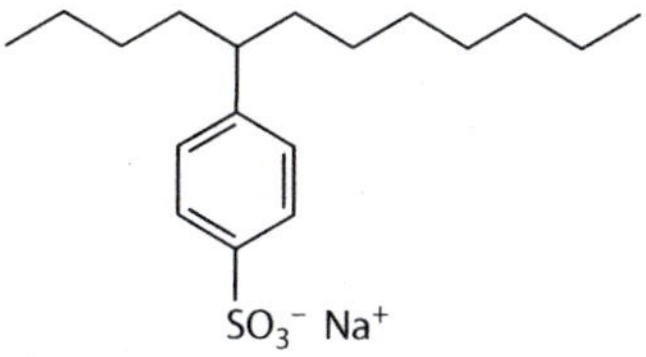

(7.3)

The detergent removes dirt and grease from hair. These are nonpolar molecules and thus mix freely with the hydrophobic, nonpolar tail—like a long-chain alkyl group—of the detergent. The hydrophilic head is water soluble and ensures that the detergent pulls the dirt and grease into the aqueous phase and off the hair as it is rinsed.

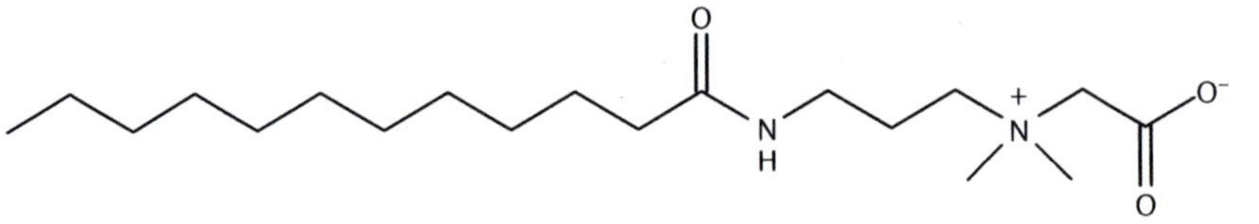

$$(7.4)$$

HAIR COLOURING AND WAVING

Hair is formed in deep pits in the epidermis; supplied with nutrients by the blood, it grows at the bottom of the follicle and cells gradually fill with keratin and harden. The keratin is made of protein α-helices twisted together to form various kinds of fibrils, which in turn associate into strands of hair. Hair is coloured by melanin, which is the same pigment found in skin. The number of melanin granules and the amount of melanin per granule determine whether hair is blond or dark. In fact, there are two melanins: Eumelanin causes shades from black to brown and phaeomelanin gives reds and yellows. Traditionally, hair was darkened using lead acetate, which reacts with sulfur-containing proteins to generate insoluble black lead sulfide, PbS, but nowadays there are concerns about the toxicity of lead compounds, so most colouring employs organic compounds.

Semipermanent rinses attach colour to the outside of hair, but for permanent dyeing, the dye must move through the cuticle and enter the cortex cells. Ammonia or hydrogen peroxide softens the cortex and allows small dye granules to get into the cuticle. Hydrogen peroxide oxidises the organic molecules to generate the colours (in the absence of dyes, hydrogen peroxide can also bleach melanin, causing lighter shades). Chemicals used in hair colouring formulations include the molecules that generate the dyes, such as 1,4-diaminobenzene (para-phenylenediamine; 7.5), and couplers (often phenols) that link with the oxidised dyestuffs to enrich the colours. These get into the cuticle, where oxidation is completed and the dyes are retained.

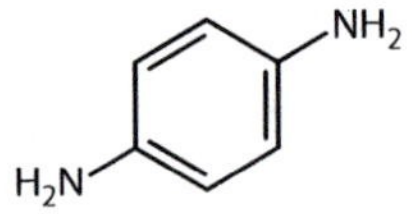

$$(7.5)$$

Permanent waving relies on the fact that the α-helices in the microfibrils contain amino acids that cross-link through the formation of disulfide bridges between cysteine groups in neighbouring helices; this confers elasticity upon the chains as well as aligning them. If these disulfide bridges are treated with a reducing agent, free cysteine–SH groups are generated as cross-links are broken; hair is then curled (e.g., with rollers), giving it a new shape, and washed before an oxidant is applied to regenerate the disulfide links, locking the new shape in place—the permanent wave (Figure 7.1).

The reducing agent often used is ammonium thioglycolate (7.6); because it is the salt of a weak acid and a weak base, it is partly hydrolysed. The solution will have a

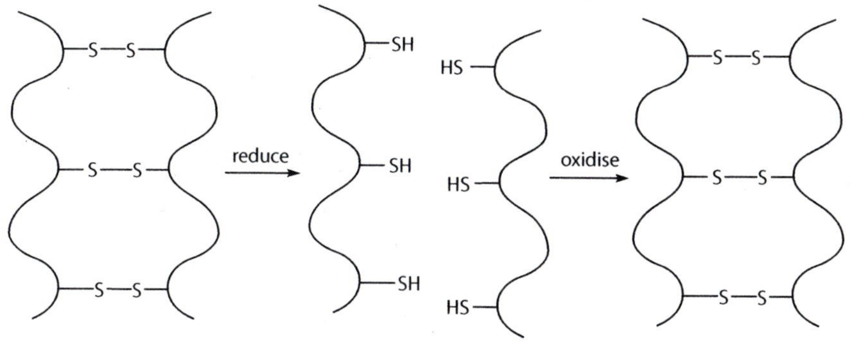

FIGURE 7.1 Breaking and reforming disulfide bridges during the formation of permanent waves.

pH of 9–9.5 and you can smell the ammonia present, which helps to make the hair permeable. The thioglycolic acid is the actual reducing agent. Dilute hydrogen peroxide is the most common oxidising agent:

$$HSCH_2COO^- + NH_4^+ \rightleftharpoons HSCH_2COOH + NH_3$$

The act of wetting hair can disrupt some of the weak hydrogen bonds between α-helices and enable the chains to slip past each other; holding the wet hair in shape until it dries out can generate temporary waves.

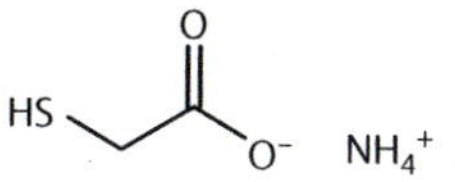

$$(7.6)$$

NAIL VARNISH

In essence, nail varnish is a simple enough substance, consisting of a solution of varnish in a volatile solvent; when the solvent evaporates, a hard protective coating is left behind. But there is a lot more to it than that. The protective coating is nitrocellulose, which was first developed as an explosive material in guncotton and used to make cordite; its use as a base for cine films was a fire hazard, but in thin films of varnish it is quite harmless. The solvents used to dissolve the nitrocellulose are commonly esters like ethyl acetate or butyl acetate, which explains why nail varnish smells nice (and also evaporates quickly). On top of that, there are pigments, if desired, so that the nails are coloured; adhesive polymers like toluenesulfonamide (tosylamide)/formaldehyde resin to help the nitrocellulose adhere to the nail; and plasticisers (butyl stearate) to keep the coating flexible.

Many nail varnishes have a sparkly appearance; this is done by adding insoluble solid particles such as mica. This in turn requires the addition to the varnish of a thickening agent, such as propan-2-ol or stearalkonium hectorite, to make the varnish more viscous so that the glitter particles do not sink and settle out. Bitrex ("the bitterest substance known", p. 58) may also be added to deter young children from consuming it. Nail varnish remover is simply a solvent, like propanone (acetone) or ethyl acetate, which dissolves the varnish. It may also contain some glycerol and lanolin to counteract any drying out that has taken place in the nail and the finger round it.

Of course, all this is not enough for some people, who want nail extensions. These are made of polymers, a polyacrylate or ABS—poly(acrylonitrile-butadiene-styrene)—copolymer. Preformed nails can be fixed to the natural nail with super-glue or other adhesives. To give the new "nail" a uniform appearance, a mixture of methacrylate monomer and some powdered polymer is painted onto the nail, on top of a fitted preformed nail or onto natural nail and fitted tip. The monomer rapidly polymerises and the mixture hardens, ready for shaping. The use of methylmethacrylate has been phased out, due to toxicity concerns, and the less toxic ethylmethacrylate is now more usual.

FACE POWDER

For many hundreds of years, there was a fashion for white cosmetic foundations; right up to the eighteenth century, this was achieved using compounds like lead acetate and lead carbonate, which proved toxic to wearers. Nowadays face powder is much less toxic; around half of it is made of talc, $Mg_3Si_4O_{10}(OH)_2$. Other insoluble substances added to it include calcium carbonate, magnesium carbonate, kaolin, zinc oxide, and zinc stearate, as well as small amounts of coloured pigments (e.g., yellow ochre, sienna, calcium lake). A sunscreen may be included. Face powder is intended to be a foundation that can be reapplied as required to make the skin less shiny and make it look smooth; the most effective colours, like beige or tan, match the skin tone. It comes as either a loose powder or as a compact material, the latter requiring a binding agent. Talc is also the main ingredient of talcum powder, also including materials like perfumes and zinc stearate, which is thought to be a water repellent and to contribute adhesive power.

LIPSTICK

It is the one thing that a girl does not want to be without. Women have been using it as far back as the time of Cleopatra, though it has only become accepted in the last century. Lipsticks are made from a mixture of ingredients, generally castor oil (up to two-thirds of the total); waxes, mainly beeswax and carnauba wax, which give it its solid form; lanolin and other emollients to make it smoother; pigments; and perfumes. Obviously, the texture can be changed by varying the quantities of the wax and oil; thus, using silicones gives a lipstick that adheres to the lips longer. Fragrances used include fruit (e.g., orange), vanilla, and floral scents like rose.

To manufacture lipstick, the solid ingredients are ground to a very fine paste, heated with the waxes, and the oil and lanolin added; when the mixture has reached homogeneity, it is poured into moulds and chilled. The cold stick is removed from the mould and given a half-second's lick with a flame to give it a shiny and glossy surface. After all that, the lipstick case may cost more than what is inside.

Red is the most popular lipstick colour, originating in carmine from insects (*Kermes vermilio*) in the Old World and cochineal (*Dactylopius coccus*) in the New World. Other animals, including nonhuman primates and birds, use red as a symbol of physiological state. Recent research indicates that red skin colour in humans influences perceived health, and that females wearing red provoke more attraction and romantic behaviour in men. Pink lipstick is obtained by dilution with some white TiO_2, and other colours including white and black enjoy periodic vogues.

MASCARA AND EYE SHADOW

Giving eyelashes and the surrounds of the eye a dark appearance goes back thousands of years to the use of kohl by the ancient Egyptians; they used insoluble black substances like soot or charcoal, as well as some minerals. By the late nineteenth century, concotions designed to make longer eyelashes included mixtures of soot and petroleum jelly, and soap and black dyes. Nowadays typical mascara is made from ingredients similar to lipstick: waxes (beeswax, carnauba wax, paraffin wax), oils like castor oil and lanolin emollient, and pigments. The main pigments used are carbon black, iron oxide (brown), ultramarine (blue), and carmine (red). Extender materials to create longer lashes are very small fibres of nylon or rayon that adhere to the eyelashes as the solvent evaporates.

SUNSCREEN AND TANNING

"Beauty is only skin deep", and people can go to great lengths to deepen it by tanning. Visible light is not dangerous to the skin, but the more energetic ultraviolet radiation (UV) is; it has the energy to break chemical bonds and, in particular, to cause damage to DNA. The ozone layer removes all UV-C (200–290 nm) and some UV-B (~290–320 nm); most UV incident on the skin is UV-A (320–400 nm), with some UV-B. UV-B does not penetrate beyond the top layers of the epidermis, whilst a high percentage of UV-A goes deep into the dermis of Caucasians, with the potential of long-term damage; thus, more people wear sunscreens and sunblocks.

The amount of UV received depends upon several variables, such as latitude, altitude, season, and time of day, as well as the amount of cloud. The best known inorganic materials are sunblocks (just think of Australian cricketers like Shane Warne and Mike Hussey); these are based on TiO_2 and ZnO (the best). They work by reflecting, scattering, and absorbing UV light; TiO_2 works particularly well for UV-B whilst ZnO has a better overall response to UV-A and UV-B. Thanks to nanotechnology, the particles used nowadays in sunblocks are smaller, so wearers do not have to look as if they have been coated in white paint. Sunscreens include some sunblock material but also incorporate UV-absorbing organic molecules, which work by converting ultraviolet energy into heat. Most of these are absorbers of UV-B (such as cinnamates and para-aminobenzoic acid) but, increasingly, materials are being sought that are good UV-A absorbers.

Absorption by octyl methoxycinnamate (OMC; 7.7) peaks at 310 nm, whilst 4-methylbenzylidene camphor (4-MBC; 7.8) absorbs UV-B well, peaking at 299 nm. Butyl methoxydibenzoylmethane (avobenzone; 7.9) is one of the most popular UV-A absorbers. OMC and 4-MBC are both reported to be endocrine disrupters, though this is disputed.

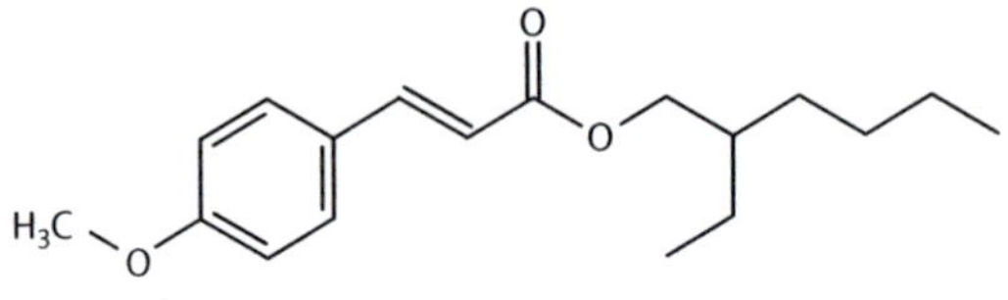

(7.7)

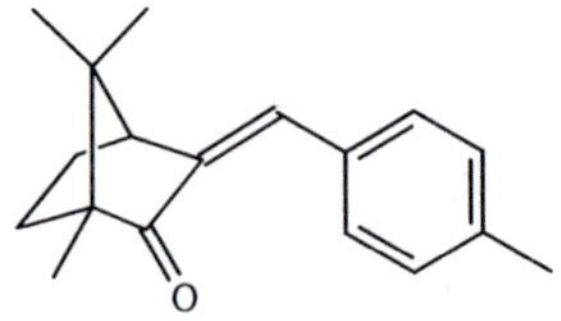

(7.8)

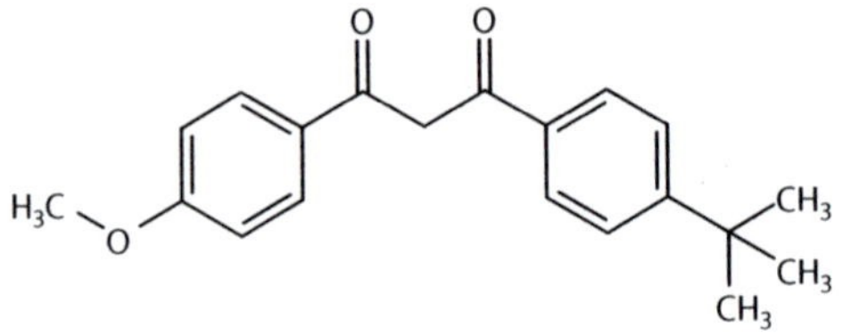

(7.9)

All sunscreens are assigned a protection factor, which is the ratio of the times taken for the skin to go red (erythema) either with or without the sunscreen. Thus, a protection factor of 10 means that a person wearing the sunscreen can spend 10 times longer in the sun and get the same reddening that he or she would without the sunscreen. In tanning, melanin near the skin is oxidised by UV radiation to a brown form, giving an initial tan; later, new melanin is produced deeper down and is moved upward, so the tan spreads. Self-tanners all use DHA (dihydroxyacetone, aka 1,3-dihydroxypropan-2-one, $HOCH_2C(=O)CH_2OH$; 7.10); this can be made by careful oxidation of glycerol and is naturally involved in glycolysis. It reacts with the terminal amine groups in molecules of protein in the skin (notably arginine) in a Maillard-type reaction. This generates melanoidins, which are molecules with similar colours to melanin. These colours do not afford protection against ultraviolet light.

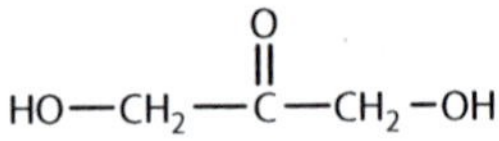

$$HO-CH_2-\overset{\overset{\text{O}}{\|}}{C}-CH_2-OH$$

(7.10)

TOOTHPASTE

Soon after they have been cleaned, teeth acquire a coating of saliva; in turn, this coating picks up bacteria that generate plaque as their living environment. Modern sugar-rich diets assist fermentation of sugars by these bacteria, especially *Streptococcus mutans*. This generates the acids—particularly lactic acid but also ethanoic, methanoic, and propanoic acids—that attack apatite ($Ca_5(PO_4)_3(OH)$) in the enamel of the teeth, leading to tooth decay. The normal pH of the mouth is around 6.8, but demineralising starts when the pH drops below 5.5, making the enamel more porous. Brushing mechanically can remove some debris from teeth, but we use toothpaste to make a better job of removing food residues, plaque, and tartar (calcified plaque).

Tetrasodium pyrophosphate, $Na_4P_2O_7$, may be present in toothpaste to remove Ca^{2+} and Mg^{2+} ions to help control tartar, but complete tartar removal may require mechanical cleaning by a professional dentist. A scouring agent like calcium phosphate ($CaHPO_4$) rubs away stains and cuts through plaque, whilst a detergent (e.g., sodium lauryl sulfate) helps disperse the plaque and "debris". The toothpaste also contains necessary substances, which include antibacterial agents like triclosan, 5-chloro-2-(2,4-dichlorophenoxy)phenol (7.11), and sodium benzoate; fluorides (NaF, Na_2PO_3F, SnF_2) to strengthen teeth; artificial sweeteners (sorbitol or saccharin) that will not promote bacterial growth, and flavourings like peppermint. Glycerol is added to toothpaste as a sweetener and also as a humectant to stop the toothpaste drying out when you do not put the cap back. If teeth still are not white enough after brushing, some people use whitening agents that are gels based on hydrogen peroxide, often as the 1:1 adduct with urea, carbamide peroxide; as a solid, this is easier to handle than liquid H_2O_2.

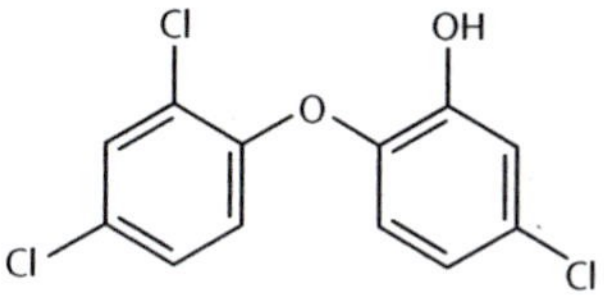

(7.11)

BAD BREATH AND MOUTHWASH

Bad breath is mainly created by bacterial breakdown of protein in the mouth, especially around the back of the tongue. It is particularly caused by molecules like H_2S, CH_3SH, CH_3SCH_3, and CH_3SSCH_3, themselves formed by decomposition of the sulfur-containing amino acids cysteine and methionine. Other compounds that may be present include the amines putrescine ($H_2N(CH_2)_4NH_2$) and cadaverine ($H_2N(CH_2)_5NH_2$) formed in the body from amino acids like lysine. Mouthwashes are intended to freshen breath and also to reduce plaque. Their antiseptic actions are due to a variety of compounds. Listerine uses thymol, eucalyptol and eugenol; these all contain phenolic groups capable of penetrating the bacterial cell wall; it also contains some ethanol as a solvent. Another compound widely used in other mouthwashes is cetylpyridinium chloride (1-hexadecylpyridinium chloride; 7.12), which also kills microorganisms, notably bacteria.

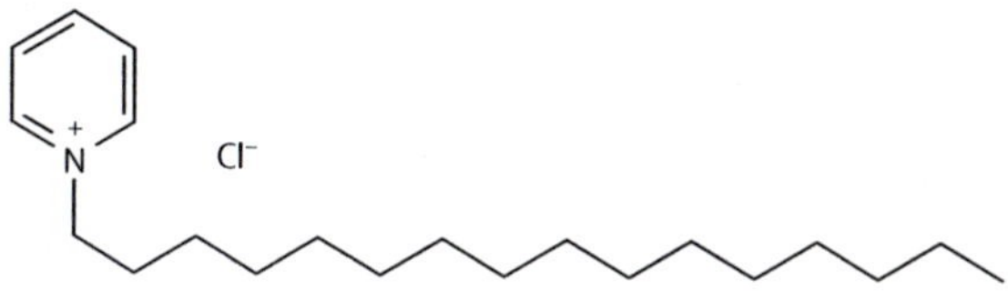

(7.12)

Scientists have studied other chemicals that might be used in chewing gum or mints to kill bacteria and freshen breath; an extract from magnolia bark has been found (2007) to be very effective due to the phenolics magnolol (7.13) and honokiol (7.14).

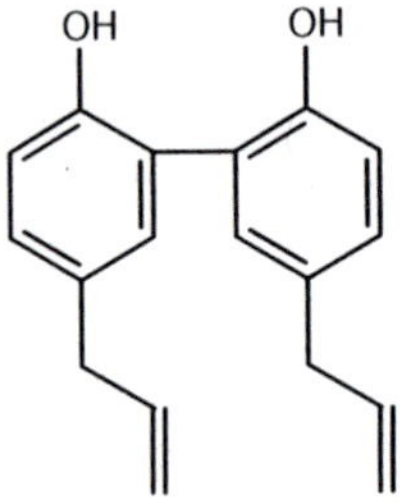

(7.13)

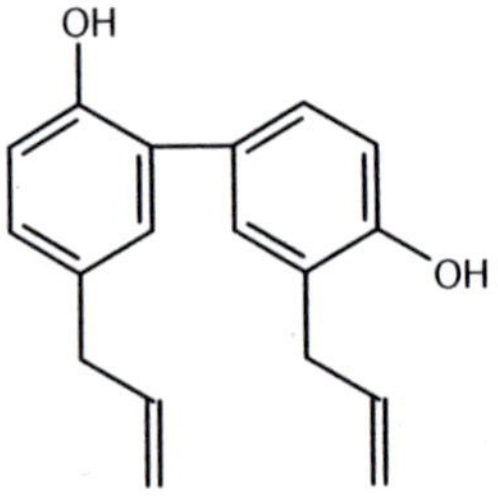

(7.14)

Around 1 person in 10,000 suffers from "fish breath syndrome" due to a genetic disorder (trimethylaminuria). Choline ($Me_3N^+CH_2CH_2OH$), found in many dietary substances like eggs, liver, legumes, and some grains, is broken down by bacteria in the gut, forming trimethylamine, $(CH_3)_3N$. An enzyme coded for by the FMO3 gene oxidises this smelly molecule to odourless $(CH_3)_3NO$, but a few people have an impaired ability to do this, leading to a very unpleasant body odour. Various treatments can reduce this embarrassing problem, such as dietary control, the use of certain antibiotics, and also using dietary supplements like activated charcoal.

BODY ODOURS AND DEODORANTS

When you have just had a bath, brushed your teeth, and maybe gargled with a mouth-wash, you smell pretty good, and not just to yourself. Soon enough, though, smelly molecules start to form on the skin. Humans have two sorts of sweat glands. The eccrine glands cover much of the body and produce a very dilute and slightly acidic aqueous solution of mainly sodium chloride and potassium chloride with lactic acid and urea. On the other hand, the apocrine glands, mainly located under the arms and groin, produce an odourless organic oil containing components that are broken down by bacteria, forming carboxylic acids (normal and branched, saturated and unsaturated), typically with 6–11 carbon atoms. Molecules like butanoic acid and 2-methylbutanoic acid have been described as having "sweaty" smells, but the key acid responsible for axillary odour is (E)-3-methyl-2-hexenoic acid (7.15). Another dominant odourant is the related (–)-(S)-3-hydroxy-3-methylhexanoic acid (7.16).

These acids are released from the eccrine glands as glutamine conjugates, which are broken down by corynebacteria.

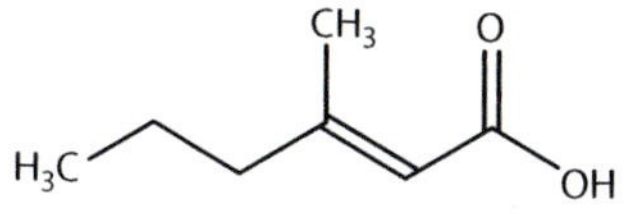

(7.15)

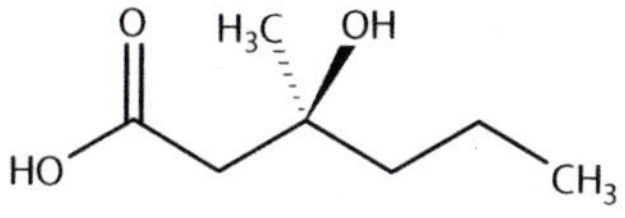

(7.16)

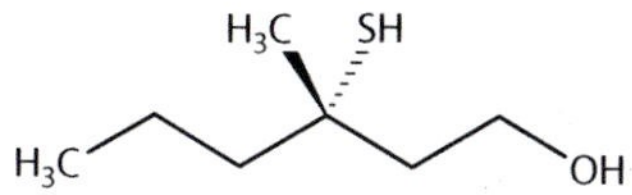

(7.17)

Another substance responsible for sweaty smell is a sulfanyl alcohol, (S)-3-methyl-3-sulfanylhexan-1-ol (7.17), described as having a sweaty and onion-like odour. It is present in sweat as a conjugate with the dipeptide Cys–Gly, from which the sulfanyl alcohol is again released by the corynebacterium. The smelly steroids 5α-androst-16-en-3-one and 5α-androst-16-en-3α-ol have also been implicated.

The first deodorant to be marketed (Mum, 1888) was a cream containing zinc oxide; because it was basic, the zinc oxide neutralised acids and was mildly anti-bacterial into the bargain. Nowadays combined deodorants and antiperspirants are most often used because they remove odour and also prevent wetness. They tend to use the aluminium chlorohydrates $Al_2(OH)_5Cl$ and $Al_2(OH)_4Cl_2$ (which are more basic than $AlCl_3,6H_2O$, so less likely to attack clothing), and some $ZrOCl_2$ is usually added. Upon hydrolysis, they form gels that block skin pores and prevent sweating. They can be applied in solutions, gels, or emulsions; antibacterial action is usually provided by triclosan, along with ethanol solvent.

PERFUMES

Perfumes have been around for thousands of years. As well as burning herbs and incense, the ancient Egyptians used various perfumed materials, as did the Greeks and Romans. Along with spices, the returning Crusaders brought perfumes to medi-aeval Europe. By the seventeenth century, a French perfume industry was well estab-lished at Grasse, in Provence, growing plants like rose, jasmine, and lavender. Few plants are grown there nowadays because of cost, but perfumes are still made using imported flowers and spices.

Materials can be extracted from plants and herbs in three ways: by pressure (expression), by steam distillation, or by solvent extraction using an organic solvent. A modern version of the latter uses supercritical CO_2 because its volatility means that residual solvent is not a problem. Until the late nineteenth century, perfumes were based on available materials that could be extracted from flowers and spices, but the growth of synthetic organic chemistry meant the availability of new compounds with new odours. Today, virtually all perfumes involve some man-made (or woman-made) molecules. In the mid-nineteenth century, perfumes were just worn by women—and rich ones at that. Nowadays, however, perfumes are available to a wide range of consumers, male and female, and also are used in all sorts of domestic materials, like soaps, detergents, and cleaning materials (even bleach).

The first use of synthetic materials was by Paul Parquet, who used coumarin in creating the fern-scented *Fougère Royale* for Houbigant in 1882; in 1889 Aimé Guerlain used vanillin to make *Jicky*, a perfume that can still be enjoyed today. Whilst *Fougère Royale* was rather conventional, *Jicky* combined a whole range of different notes and moved perfumes beyond mere flowery odours. Its initial impact was due to lavender, lemon, and bergamot, followed by a middle featuring jasmine, vetiver, patchouli, and rose, supported by a warm base of amber, civet, coumarin, and, above all, vanilla. It laid down the formula followed by perfumes to this day: a top note (head note)—the initial impact of a fragrance, which appears within the first few minutes—leading to the middle note (heart note)—the key fragrances and personality of the perfume that lasts for several hours. At the end, after a few hours, the middle note evaporates and is replaced by the bottom note (base note, end note, or drydown), which is the residual smell, the heavy molecules.

Typically, the top note involves aldehydes (citrus or lavender); the middle note is floral, with rose, jasmin, lilac, or lily of the valley, whilst the bottom note is musk or amber. Perfumes are a triumph of the chemist, bearing in mind that only around 3% of the shop price is the cost of the chemicals, the rest being bottling and packaging, advertising, and profit. In the following sections, four perfume-associated materials are discussed: rose, musk, vanillin, and aldehydes.

ROSE

Roses use small, fragrant molecules, synthesized in the petals, to attract pollinators. These compounds are formed by enzymatic oxidation of the carotenoid pigments, which give the petals their colour. Huge numbers of rose petals are used to make rose oil, by steam distillation, solvent extraction, or extraction using supercritical carbon dioxide; it is reckoned that you need around 1,400 flowers to get 1 g of the rose oil. Most of the rose oil is made from *Rosa damascena* (Bulgaria, Turkey) or from *Rosa centifolia* (France, Morocco). Though the composition of the scent can vary, analysis of Bulgarian rose oil obtained from *Rosa damascena* showed that over 60% of it was made of just four molecules: geraniol (7.18: 14%), (–)-citronellol (7.19: 38%), nerol, the Z-isomer of geraniol (7.20; 7%), and 2-phenylethanol (7.21; 2.8%).

(7.18)

(7.19)

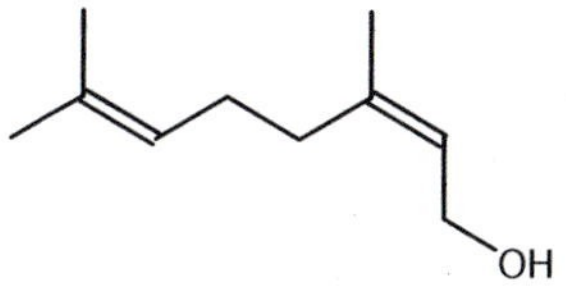

(7.20)

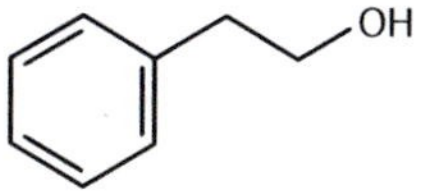

(7.21)

However, these molecules are not very important to the overall rose smell. They have rather high odour thresholds (the level at which they can be detected). The importance of the contribution of an odorant to the smell of a perfume can be assessed in terms of odour units, defined as

$$\text{odour units} = \text{concentration of molecule/odour threshold}$$

Therefore, the greater the value of odour units is, the more important is that molecule's contribution (see Table 7.1).

It was not until the 1960s and 1970s that the key odour-impact molecules essential to a rose fragrance were identified. Though molecules like β-damascenone (0.14%), β-ionone (0.03%), and damascone are present in extremely low abundance, individual molecules have very strong smells (low odour thresholds) and thus they contribute most to rose aroma.

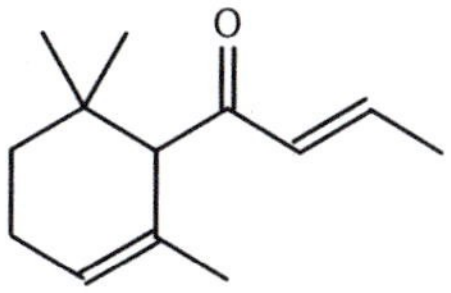

(7.22)

TABLE 7.1

Contributions of Odorants to the Smell of Rose Oil

Component	Percent of Oil	Threshold (ppb)	Odour Units ($\times 10^{-3}$)	Rel. % of Odor Units
(−)-Citronellol	38	40	9,500	4.3
Geraniol	14	75	1,860	0.8
Nerol	7	300	233	0.1
Phenylethanol	2.8	750	37	0.016
β-Damascenone	0.14	0.009	156,000	70.0
β-Ionone	0.03	0.007	42,860	19.2
(−)-Rose oxide	0.46	0.5	9,200	4.1

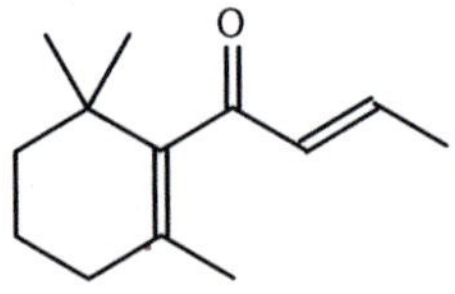

(7.23)

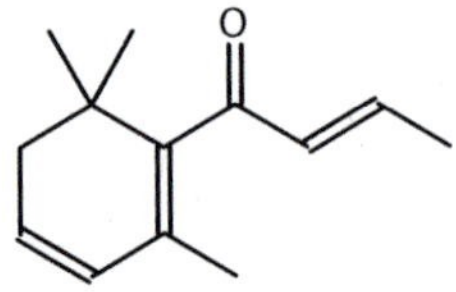

(7.24)

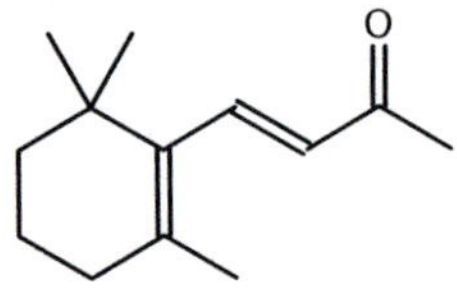

(7.25)

The damascones have slightly different smells; α-damascone (7.22) has a rose–apple note and β-damascone (7.23) has a blackcurrant–plum note. Beta-damascenone (7.24) is described as brilliant and damascone-like and β-ionone (7.25) in dilution is violet-like. Molecules like β-damascenone and β-damascone are important perfume materials, notably in *Poison* (Christian Dior, 1985). Rose oxide exists as *cis*-(Z-) and *trans*-(E-) isomers, each having (+)- and (−)-stereoisomers (7.26–7.29). It is the (−)-*cis* isomer (7.26) that is responsible for the floral green rose fragrance; its aroma threshold at 0.5 ppb is at least 100 times lower than those of the other isomers.

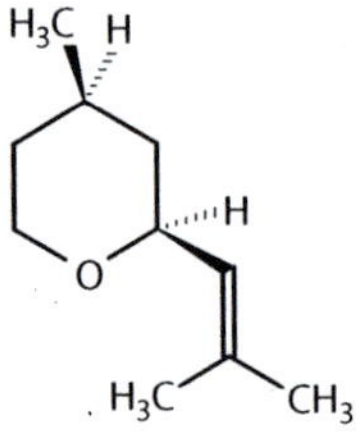

(7.26)

(4R, 2S)-(−)-*cis*-rose oxide

(7.27)

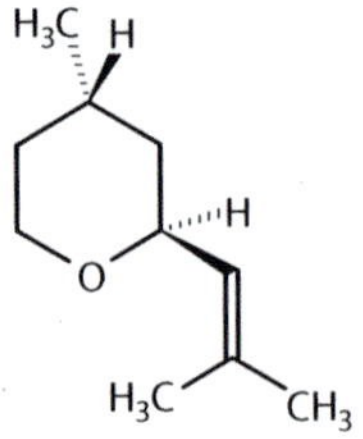

(4S, 2S)-(+)-*trans*-rose oxide

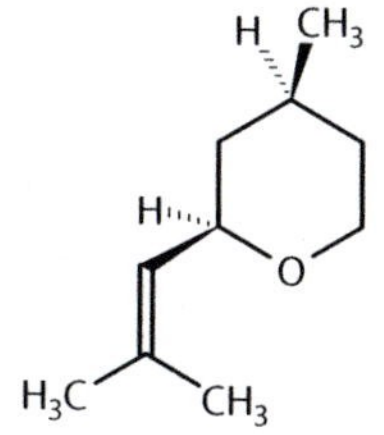

$$(7.28)$$

(4S, 2R)-(+)-cis-rose oxide

$$(7.29)$$

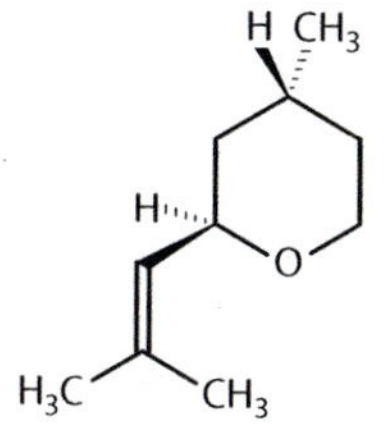

(4R, 2R)-(–)-trans-rose oxide

European roses have "floral" scents, based on the intensely smelly damascenone, damascone, and ionone, with contributions from monoterpenes such as nerol, geraniol, and citronellol; on the other hand, Chinese roses and "hybrid tea roses" (crosses from Chinese roses and European roses) often have the odour of black tea. The most important contributor to the smell of Chinese roses is the aromatic molecule 3,5-dimethoxytoluene.

Chinese rose plants use two very similar orcinol O-methyltransferase enzymes, OOMT1 and OOMT2, to turn orcinol into 3,5-dimethoxytoluene (Figure 7.2). Research carried out by Philippe Hugueney and his colleagues at the University of Lyon in France has shown that European roses do not possess the OOMT-2 gene in their petals (where the scent is generated) and do not have OOMT1 at all, so they cannot make 3,5-dimethoxytoluene. It was discovered that there is only a very small difference—one amino acid—between the active sites of OOMT1 and OOMT2 enzymes. It seems that the OOMT1 gene arose when an OOMT2-type gene duplicated and one copy underwent a mutation, a fortunate event resulting in the characteristic smell of Chinese roses.

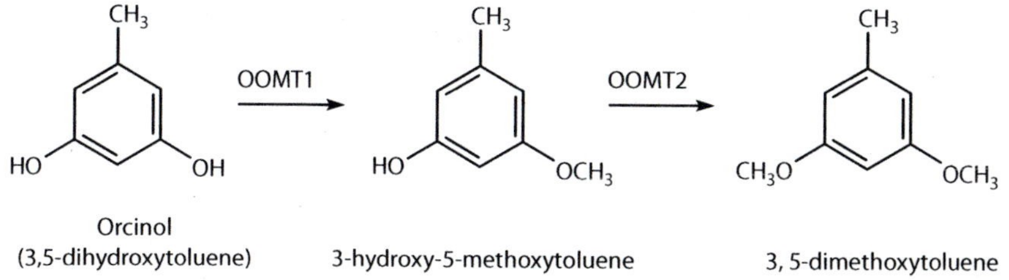

Orcinol
(3,5-dihydroxytoluene)

3-hydroxy-5-methoxytoluene

3, 5-dimethoxytoluene

FIGURE 7.2 Biosynthesis of 3,5-dimethoxytoluene.

MUSKS

Musk has a peculiar animalic and sensory—even erotic—smell and has been used in fragrance and perfume applications for well over a thousand years, especially as an involatile fixative that retains the odour of the other molecules. The male musk deer (*Moschus moschiferus*) secretes a smelly substance from an anal gland to attract the female (musk is derived from the Sanskrit word for testicle, *muská*). The secretion is a mixture that includes steroids, but the most important component is muscone (7.30). The deer became a protected species in 1979, but is still endangered due to both habitat loss and poaching. The African civet cat (*Civettictis civetta*) similarly yields a musky substance, civetone (7.31), from its anal glands, though in this case the compound can be extracted without killing the animal. As obtained, civetone has a faecal smell, owing to the presence of skatole. Modern syntheses of muscone are available, often using Grubbs's catalysts in ring-closing metathesis—for example, starting from citronellol, using $[RuCl_2(PCy_3)_2(=CPh)]$ catalyst.

(7.30)

(7.31)

Chemists knew that muscone was a ketone with molecular formula $C_{16}H_{30}O$ and that civetone was $C_{17}H_{30}O$, but it was not until the mid-1920s that Lavoslav (Leopold) Ružička (1887–1976), working at ETH Zurich, discovered their cyclic structures. Until then, it was believed that rings with more than eight atoms were too unstable to exist. He also developed the Ružička large ring synthesis (Figure 7.3) involving pyrolysis of heavy metal salts of a long-chain dicarboxylic acid; this was used particularly to make the musk exaltone (cyclopentadecanone), one of the ingredients of the musky secretion from the Louisiana muskrat.

Nitromusks

Nitromusks, the first synthetic musks, were discovered (as so often happens) by a scientist looking for something else. In 1888, Albert Baur synthesised 3-*tert*-butyl-toluene by a Friedel–Crafts reaction of toluene with isobutyl bromide using $AlCl_3$ catalyst, then nitrated it. He hoped to make an even better explosive than TNT, but 2-*tert*-butyl-4-methyl-1,3,5-trinitrobenzene was not explosive; however, it had an outstanding musk smell and became known as "musk Baur" (7.32). He went on to make molecules with even more effective musk smells, known as musk xylene, musk ketone (7.33; believed to be the nearest to natural musk), and musk ambrette (7.34). These were widely used in perfumes until around 1980, when concerns

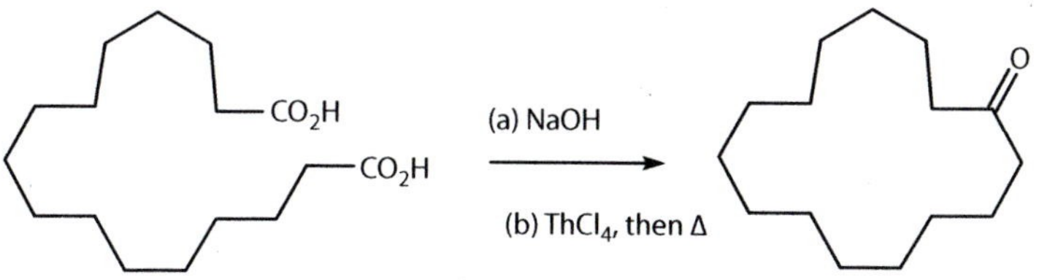

FIGURE 7.3 Ružička large ring synthesis of cyclopentadecanone.

emerged about their neurotoxicity and phototoxicity, as well as their poor biodegradability. Ernest Beaux used over 10% nitromusks, especially musk ketone, in *Chanel No. 5* (1921); musk ambrette was used by Francis Fabron in *L'air du Temps* (1948).

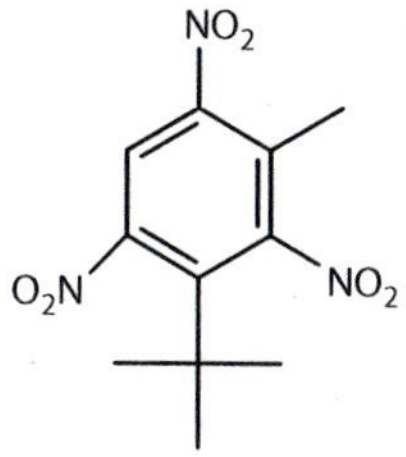

(7.32)

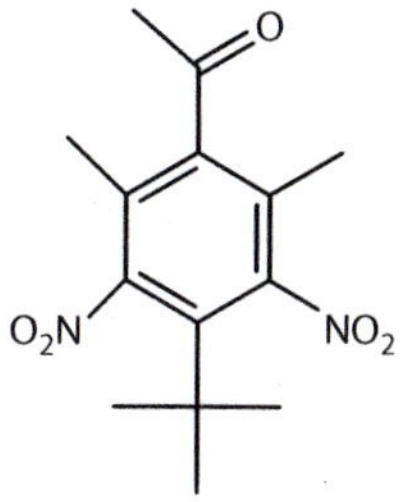

(7.33)

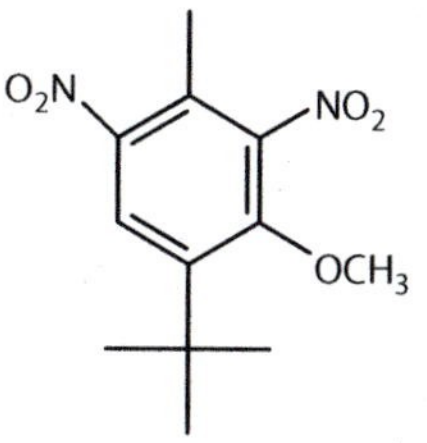

(7.34)

Polycyclic Musks

Because of their instability in alkaline solution, as well as their sensitivity to light, nitromusks were not ideal for applications in detergents and washing powders. The first polycyclic musks emerged after World War II. In 1951, Kurt Fuchs discovered phantolide (7.35); its name reflects the fact that its structure was not known initially.

It was followed by products with names such as celestolide, fixolide, tonalide, and galaxolide (1965). Galaxolide (7.36) was used in considerable amounts in fabric softeners (up to 40%) and detergents, as well as in the original formulation of the perfume *White Linen* (Estée Lauder, 1978). With its clean, metallic, musky smell, it typifies the idea of a "white musk".

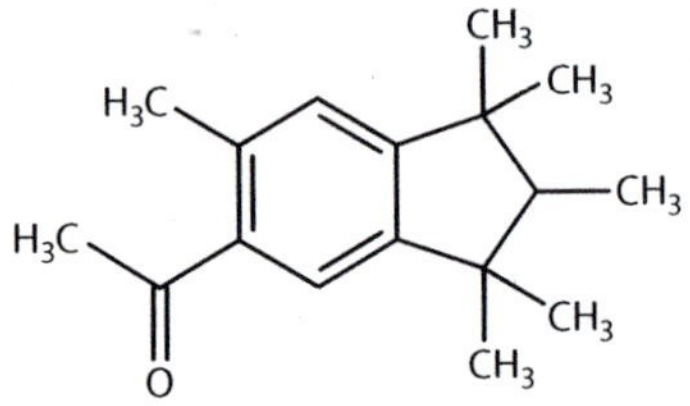

(7.35)

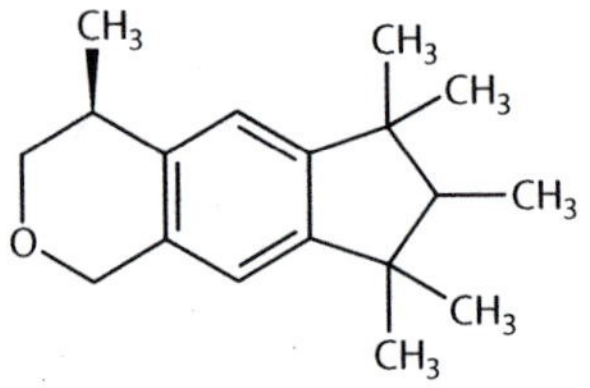

(7.36)

Polycyclic musks were very important commercially into the 1990s, but concerns about their lack of biodegradability (though they are not toxic) have shifted the market toward macrocyclic musks.

Macrocyclic Musks

Kerschbaum identified the fragrance ambrettolide (7.37) from the oil of the ambrette plant; it is a lactone, $C_{16}H_{28}O_2$. Cyclopentadecanolide (7.38), known as Exaltolide® and Thibetolide®, was discovered in angelica root oil in 1927. Animal musks are ketones, whereas musks from plant sources are lactones.

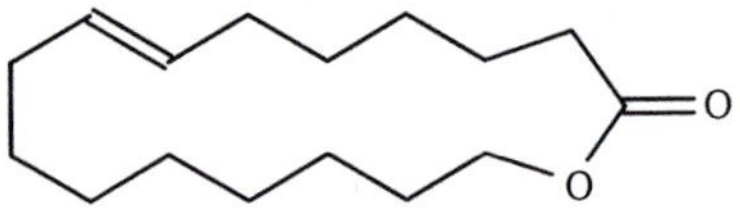

(7.37)

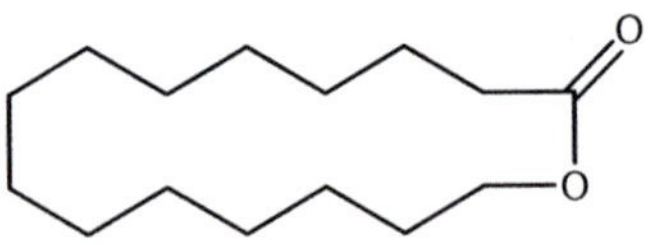

(7.38)

Other important macrocycles include globalide (7.39; habanolide), used to advantage in *White for Her* (Giorgio Armani, 2001) and *Glow by J. Lo* (2002). Globalide is a mixture of the isomeric 11- and 12-pentadecen-15-olides. Catalytic hydrogenation of globalide (Raney Ni catalyst) affords exaltolide. Nirvanolide® is reported to have a very similar smell to Baur's musk ketone, making up 6%–7% of *Forever Elizabeth*

(David Apel, 2002). (12R,9Z)-Nirvanolide (7.40) is described as having an intense musky, fruity, powdery odour whilst the (12S,9Z)-isomer is odourless.

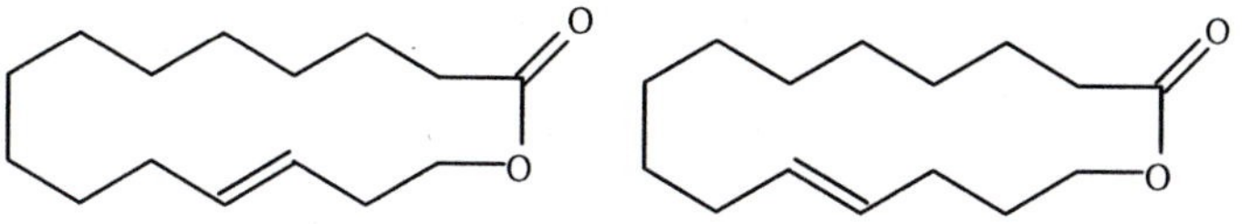

(7.39)

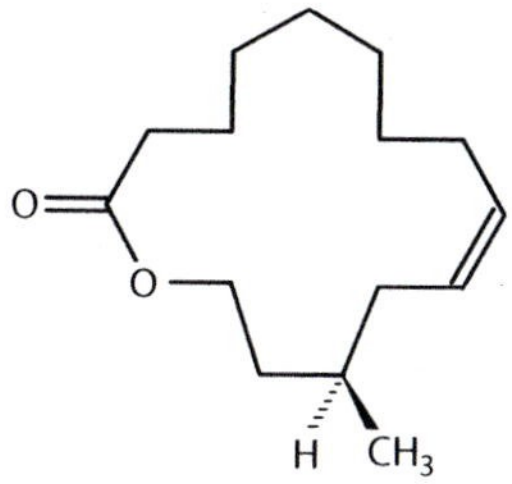

(7.40)

VANILLIN

Years before vanilla was brought to Europe, pre-Columbian peoples in Mexico were adding it, along with chilli peppers and honey, to chocolate drinks, particularly the royal drink xocolatl.

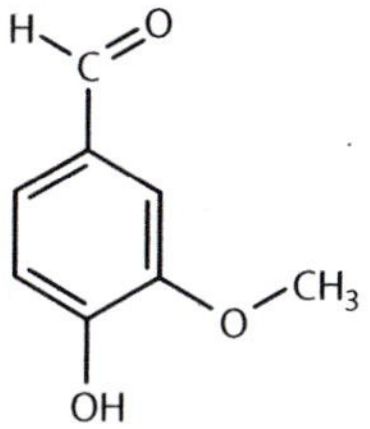

(7.41)

The *Vanilla planifola* orchid probably makes vanillin (7.41) starting from phenylalanine and proceeding via *trans*-4-hydroxycinnamic acid and ferulic acid (3-methoxy-4-hydroxycinnamic acid); in the plant, the vanillin is present linked to glucose as a β-D-glycoside (7.42). Enzymatic breakdown liberates free vanillin molecules in a curing process during which the green beans turn brown and shrivel up. Most vanillin today is used to flavour foods, mainly chocolate and ice cream. Other applications include fragrances, perfumes, and medicines, as well as making other chemicals. A lesser application, though important for laboratory chemists, is as a developing agent for TLC plates.

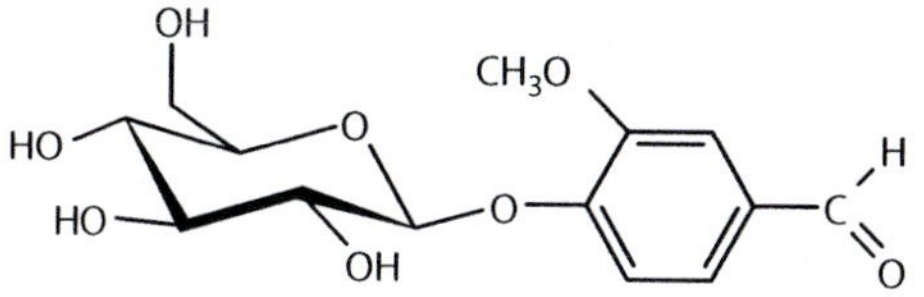

(7.42)

However, demand for vanilla is about 20 times greater than that which can be supplied by harvesting vanilla pods; this makes genuine vanilla extract around 200 times more expensive than counterfeit vanilla essence that uses synthetic vanillin. Detection of fake vanilla is thus important. Since natural vanilla extract contains many "impurity" chemicals detectable by GLC (gas liquid chromatography), such as 4-hydroxybenzaldehyde, vanillic acid, and 3,5-dihydroxybenzoic acid, it can be difficult for someone to reproduce this closely enough, but fakers are more ingenious than that. There has been a race between the analytical chemists on the one side and the merchants selling fake vanilla extract on the other, with each side striving to stay ahead.

Therefore, analysts did radiocarbon tests because vanillin obtained from fresh vanilla pods had higher radiocarbon (^{14}C) levels than vanillin from petrochemical sources; the fakers responded by adding extra ^{14}C to their vanillin. Analysts then checked the ^{13}C:^{12}C ratios in the samples by mass spectroscopy. The vanilla orchid has an unusual biosynthetic pathway for vanillin, so more ^{13}C than usual is incorporated; "fake" vanilla extract will originate in material with a lower ^{13}C:^{12}C ratio. In their turn, the fakers added ^{13}C to their molecules, usually in just the CHO and CH_3 groups. The analytical chemists then used site-sensitive NMR (nuclear magnetic resonance) techniques (SNIF-NMR) to see if the ^{13}C:^{12}C ratio was the same throughout the molecules.

Synthetic vanilla is made by various routes (Figure 7.4). One classic route relies on the isomerisation of eugenol, sourced from cloves, nutmeg and cinnamon, in alkaline solution; this forms isoeugenol, which undergoes gentle oxidation by nitrobenzene to vanillin. In the Reimer–Tiemann method (1876), guaiacol (2-methoxyphenol), obtained from sources like pine tar, gains an aldehyde group when refluxed with an alkaline solution of chloroform; a method widely used today reacts guaiacol with glyoxylic acid. Low-temperature combustion of wood breaks its polymeric structure down to small molecular units including vanillin, which is partly responsible for the smell and taste of smoked food; similarly vanillin leached from oak casks can contribute to the flavour of stored wines. Until recently, much of the world's vanillin was obtained from wood-derived lignin in the waste water from paper mills.

Vanillin had a key role in the perfume *Jicky* (Aimé Guerlain, 1889), the first perfume to make use of synthetic materials rather than just plant-sourced materials, thus moving away from floral smells. Apart from pioneering the use of vanillin, this perfume also introduced the now-traditional makeup of a top note, a middle note, and a bottom note (p. 78). Jacques Guerlain, nephew of Aimé, used the synthetic ethyl vanillin (7.43) to create *Shalimar* in 1925; ethyl vanillin does not occur in nature and has a much stronger vanilla smell than vanillin. The odour threshold of ethyl vanillin is given as 0.000007 mg/m^3 of air.

(7.43)

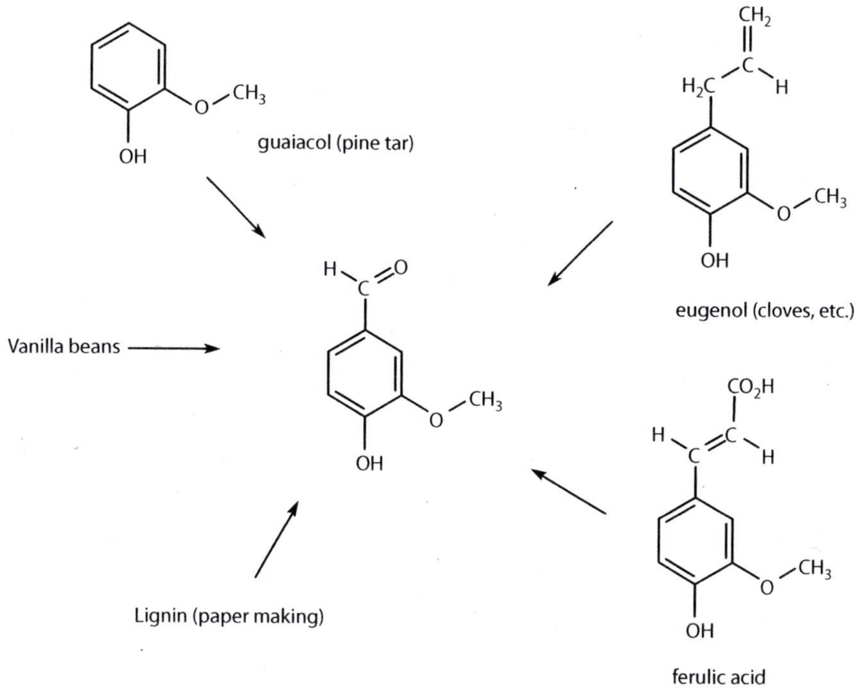

FIGURE 7.4 Synthetic vanilla is made via various routes.

ALDEHYDES AND 2-METHYLUNDECANAL

Aldehydes are important odorant molecules. Ethanal is responsible for providing the fresh note in fruit juices like orange and grapefruit, whilst octanal and nonanal are also important in orange flavour (decanal is important in grapefruit). The unsaturated aldehyde citral (a mixture of the isomers neral and geranial) contributes to lemon flavour. Short-lived *cis*-3-hexenal causes the characteristic aroma of freshly cut grass.

There is no single characteristic smell for aldehydes. Short-chain alkanals tend to have sharp, sometimes unpleasant smells (butanal being described as "sweaty" and "cheesy"), but as the number of carbon atoms increases, the odour does become less offensive. The C_8–C_{12} aldehydes show floral or fruity characteristics, with some fatty character, but after C_{13}, the odours become much fainter. Introducing a methyl side chain at carbon 2 makes the odours stronger and more pleasant, but once this side chain is longer than two carbons, the aroma gets much weaker; the position of the side chain also affects smell. 12-Methyltridecanal with a meaty, tallow smell is found in beef, from which it is slowly released on stewing (but not frying or roasting).

The characteristic smells of C_7–C_{14} aldehydes, some of importance in perfumes, are summarised in Table 7.2.

TABLE 7.2
Characteristic Smells of C_7–C_{14} Aldehydes

Aldehyde	Odour Characteristics	Aldehyde	Odour Characteristics
Heptanal	Herbal, green		
Octanal	Fatty-orange	2-Methyloctanal	Orange-floral, less fatty than nonanal
Nonanal	Fatty-waxy, floral note	2-Methylnonanal	Green, fatty-floral
Decanal	Fatty-orange	2-Methyldecanal	Fresh, orange
Undecanal	Green, fatty, floral-orange	2-Methylundecanal	Herbaceous-orange-ambergris like
Dodecanal	Fatty, floral-orange	2-Methyldodecanal	Fatty-floral with ambergris note
Tridecanal	Intensive, fatty with fresh citrus note	2-Methyltridecanal	Pleasant, mild, ambergris-like
Tetradecanal	Faint, fatty, citrus and ambergris-like notes		

The odour thresholds in humans of some C_3–C_9 straight chain aldehydes are as follow:

C_2H_5CHO: 2.0 ppm
C_3H_7CHO: 0.46 ppm
$C_5H_{11}CHO$: 0.33 ppm
$C_7H_{15}CHO$: 0.17 ppm
$C_8H_{17}CHO$: 0.53 ppm

The lowest odour detection threshold is thus in octanal.

Several research groups have examined the way in which rodents respond to the smell of aliphatic aldehydes. The I7 odorant receptor has been studied in particular. In mice, this gives its strongest response with heptanal. An MRI study of the mouse olfactory bulb showed that each one of the C_4–C_8 aldehydes caused different globular activity patterns, corresponding to different "smells". The rat I7 odorant receptor differs from the mouse I7 receptor by a single isoleucine-to-valine substitution and gives its strongest response to octanal. Ca^{2+} imaging studies of the rat OR-I7 receptor with straight-chain aldehydes indicate that it only responds to C_7–C_{11} aldehydes and not to longer or shorter chains. Short-chain aldehydes appear to bind to the receptor but not to activate it; longer chains are too large to fit in the binding pocket. This receptor shows no response to many odorants, including ketones and esters, and likewise does not respond to C_8 alcohols or acids.

Chanel No. 5 (Chanel, 1921), probably the most famous perfume in the world, is generally thought of as the first perfume to make real use of a totally synthetic substance—2-methylundecanal (also known as 12-MNA, methyl nonyl acetaldehyde, 7.44)—though this is a slight simplification.

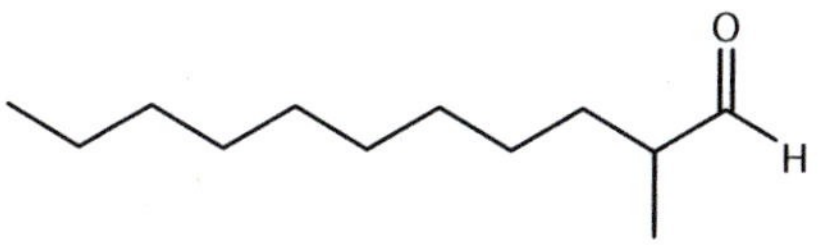

$$(7.44)$$

Perfumers started to use aldehydes in the first decade of the twentieth century. 2-Methylundecanal was first reported by Darzens in 1904, and soon it was used in perfumes like *Floramye* (1905). Within a decade of the discovery of the Darzens reaction, Robert Bienaimé used 2-methylundecanal in the important perfume *Quelques Fleurs* for Houbigant (1912). In 1914, Ernest Beaux, then working for Alphonse Rallet & Co, the leading Russian perfume house, used aldehydes in his new perfume, *Rallet No. 1*, whose composition is known thanks to analysis of surviving sealed bottles.

Beaux used undecanal, dodecanal, and undec-10-enal, together with neroli and ylang-ylang, for the top note; jasmine absolute, rose de mai, orris, and lily of the valley for the middle note; and sandalwood, vetiver, styrax, vanillin, and nitromusk in the base note. In 1920, working in France, he composed *Chanel No. 5* for Coco Chanel, again employing $C_{10}–C_{12}$ aldehydes, neroli, and ylang-ylang for the top note, along with jasmine, rose, lily of the valley, and iris for the middle note, and sandalwood, vetiver, musk, vanillin, civet, and oak moss for the base note. He achieved great impact by using 2-methylalkanals—particularly 2-methylundecanal—in much larger amounts than had been done hitherto. In a famous comment, Beaux compared the effect of aldehydes to "lemon juice on strawberries". 2-Methylundecanal has a chiral carbon, but the two optical isomers have very similar smells.

8 Natural Killers

CURARE

Native South Americans have long used arrow poisons for both hunting and defence; they are generically referred to as curare. (+)-Tubocurarine chloride (8.1), which is obtained by extracting the roots and stem of *Chondrodendron tomentosum* with boiling water and was once used in anaesthesiology, is the best representative of this family. It operates as an antagonist at the nicotine receptor, stopping acetylcholine binding and preventing muscular contraction, so the diaphragm is paralysed.

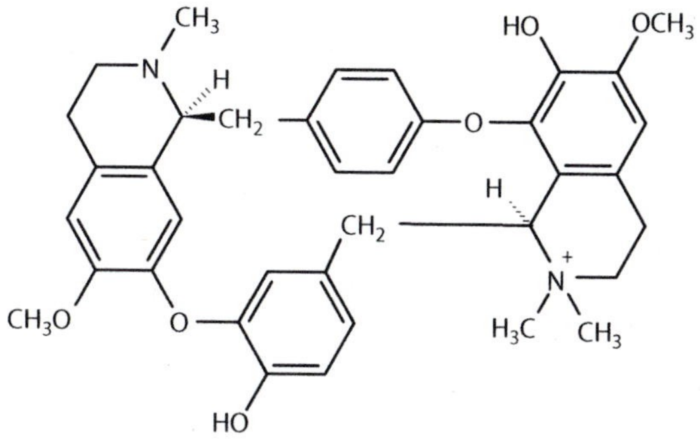

(8.1)

These molecules are only effective by injection (by hypodermic or poisoned arrow) and cannot be absorbed from the digestive system—a point important to any native Indian eating a monkey that he has just killed with a curare-tipped dart.

FROG VENOMS

Much more toxic molecules are produced by tiny, brightly coloured frogs, which excrete them from their skins when threatened. Such frogs, traditionally known as "poison dart frogs", are associated especially with tropical South America, but are also found in other regions, including Australia and Madagascar. These molecules exhibit very different structures. In their poison darts, the Chocó Indians of western Colombia use the most toxic molecule known: **batrachotoxin** (8.2), which has a steroid-related structure. Together with the similar homobatrachotoxin, batrachotoxin is associated especially with dendrobatid frogs of the species *Phyllobates terribilis*, which produce their venom from glands in their back and from behind their ears—the most toxic venom known.

(8.2)

An average *P. terribilis* frog contains 1100 μg of batrachotoxin; the lethal quantity is ca. 1–2 μg/kg in a human—around 100 μg in an 11-stone person, making batrachotoxin at least 10 times as toxic as curare or tetrodotoxin. It is said that a person with a cut hand could obtain a lethal dose from touching one of these frogs. The frogs are immune to batrachotoxin because they have a modified sodium channel protein in their nerves and muscles to which batrachotoxin cannot bind (an analogous situation to fugu fish and tetrodotoxin, p. 94).

The frogs do not make the batrachotoxin themselves; rather, it comes from their diet. *P. terribilis* born in captivity do not have any batrachotoxin in their skin, though if they are fed it, it accumulates there. Wild-caught frogs gradually lose their batrachotoxin the longer they are kept. New Guinea song birds of the genus *Pitohui* have batrachotoxin in their plumage (as do *Ifrita kowaldi*); their diet includes melyrid beetles (*Choresine pulchra*), which also contain batrachotoxin, but where the beetles get it from is unclear. These beetles are also found in Colombia and thus could be part of the frog's diet. Batrachotoxin is poisonous because it binds to the voltage-gated Na⁺ channel in the outer membrane of nerve and muscle cells and makes it very permeable to sodium ions. The channel cannot close, so an action potential cannot develop and nerve impulses do not get transmitted. Ultimately, this results in cardiac failure.

The toxic molecule in frogs that has attracted the most attention is not a steroid but rather an alkaloid. **Epibatidine** (8.3) was first isolated from the rare *Epipedobates tricolor* in 1974, but so little (1 mg) was available for study that its structure was not elucidated for another 20 years, until the arrival of more sophisticated and sensitive spectroscopic methods. When John Daly (the leading researcher in poison-dart frogs) injected it into mice, he got a very strong Straub-tail reaction (where the tail stands up and arches over the mouse's back)—a key indication of opiate behaviour. It turned out that epibatidine was an extremely strong nicotine agonist, some 200 times more potent than morphine. Since it did not act through the opioid receptor, hopes of its being useful as a nonaddictive painkiller were raised. Synthetic chemists soon made epibatidine, though it is too toxic to be used on humans (the lethal dose is around 2 mg); experiments with synthetic analogues have so far not yielded a useable drug.

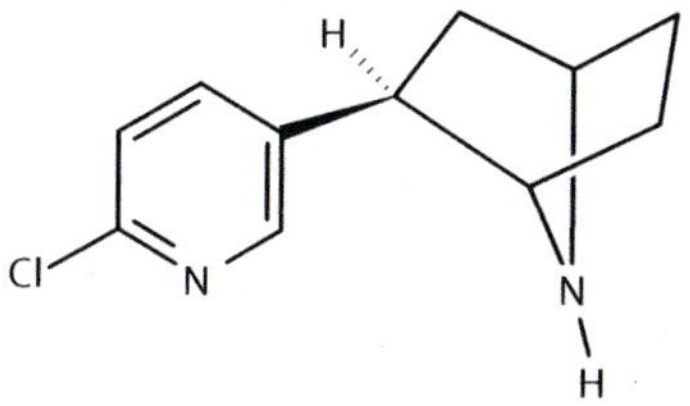

(8.3)

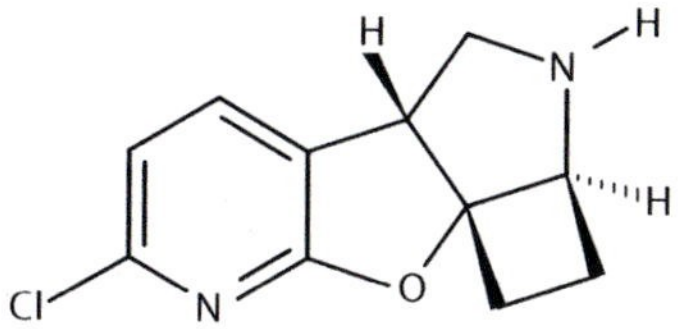

(8.4)

The frog's source of the epibatidine is unknown; captive frogs do not make it. One possibility is arthropods on the forest floor, though because the structure resembles nicotine, a plant source is not out of the question. More recently (2010), a tetracyclic analogue named phantasmidine (8.4) has been found in *Epipedobates anthonyi* frogs; this also has the unusual feature of a chloropyridine unit.

Pumiliotoxins (8.5) are the most widely spread of these toxic molecules. They were first identified in the Central American strawberry poison frog, *Dendrobates pumilio* (now known as *Oophaga pumilio*), but are also found in Madagascar frogs (genus *Mantera*) and in Australia (*Pseudophryne* frogs). Pumiliotoxins are some 100 to 1,000 times less toxic than batrachotoxin; pumiliotoxin A is the most poisonous, acting particularly upon sodium channels.

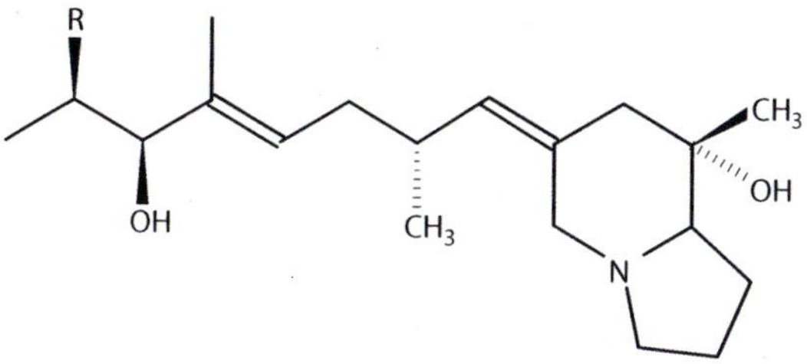

(8.5)

R = H, Pumiliotoxin A
R = OH, Pumiliotoxin B

The frogs probably get the pumiliotoxins from certain insects that are known to contain them, including some formicine ants and oribatid mites. Australian mycobatrachid frogs of the genus *Pseudophryne* contain not only pumiliotoxins but also **pseudophrynamines** (8.6A, 8.6B), which are potent noncompetitive blockers of nicotinic channels. Pseudophrynamines are at present unique because they are synthesised by the frog, rather than obtained in the diet.

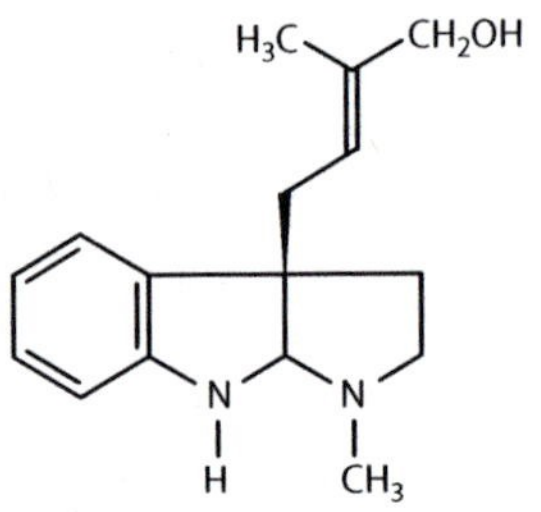

Pseudophyrinaminol

(8.6A)

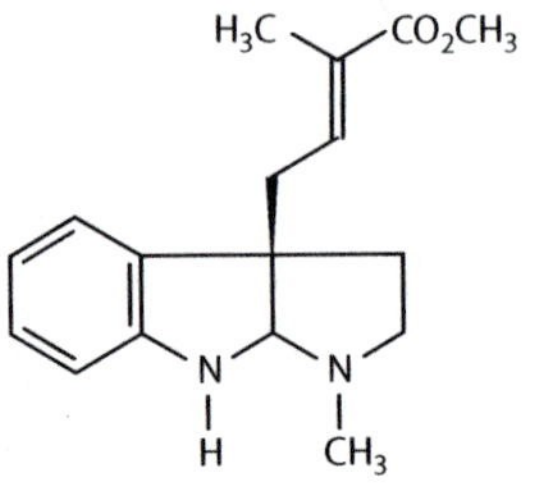

Pseudophyrinamine

(8.6B)

Another unusual alkaloid, found in *Dendrobates histrionicus,* a Columbian poison frog, is **histrionicotoxin** (8.7), a quite singular spiropiperidine alkaloid featuring two terminal alkyne groups. The food source is not known. An LD50 of around 500 μg/kg has been quoted.

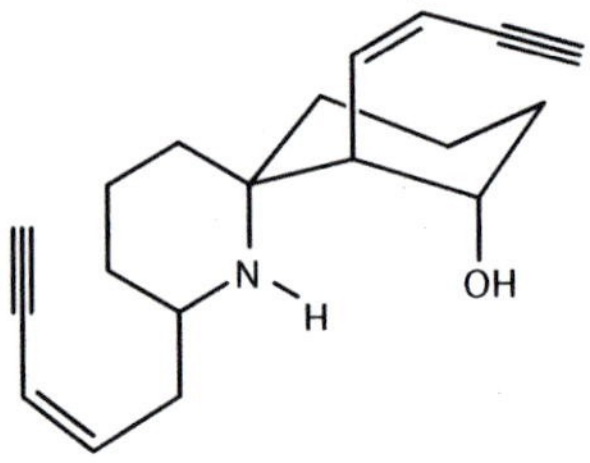

(8.7)

TETRODOTOXIN

Tetrodotoxin (TTX; 8.8) is a potent marine poison, a neurotoxin for which as yet there is no antidote. It is found in various species, including the blue-ringed octopus, some poisonous frogs, the rough-skinned newt, a flatworm, and certain fish, most famously the pufferfish of the family Tetraodontidae. The fish uses it as a defensive toxin, but the octopus appears to employ it as a predatory venom. The $=NH_2^+$ group of tetrodotoxin binds to the voltage-gated Na^+ channels in nerve cell membranes, blocking the pore opening and causing paralysis; the victim ultimately dies from asphyxiation. The fish has a modified sodium channel, due to a single amino acid mutation, and thus is immune to its own poison. First symptoms are a feeling of warmth, followed by a tingling round the mouth; however, this is succeeded by less pleasant spreading paralysis. It does not affect the brain, so the victim is likely to remain conscious. Death usually occurs within a few hours; any victim who lasts for 24 hours is likely to survive.

$$(8.8)$$

It was once claimed that tetrodotoxin was used in the Haitian voodoo culture to create "zombies", but that idea is now discredited (it was used in this context in a 1985 episode of *Miami Vice*). *Fugu* (cooked pufferfish meat) is an important Japanese delicacy; because the tetrodotoxin molecule is too robust to be destroyed by cooking, the tetrodotoxin-containing organs, such as the liver, ovaries, and intestines, have to be removed by the chef. *Fugu* chefs face 3 years' training and then sit an examination, both written and practical, before they are allowed to sell it to the public. Some of the very best chefs can leave just a trace of the toxin in the fish so that their customers can enjoy that delicate tingle without risk. Despite that, around 100 Japanese die each year from tetrodotoxin poisoning.

Captain James Cook, the explorer, ate pufferfish in 1774 in Polynesia and survived, as did James Bond, a victim of Rosa Klebb's TTX-coated steel-tipped shoe in *From Russia with Love*. It is revealed in the following Bond book, *Dr. No*, that Bond had been successfully treated with an antidote for curare poisoning—a mistake by author Ian Fleming because, even now, there is no antidote for tetrodotoxin.

The common garter snake (*Thamnophis sirtalis*) is engaged in an "arms race" with the rough-skinned newt (*Taricha granulosa*) as it builds up resistance to the newt's TTX. TTX is produced by certain symbiotic bacteria, possibly acquired by the fish from its diet, as pufferfish raised in captivity have much lower levels of TTX.

CONE SNAIL VENOMS (CONOTOXINS)

Conotoxins, or conopeptides, are chemicals made by marine gastropods of the genus *Conus* comprising snails that live in shallow waters of the Atlantic, Pacific, and Indian Oceans. These snails have very collectable shells but compensate for their immobility by harpooning their prey, then injecting the venom, which is a complex cocktail of chemicals containing up to 100 different compounds. This paralyses the prey, which is other molluscs, worms, and fish. Human fatalities have resulted from mishandling the snails. Each species seems to come up with different types of conotoxin, most of which are small peptides containing between 10 and 40 amino acids, rather smaller than the neurotoxins with 60–70 amino acids that snakes use in their venom. Several kinds are known:

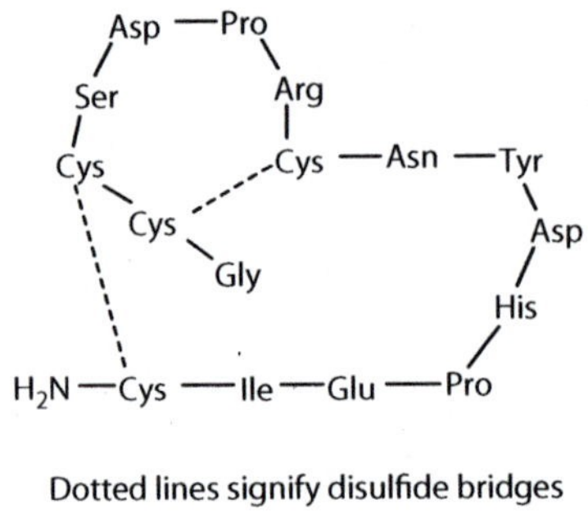

Dotted lines signify disulfide bridges

FIGURE 8.1   Alpha-conotoxin.

- Alpha (α)-conotoxins block nicotinic acetylcholine receptors.
- Mu (μ)-conotoxins block sodium ion (Na^+) channels in muscle and prevent muscle contractions.
- Omega (ω)-conotoxins block calcium ion (Ca^{++}) channels in neurons. Many different ω-conotoxins, from different species of cone snails, affect different subtypes of Ca^{++} channels. Another kind, delta (δ)-conotoxins, also targets Na^+ channels.

Conotoxins could be potent new medicines. Alpha-conotoxin Vc1.1 (Figure 8.1), from the mollusc hunter *Conus victoriae,* a polypeptide composed of 16 amino acid residues, is being tested as an analgesic for a wide range of neuropathic conditions.

Omega-conotoxin MVIIA, from the fish hunter *Conus magus,* is a polypeptide of 25 amino acid residues (Figure 8.2) that selectively blocks N-type calcium-ion channels. Under the name PRIALT™ (it has also been known as Ziconotide and SNX-III), it is a powerful pain therapeutic that successfully completed phase III clinical trials and received FDA approval for the management of severe chronic pain in malignant diseases such as cancer and AIDS. The patient wears a pump that injects the drug straight into the fluid around the spinal cord. PRIALT is said to be perhaps 1,000 times stronger than morphine, but, because it targets a different receptor in the body, it is free of the addictive properties of morphine.

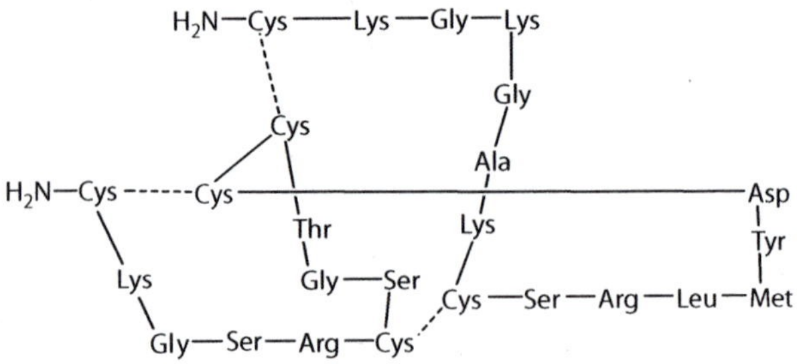

FIGURE 8.2   Omega-conotoxin MVIIA.

SPIDER VENOMS

Spider venoms are complex multicomponent mixtures of polyamines and peptides. Most spiders are venomous; their venom is usually aimed at insects but a few can be very toxic to humans. The main active component of the female black widow spider is the peptide α-**latrotoxin.** The structure of α-latrotoxin from the American black widow spider, *Latrodectus mactans,* shows four α-latrotoxin molecules assembled into a tetramer that forms pores in cell membranes leading to calcium release. It causes an enormous release of neurotransmitters, particularly acetylcholine, which block the neural synapse, resulting in impairment of cardiovascular and neuromuscular functions and leading to pain and cramps. Normally, the victim just has flu-like symptoms, but the paralysis extends to the respiratory system and can occasionally prove fatal to humans.

The male Sydney funnel web spider, *Atrax robustus,* has caused fatalities in humans, though an antivenom is now available. A key toxin is a 42 amino acid peptide, **robustoxin,** which targets sodium channels; a very similar peptide, **versutoxin,** is found in *Atrax versutus* from the hills west of Sydney. These are δ-atracotoxins, which bind to the outside of tetrodotoxin-sensitive sodium channels, causing massive release of the neurotransmitters acetylcholine, norepinephrine, and epinephrine. This leads to muscular paralysis. Robustoxin has the following amino-acid sequence:

Cys-Ala-Lys-Lys-Arg-Asn-Trp-Cys-Gly-Lys-Asn-Glu-Asp-Cys-Cys-Cys-
Pro-Met-Lys-Cys-Ile-Tyr-Ala-Trp-Tyr-Asn-Gln-Gln-Gly-Ser-Cys-Glu-Thr-
Thr-Ile-Thr-Gly-Leu-Phe-Lys-Lys-Cys

Versutoxin differs in eight of the amino acids and has the following sequence:

Cys-Ala-Lys-Lys-Arg-Asn-Trp-Cys-Gly-Lys-Thr-Glu-Asp-Cys-Cys-Cys-Pro-
Met-Lys-Cys-Val-Tyr-Ala-Trp-Tyr-Asn-Glu-Gln-Gly-Ser-Cys-Gln-Ser-Thr-
Ile-Ser-Ala-Leu-Trp-Lys-Lys-Cys

(Ala = alanine; Arg = arginine; Asn = asparagine; Asp = aspartic acid; Cys = cysteine; Gln = glutamine; Glu = glutamic acid; Gly = glycine; His = histidine; Ile = isoleucine; Leu = leucine; Lys = lysine; Met = methionine; Phe = phenylalanine; Pro = proline; Ser = serine; Thr = threonine; Trp = tryptophan; Tyr = tyrosine; Val = valine)

The main toxic component of the brown recluse (fiddleback) spider, *Loxosceles recluse,* is a phospholipase D enzyme; after a painless bite, swelling and pain develop in hours, and there can be severe tissue damage. There may also be damage to blood vessels; a secondary bacterial infection; occasionally, kidney failure caused by coagulation of the blood; and death.

Tarantulas, spiders of the family Theraphosidae, have a fearsome reputation but, at worst, cause symptoms ranging from local pain to muscle cramps lasting for some weeks, though bites from Australian tarantulas have proved fatal to dogs. Tarantula venoms contain a mixture of chemicals, with salts, nucleotides, and neurotransmitters in addition to polyamines, peptides, proteins, and enzymes. Most of the toxic

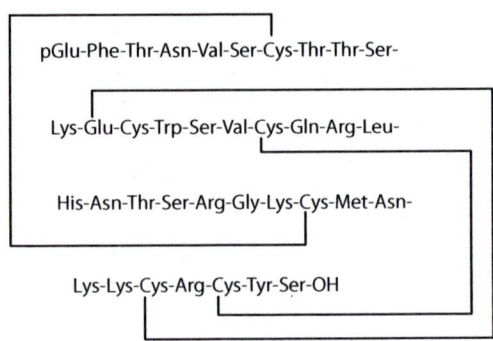

FIGURE 8.3   Structure of charybdotoxin.

peptides contain 33–41 amino acids cross-linked with three disulfide bridges. The venom of the West Indian tarantula, *Psalmopoeus cambridgei,* contains peptides known as vanillatoxins, which affect the same TRPV1 "vanilloid" receptor as capsaicin, the hot ingredient of chilli peppers. This is what produces the pain and inflammatory response when someone is bitten by this spider.

SCORPION VENOMS

Scorpion venoms often contain a mixture of a hundred or more neurotoxic polypeptides of two main types: long-chain toxins of 60–80 amino acid residues (tending to act upon Na^+ and Ca^{2+} channels) and short-chain toxins containing 30–40 amino acid residues (tending to act on K^+ and Cl^- channels). Perhaps the best known is **charybdotoxin,** a 37 amino acid polypeptide with three disulfide bridges, from the scorpion *Leiurus quinquetriatus hebraeus.* This blocks the external mouth of calcium-activated potassium channels, interfering with their ability to pass K^+ ions and transmit nerve messages (Figure 8.3).

In contrast, **chlorotoxin** is a 36 amino acid peptide with four disulfide bridges. It is also found in the venom of *Leiurus quinquestriatus* and blocks chloride channels. Glioma is a highly invasive form of brain cancer; it has been found that chlorotoxin binds specifically to the surface of glioma cells, and it is under investigation for treatment of these tumours. For example, a near-IR emitting conjugate of chlorotoxin and a fluorescent substance identifies tumours and may make it possible to remove the cancerous cells without affecting healthy tissue. ^{131}I-chlorotoxin has given encouraging results in phase 1 clinical trials.

The most deadly scorpion in the United States, the Arizona bark scorpion (*Centruroides sculpturatus*), uses a mixture of "long" neurotoxins with 60–70 amino acid residues with four disulfide bridges; thus, the *Centruroides sculpturatus* Ewing variant 2 neurotoxin has 66 amino acids.

SNAKE VENOMS

In contrast to plants or even poison-dart frogs, which generally use one or two toxic molecules, snake venoms are tremendously complicated. They tend to be cocktails of chemicals; some are cardiotoxins, some neurotoxins, and some enzymes that cause haemorrhages or destruction of flesh. Snakes cannot chew their prey, so they swallow their victims whole and digest them. When they bite the victim, the venom runs down the fangs into the wound, and enzymes get into the circulatory system. These enzymes include proteolytic enzymes, which break down cells and tissue starting at the site of the bite; phospholipases, which destroy muscles and nerves; haemorrhagins, destroyers of capillary walls, leading to haemmorhages; other compounds that prevent clotting; and haemolytic enzymes that destroy red blood cells. But worse damage may be done by smaller molecules, usually polypeptides, which have more specific actions, such as cardiotoxins that affect the heart and neurotoxins that stop nerve impulses being sent to muscles.

The best known family of polypeptides are the "three finger" toxins; the first structure of this type to be discovered was that of α-**bungarotoxin,** found in the venom of the Taiwanese banded krait (*Bungarus multicinctus*). It is a polypeptide, containing 74 amino acids, with a molecular mass of about 8,000. It is a "long" α-neurotoxin, with three loops ("fingers") extending from a globular core, cross-linked by five disulfide bridges. Bungarotoxin is a competitive antagonist, binding irreversibly to the nicotinic acetylcholine receptor, stopping acetylcholine binding, and blocking the channels; initial paralysis of limbs spreads throughout the body, causing respiratory failure. Death can happen within minutes, as with the related **cobratoxin** (a 71 amino acid polypeptide with five disulfide bridges). Venom from the sea snake, *Laticauda semifasciata,* contains three closely related neurotoxins: **erabutoxins a**, **b**, and **c**. These are "short" α-neurotoxins with 62 amino acids and four disulfide bridges. However, other "three finger" toxins attack different receptors, as with the cardiotoxins used by cobras, such as **cardiotoxin V^{II}4** from *Naja mossambica mossambica,* a peptide containing 60 amino acids and four disulfide bridges. This does not bind to the nicotinic acetylcholine receptor.

Mambas use a mixture of toxins, which work synergistically. **Fasciculins** have three-finger structures, like the α-neurotoxins. These are strong acetylcholinesterase inhibitors that result in the generation of an overwhelming number of nerve impulses. **Dendrotoxins**, peptides ~ 60 amino acid residues long with three disulfide bridges, bind to residues near the pores of voltage-gated potassium channels, blocking nerve impulse by paralysing certain muscles. The burrowing asp, *Atractaspis engaddensis,* has a highly toxic venom in which the most dangerous ingredient is **sarafotoxin,** which binds to the endothelin (ET) receptor, leading to constriction of blood vessels and causing a sharp rise in blood pressure. This can lead to aortic rupture and death in some cases. Sarafotoxins are relatively small peptides with 21 amino acid residues, with two disulfide bridges.

The pit viper *Bothrops asper* is responsible for the majority of snakebites in Central America. One component of its venom is **BaP1**, a zinc-dependent metalloproteinase containing a single chain of 202 amino acids with six cysteine residues involved in three disulfide bridges. Its active site contains a Zn^{2+} ion bound tetrahedrally to three histidine residues and a water molecule. This enzyme causes severe tissue damage, haemorrhaging, and oedema.

But the study of snake venoms has also yielded dividends in medicinal chemistry. The venom of the South American pit viper, *Bothrops jararaca,* produces a big drop in blood pressure. Scientists isolated teprotide, a nonapeptide (Pyr-Trp-Pro-Arg-Pro-Gln-Ile-Pro-Pro), and this led to the development of Captopril, a very effective oral treatment for high blood pressure. Echistatin is an antiplatelet compound in the venom of the saw-scaled viper, *Echis carinatus;* it is a polypeptide containing 49 amino acids and four cysteine bridges:

Glu-Cys-Glu-Ser-Gly-Pro-Cys-Cys-Arg-Asn-Cys-Lys-Phe-Leu-Lys-Glu-Gly-
Thr-Ile-Cys-Lys-Arg-Ala-Arg-Gly-Asp-Asp-Met-Asp-Asp-Tyr-Cys-Asn-
Gly-Lys-Thr-Cys-Asp-Cys-Pro-Arg-Asn-Pro-His-Lys-Gly-Pro-Ala-Thr

It was the lead compound for the anticoagulant drug tirofiban (Aggrastat).

STRYCHNINE

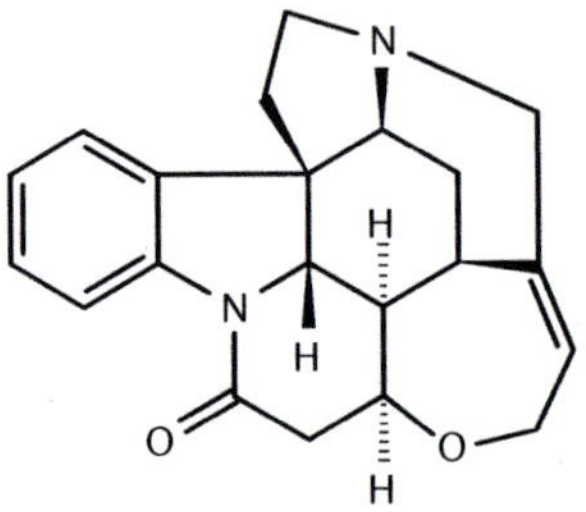

(8.9)

Strychnine (8.9) will suffice as another natural killer that is not from an animal. A famously bitter-tasting indole alkaloid, strychnine is usually obtained from the *Strychnos nux vomica* tree; the purpose of this secondary metabolite in the plant, bio-synthesised from the amino acid tryptophan, is unknown, though it could be to deter predators. Poisoners have often masked its taste with alcoholic drinks, coffee, or hot chocolate (e.g., some Agatha Christie novels), and several famous nineteenth century murderers had recourse to it—notably, two doctors, William Palmer of Rugeley and Thomas Neill Cream, the "Lambeth Poisoner". R. B. Woodward remarked in 1955, "Certain is it that it has no competitor as the lethal weapon of choice in the literary sphere", and Agatha Christie used strychnine as the vehicle of the murder of Emily Inglethorp in her first book, *The Mysterious Affair at Styles* (1920).

The symptoms of strychnine poisoning are unmistakeable: The patient has violent convulsions and contortions, notably arching the back, whilst the jaw muscles can contract, leaving the victim with a fixed grin (*Risus sardonicus*), as noted by Conan Doyle in *The Sign of Four*. Strychnine operates by blocking a glycine receptor that stops motor nerves in the spinal cord from operating normally. It heightens sensitivity to stimuli, resulting in excessive muscular contractions. Death is a result of asphyxia from paralysis of breathing or exhaustion. Strychnine does not cross the blood–brain barrier, so the victim remains conscious.

Despite its toxicity, strychnine was used medicinally as a stimulant and tonic into the early twentieth century, most famously by the American marathon runner Thomas Hicks in the St Louis Olympic Games of 1904. In Victorian times, it was even used as a reputed aphrodisiac. The dose would be up to a milligram; the lethal oral dose was in the range of 50–100 mg. It is much lower when strychnine is administered intravenously, and it has been suggested that this permits the use of strychnine in poison darts by hunters in Africa and South America. A traditional use of strychnine was as a pesticide for small rodents and birds, though its use for poisoning moles was made illegal in the European Union in 2006. There are still some states in the United States, such as West Virginia, where a few Christian groups take Mark 16:17–18 literally, both handling rattlesnakes and drinking strychnine, though a wit has remarked that "this is a very small sect and getting smaller all the time".

Strychnine was first isolated by ether extraction from the beans of *Strychnos ignatii* by P. J. Pelletier and J. B. Caventou in 1818. Chemists argued over its structure for years and it was eventually determined by Sir Robert Robinson in 1952, who remarked that "for its molecular size it is the most complex substance known". Its structure, involving seven fused rings and six chiral centres, presented R. B. Woodward with one of his greatest challenges before he reported its total synthesis in 1954.

9 Unnatural Killers

DIMETHYLMERCURY

$$H_3C - Hg - CH_3 \quad (9.1)$$

Dimethylmercury (9.1) was first synthesized in 1858 by George Buckton, who worked at the Royal College of Chemistry (now Imperial College) as assistant to A. W. von Hofmann. $(CH_3)_2Hg$ was soon known to be poisonous. Two English chemists were fatally poisoned by it in 1865; in 1943, two Canadian secretaries were killed by fumes of the similar diethylmercury and, in 1971, a Czech chemist who had been making dimethylmercury died within a month. But no one knew *quite* how toxic this colourless, volatile liquid (b.p. 92°C) was until 1997.

Mercury and some of its compounds have been known for over 2,000 years; the ancient Romans mined cinnabar (HgS) at Almaden in Spain. They knew how toxic it was; criminals sentenced to work in the mines treated it as a death sentence. For the next millennium or more, mercury compounds were used in medicine; later on, mercury nitrate was used to preserve the felt used in hat-making, as well to soften the hairs. Workers absorbed the toxic mercury through their skins; the Mad Hatter in *Alice in Wonderland* was no exaggeration of Lewis Carroll's mind, though the character exhibits none of the classic symptoms of mercury poisoning.

In the twentieth century, organomercury compounds were widely employed as fungicides on seed grains. Epidemics of mercury poisoning in West Pakistan (1961), Guatemala (1965), and Iraq (1956, 1960, 1971–1972) were caused by people eating bread made from the wheat grain or the grain itself.

It was two episodes of mercury poisoning that brought the danger home to chemists across the world. Over 1,700 people were poisoned in Minamata, Japan, when a chemical company discharged mercury wastes into the sea between 1932 and 1968. This "inorganic" mercury was converted into the methylmercury ion, $CH_3Hg^+(aq)$, a powerful neurotoxin; absorbed by plankton; and passed up the food chain through fish to humans. The second episode involved just one fatality, Professor Karen Wetterhahn of Dartmouth College in New England. On 14 August 1996, she put on protective gear—lab coat, safety glasses, and latex gloves—went to a fume cupboard, and pipetted a very small amount of dimethylmercury into an nuclear magnetic resonance (NMR) tube, using it as her standard in [199]Hg NMR spectra. She spilled a drop or two of the dimethylmercury on her gloves, but thought no more of it; having cleaned up, she got on with running the spectra.

In January 1997, Wetterhahn noticed symptoms that rapidly worsened: She slurred her speech and had tingly fingers and toes, as well as problems with her eyesight and balance. Mercury poisoning was diagnosed; she had a blood mercury level of 4000 μg L^{-1}, over 500 times the normal level of 1–8 μg L^{-1} and at least 20 times the toxic threshold. Nothing could be done for her; she slipped into a coma on 7 February and died on 8 June 1997. Tests later showed that dimethylmercury could penetrate ordinary laboratory rubber gloves in seconds; scientists must now wear highly resistant laminated gloves underneath a pair of heavy duty gloves when using dimethylmercury, which itself is strongly discouraged.

Like many heavy metals, mercury is a "soft" Lewis acid, binding to "soft" bases, with very polarisable donor atoms; this means that it has a high affinity for sulfur-containing ligands and will therefore bind to the –SH groups in enzymes and inhibit them. Dimethylmercury is a deadly neurotoxin. It appears that once it is in the body, it is metabolised to $CH_3Hg^+(aq)$, which can complex with cellular proteins. It probably crosses the blood–brain barrier as a methylmercury–cysteine complex and causes severe brain damage. By the time that the effects of dimethylmercury are noted, it is too late to do anything about it. The scientists and doctors who treated Karen Wetterhahn concluded:

> Dimethylmercury appears to be so dangerous that scientists should use less toxic mercury compounds whenever possible. Since dimethylmercury is a "supertoxic" chemical that can quickly permeate common latex gloves and form a toxic vapour after a spill, its synthesis, transportation, and use by scientists should be kept to a minimum, and it should be handled only with extreme caution and with the use of rigorous protective measures.

And, in case you wondered, George Buckton lived to the age of 87, dying in 1905. For the last 40 years of his life, he was exclusively devoted to entomology.

TETRAETHYLLEAD

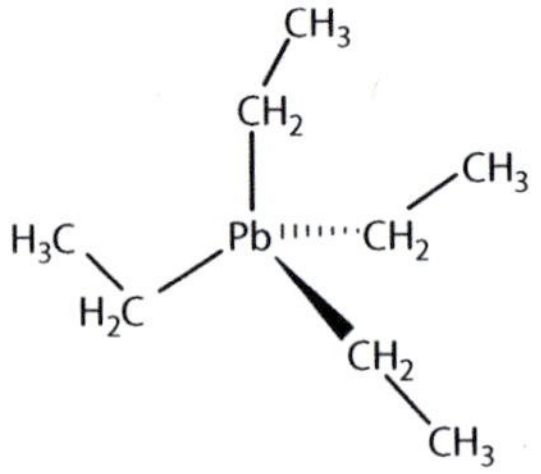

$$(9.2)$$

As a soft and unreactive metal that is easy to extract from its ores, lead has been mined from antiquity. The ancient Romans had many uses for it: in pipes, paint, cooking vessels, and even in the sweetener they added to food—"sugar of lead" (lead ethanoate). They did not appreciate the toxic nature of lead, which interferes with the synthesis of haemoglobin and also inhibits enzymes with –SH groups. As time went on, more uses were found for lead and its compounds that accompanied changes in society and science: lead roofs, lead printing type, and lead bullets.

Tetraethyllead (9.2) is a molecule that was of worldwide importance for about 50 years. It was first synthesised in 1853 by Löwig, and George Buckton achieved its characterisation in 1859. Löwig's synthesis from Pb–Na amalgam and iodoethane was to be adopted as the usual route to large-scale production nearly a century later. Tetraethyllead was for many years a molecule without an application, until the advent of the motor car. Problems of preignition and knocking, caused by some of the fuel–air mixture not burning at the same time as the rest, got worse as more powerful engines were developed.

Research led by Thomas Midgley at the Research Division of General Motors Company examined a range of organic and organometallic compounds until, on 9 December 1921, a solution of $Pb(CH_2CH_3)_4$ in kerosene was found to give knock-free combustion in an engine, down to a level of 0.025%. Residues of lead oxide on the engine were a problem until it was found that when 1,2-dibromoethane was added to the petrol, the lead content was converted to volatile $PbBr_2$, thus removing the lead from the engine. Public sales of "leaded petrol" began in 1923, with the tetraethyllead eventually being made from a 1:1 Na: Pb alloy and chloroethane:

$$4 \text{ Na/Pb} + 4 \text{ CH}_3\text{CH}_2\text{Cl} \rightarrow (\text{CH}_3\text{CH}_2)_4\text{Pb} + 4 \text{ NaCl} + 3 \text{ Pb}$$

From the start, there were problems arising from the toxicity of $Pb(CH_2CH_3)_4$ due to its covalent nature; its volatility meant that the vapour was readily inhaled, whilst the lipid-soluble liquid could be absorbed through the skin. Midgley himself had to have a month's holiday in Florida in February 1923; worse than that, one worker died of tetraethyllead poisoning in 1922. There were four fatalities in just the month of September 1924 at a Standard Oil plant in New Jersey, as well as eight more in 1923–1925 at plants run by DuPont. Safety precautions were negligible and ventilation standards poor.

The state of New Jersey, and New York and Philadelphia, all suspended sales of leaded petrol, but in May 1926 leaded fuels were back on the market because the manufacturers had introduced new safety regulations. At the time, many health and environmental campaigners were unconvinced of the claim by the manufacturers that tetraethyllead was harmless or that lead emissions into the atmosphere from the exhaust were safe. These fine dusts of lead halides (and others) not only accumulated near roads, but also could even be transported for many hundreds of miles and were detectable in areas like Greenland. It was years before the risks associated with leaded fuel could be proved.

It was the American scientist Clair Cameron Patterson (1922–1995) who discovered that much more lead was entering the environment than in preindustrial times. Patterson realised that many scientists had grossly underestimated the problem of lead pollution, owing to their using contaminated "blank" background samples. In 1969 he showed that lead concentrations in Greenland ice had increased by a factor of at least 200 between 800 BC and AD 1969. The sharpest rise happened after 1940, and he attributed this to the use of lead alkyls in petrol. A later study (1994) went back nearly 8,000 years and showed that lead levels in ice during the times of the ancient Greek and Roman civilizations were much greater than natural values, showing how much lead was mined in the Greek and Roman Empires. At the same time, Herbert Needleman (1927–) linked high bone lead levels with lower I.Q. and with increased inattention, aggression, and delinquency.

In the wake of Rachel Carson's *Silent Spring,* concerns about leaded fuels resurfaced and, together with the introduction of catalytic converters to cars in order to meet new air-quality regulations, this ensured the phasing out of leaded fuels from an increasing section of the world. Other campaigns were concerned with removing lead from canning and paints. In 1920–1921, the General Motors researchers had discovered that benzene and ethanol could be used as antiknock agents (at much higher levels than tetraethyllead), but factors such as the large-scale production of ethanol during the Prohibition era led to its abandonment. Nowadays both these chemicals are used to enhance the octane rating of lead-free petrol.

NERVE AGENTS

In the beginning, no one set out to make nerve agents; following World War I, chemists were trying to improve crop yields. Willy Lange at the Friedrich-Wilhelms University of Berlin was looking for cheaper synthetic alternatives to nicotine as insecticides. In 1932, he and his PhD student, Gerde von Krueger, reacted Ag_2FO_3P, silver monofluorophosphate, with alkyl halides and prepared $(RO)_2P(=O)F$ (R = Me, Et), dimethyl- and diethyl monofluorophosphate, in very low yield. They reported that the vapours caused breathing difficulties and that their vision was affected.

> The fumes of these compounds have a pleasant, slightly aromatic odour. But a few minutes after inhalation there's a feeling of pressure to the larynx and difficulty in breathing. Then a disturbance of consciousness develops, as well as blurred vision and a painful oversensitivity of the eyes towards light. Only after several hours [do] the problems wear off.

In unrelated research, Gerhard Schrader of I. G. Farben in Frankfurt was studying sulfur- and phosphorus-containing compounds as insecticides. On 23 December 1936, he obtained a pure sample of a clear, colourless liquid, which in even very dilute solution (5 ppm) killed 100% of leaf lice on contact. This was accompanied by unwelcome consequences for Schrader and his assistant; even a minute drop spilled in the laboratory gave a vapour of which

> the first symptom noticed was an inexplicable action causing the power of sight to be much weakened in artificial light. In the darkness of early January it was hardly possible to read by electric light, or after working hours to reach my home by car.

$$CH_3CH_2O - \overset{\overset{\displaystyle O}{\|}}{\underset{\overset{\displaystyle |}{\underset{\underset{\displaystyle N}{\overset{\displaystyle \||}{C}}}{}}}{P}} - N \overset{\diagup CH_3}{\diagdown CH_3} \tag{9.3}$$

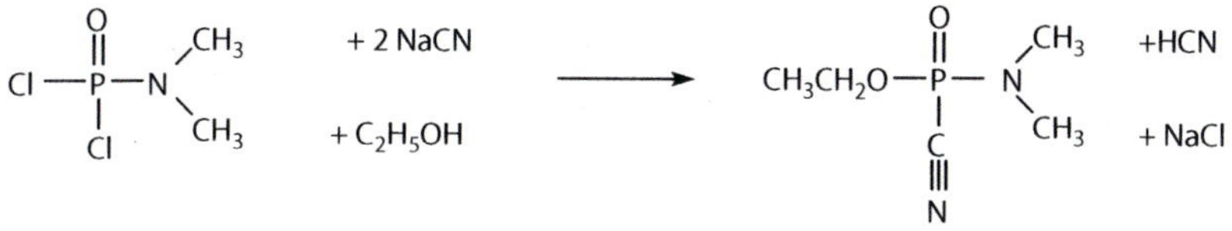

FIGURE 9.1 Making tabun.

He had observed meiosis (contraction of the pupils of the eyes), as well as short-ness of breath. He and his assistant had to stop work for 3 weeks. They had syn-thesised [(EtO)P(=O)(CN)(NMe$_2$)], now known as tabun (GA; 9.3). Animal testing revealed that it was far too toxic to mammals to be marketed as a commercial insec-ticide. Tabun can be made as shown in Figure 9.1.

Schrader had to report this discovery to the Wehrmacht, who began field trials within weeks. Construction started on a plant near Wroclaw for the manufacture of tabun in December 1939; it took 2½ years because of the hazardous nature of tabun and chemicals used in the process, and at least 10 workers were killed during the plant's operation. Meanwhile, in 1938, Schrader synthesised [MeP(=O)F(OCHMe$_2$)], sarin (GB; 9.4), which is significantly more toxic than tabun. The Germans built a pilot plant at Dyhernfurth and a large-scale plant was begun at Falkenhagen near Frankfurt (still incomplete at the end of the war). The Russians captured the plants at Dyhernfurth and Falkenhagen, so they had nerve agent capability. A synthesis of sarin is shown in Figure 9.2.

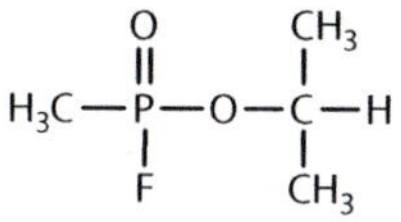

$$(9.4)$$

Evidence indicates that Saddam Hussein used sarin against Kurdish citizens in Halabja on 16–17 March 1988, leaving an estimated 5,000 dead. Tests in 2004 on an exploded Iraqi 155 mm shell (used as an improvised roadside bomb) showed it was a binary weapon containing the precursors for sarin. On 20 March 1995, 10 members of the Japanese *Aum Shinrikyo* cult, travelling on the Tokyo subway system, used

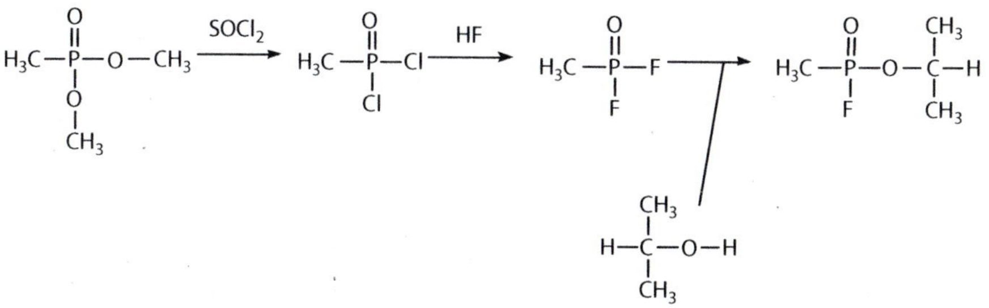

FIGURE 9.2 Synthesis of sarin.

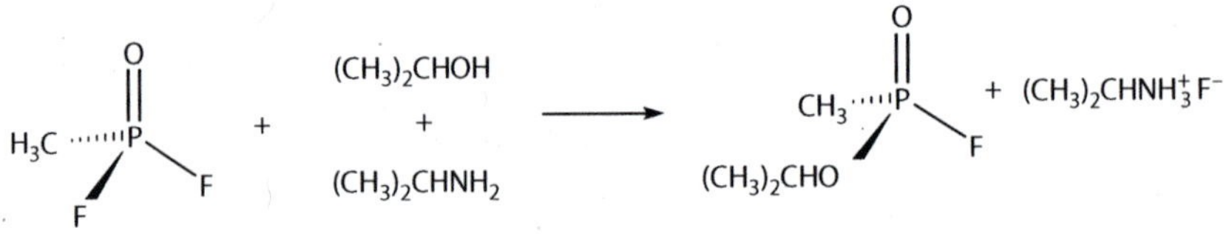

FIGURE 9.3 Synthesis of sarin in a binary weapon.

umbrellas with sharpened tips to puncture bags containing sarin. Many more people would have been killed than the 13 victims, had not the cult members been using very impure sarin.

Sarin tends to decompose upon keeping:

$$MeP(=O)F(OCHMe_2) \rightarrow MeP(=O)F(OH) + MeCH=CH_2$$

It is thus an ideal candidate for use in binary weapons. Separate containers of methylphosphoryldifluoride (DF) and a mixture of propan-2-ol and isopropylamine (OPA) are placed in a shell or rocket (Figure 9.3); when the shell or rocket is fired, the containers are fractured, and the reactants mix in the spinning shell, forming sarin.

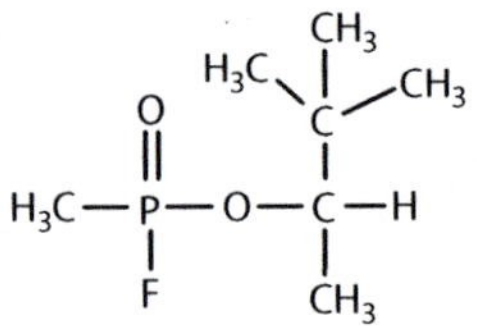

(9.5)

The third of the German nerve gases, [MeP(=O)F(OCHMe(ᵗBu))] (soman; GD; 9.5), was discovered in 1944 and is prepared in a way similar to that for sarin:

$$MeP(O)Cl_2 + MeP(O)F_2 + (CH_3)_3CCH(CH_3)OH \rightarrow$$
$$MePO(F)OCH(CH_3)C(CH_3)_3 + HCl$$

Cyclosarin (GF; 9.6) was also synthesised during World War II. It is more expensive than sarin, but less volatile and thus more persistent. Cyclosarin also demonstrates greater toxicity than sarin (GB) in humans. It was used by Iraq as a mixture with sarin, possibly because it was easier to obtain the cyclohexanol precursor.

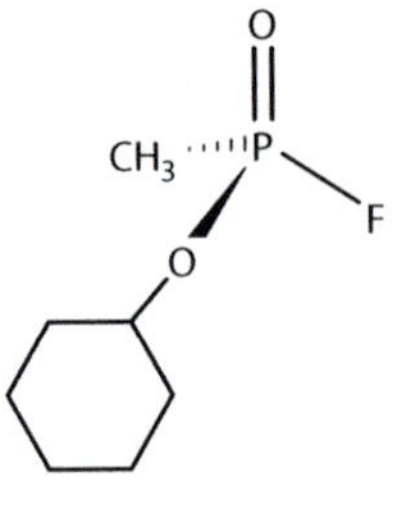

(9.6)

All these agents feature a P=O bond, as well as a halide (or pseudohalide) leaving group.

By the summer of 1943, the British were aware that the Germans were making nerve agents, but already B. C. Saunders of the University of Cambridge knew of Lange's published work and had synthesised (1941) some esters of monofluorophosphoric acid, notably [(Me$_2$CHO)$_2$P(=O)F], diisopropyl fluorophosphate (DFP; 9.7). Sanders and a volunteer sat in a sealed room and put a drop of DFP on a dish. After 1 hour, Saunders admitted that he felt strange and noticed that the room had gone very dark; he found that meiosis (pupil constriction) could last for up to 7 days. DFP was less toxic than sarin or tabun and was considered to offer no advantages over mustard gas and phosgene.

$$\text{H}-\underset{\underset{\text{CH}_3}{|}}{\overset{\overset{\text{CH}_3}{|}}{\text{C}}}-\text{O}-\underset{\underset{\text{F}}{|}}{\overset{\overset{\text{O}}{\|}}{\text{P}}}-\text{O}-\underset{\underset{\text{CH}_3}{|}}{\overset{\overset{\text{CH}_3}{|}}{\text{C}}}-\text{H} \tag{9.7}$$

$$\text{CH}_3\text{CH}_2\text{O}-\underset{\underset{\text{CH}_3}{|}}{\overset{\overset{\text{O}}{\|}}{\text{P}}}-\text{SCH}_2\text{CH}_2\text{N} \tag{9.8}$$

The first new postwar agent, VX (9.8), was made by Dr Ranajit Ghosh, in 1952, while he was trying to make pesticides at ICI. It was one of a family of "V-agents". When its toxic potential was discovered, details were passed on to the Porton Down Chemical Weapons Research Centre. It was independently and simultaneously produced by the Swedish scientist Lars-Erik Tammelin. In 1958, the UK traded knowledge of VX with the United States for information on thermonuclear weapons, and the United States started to make VX in 1961. VX is less volatile but more toxic than the preceding agents, so aerosols of VX are very dangerous; it is also absorbed through the skin. The Soviets prepared an isomeric version of VX (9.9), possibly through incomplete intelligence. On 13 March 1968, an escape of VX from the chemical weapon proving ground at Dugway killed more than 3,000 sheep grazing 27 miles away in the Skull Valley area of Utah. The only known human fatality arose when two members of the *Aum Shinrikyo* cult used VX to assassinate a former number of their sect in Osaka in 1994.

$$(\text{CH}_3)_2\text{CHCH}_2\text{O}\diagdown\overset{\overset{\text{O}}{\|}}{\underset{\text{CH}_3}{\text{P}}}\diagdown\text{SCH}_2\text{CH}_2\text{N}(\text{CH}_2\text{CH}_3)_2 \tag{9.9}$$

Subsequent Russian research in the 1970s and 1980s led to the development of the новцчок (novichok = newcomer) agents, some of which were claimed to be up to an order of magnitude more potent than VX.

These are organophosphorus compounds that have a substituted oxime group (9.10, 9.11).

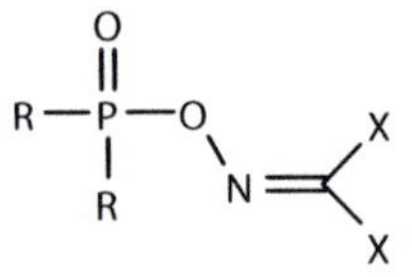

(9.10)

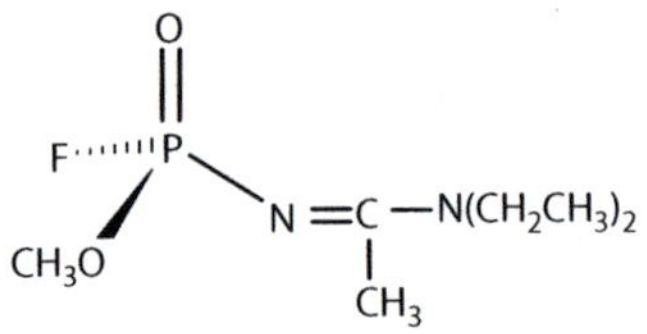

(9.11)

A-232, aka Novichok-5

Nerve agents work by disrupting the central nervous system. When nerve impulses reach a synapse (gap between cells), release of the neurotransmitter acetylcholine (ACh) is triggered. ACh migrates to the other end of the synapse and docks in a receptor in the new cell, triggering a nerve impulse in the second cell. The receptor must now release (and destroy) the acetylcholine, as if the ACh builds up there will be continuous new impulses and the nervous system will be overstimulated.

Acetylcholine is broken down by the enzyme acetylcholinesterase (AChE). In the diagram in Figure 9.4, acetylcholine binds to a serine residue, transferring the acetyl group. Upon hydrolysis, the serine group is freed, regenerating the enzyme.

Nerve gases such as sarin attack AChE (shown in Figure 9.5). All nerve agents feature a bond like P–F that is easily broken. The initial adduct rapidly eliminates HF and "ages", forming a new adduct that cannot be regenerated. This means that the active site of the enzyme is blocked and it cannot function (Figure 9.5).

At present, soldiers going into battle have limited options because the body's natural defensive paroxonase enzymes, which can regenerate AChE, are likely to be overwhelmed. If pyridostigmine bromide (PB) is taken in advance, it inhibits AChE and thus protects some of the body's AChE from the nerve agent. After the attack, because PB is a reversible inhibitor, it releases itself, freeing up AChE. Alternatively, a mixture of atropine (an alkaloid) and PAM (2-pyridinealdoxime methiodide), is taken as an injection. The atropine blocks the ACh receptor so that no message can be transmitted; there is nothing for the ACh to bind to. The PAM displaces the nerve agent from the AChE so that the enzyme can work again. PAM only works with sarin and VX, and it has to be administered quickly, before the adduct "ages". (With soman and tabun, obidoxime is used.)

Rodents are protected against nerve agents because they have high levels of carboxylesterase enzymes, which act on the nerve agent before it can bind to the enzyme and inhibit it. Human carboxylesterase 1 (hCE1) is a broad-spectrum enzyme that

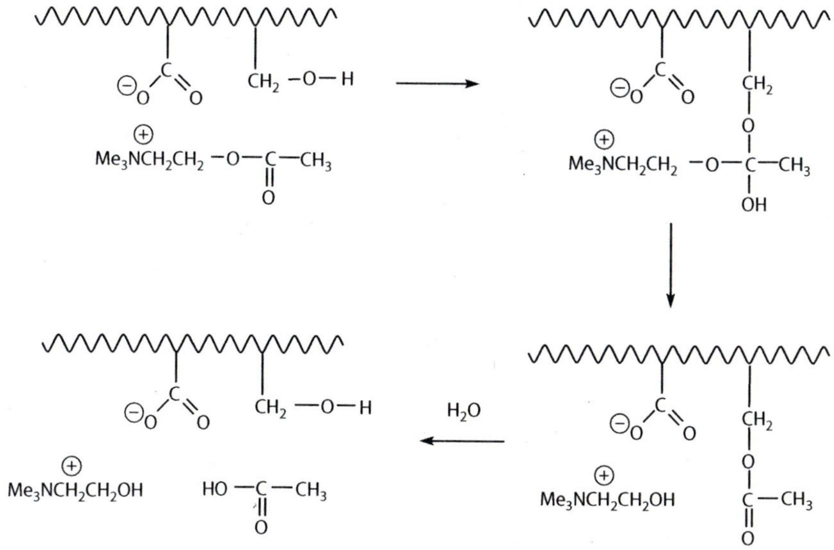

FIGURE 9.4 Acetylcholine binds to a serine residue; the serine residue can be regenerated by hydrolysis.

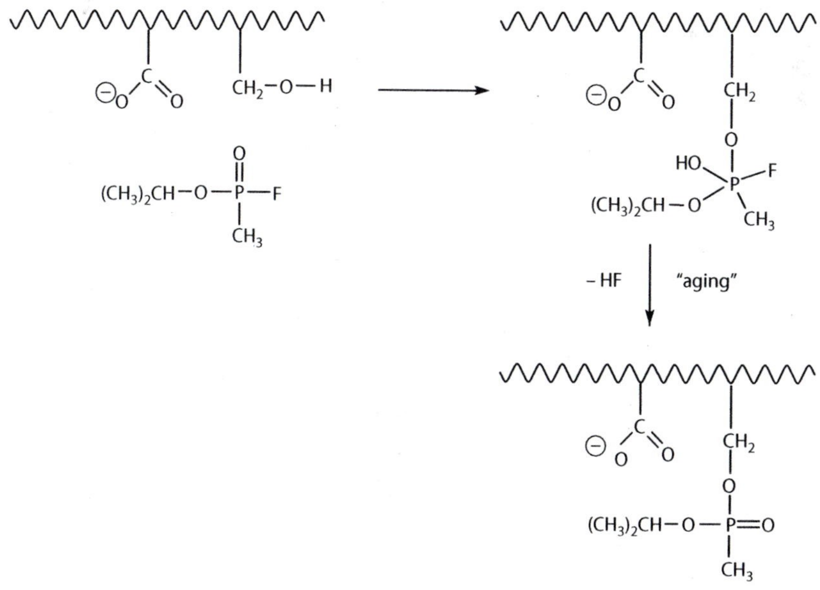

FIGURE 9.5 The active site of the enzyme is blocked by sarin.

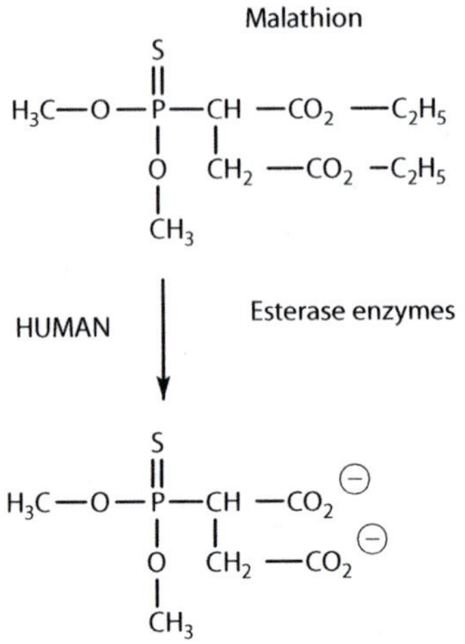

FIGURE 9.6 In insects, malathion changes to a P=O compound, a very toxic acetylcholinesterase inhibitor.

metabolises a wide range of drugs; it has only weak activity against nerve agents, so scientists are seeking to modify it so that it could be an effective enzyme to remove a nerve agent catalytically.

Organophosphorus insecticides work in the same sort of way as nerve gases, though they are generally less toxic. Some combine toxicity to insects with safety to humans, as with **malathion;** like many others, it has a P=S group rather than a P=O. Humans contain sufficient esterase enzymes to hydrolyse the ester linkages and convert it to a nontoxic metabolite, whereas insects lacking esterase enzymes instead oxidise it to a P=O compound, a very toxic acetylcholinesterase inhibitor (Figure 9.6). Thus, the ACh builds up in the fly, exciting it more and more, until it falls to the ground with its legs twitching in the air, having been excited to death.

Parathion (9.12), though, is another matter. The P=S group is oxidised inside the insect, forming the very toxic paraoxon (9.13). Whereas malathion has a toxicity of well over 1000 mg/kg in rats but only 17 mg/kg in flies, parathion has a similar toxicity in rats of around 10 mg/kg. It has killed hundreds of agricultural workers and is now little used as an insecticide because of the risk it poses to humans. Parathion has been responsible for deaths through contamination of foodstuffs (e.g., flour) and has also been used to commit suicide.

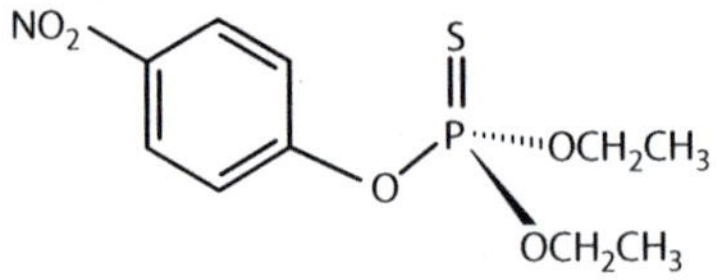

(9.12)

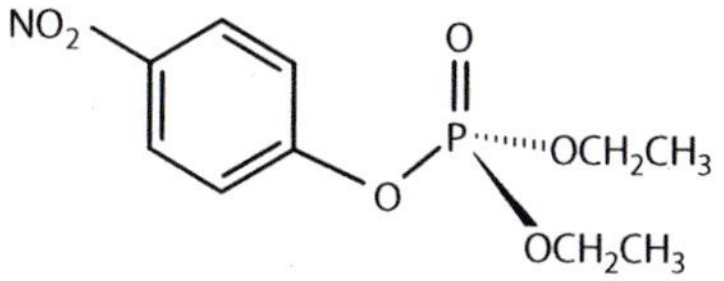

$$\text{(structure: NO}_2\text{-phenyl-O-P(=O)(OCH}_2\text{CH}_3\text{)OCH}_2\text{CH}_3\text{)}$$

(9.13)

Certain organophosphate compounds like **phosmet** (9.14) were used in sheep dips on UK farms, until 1992, to kill insect pests such as the warble fly; a number of farmers believe that they were poisoned by these compounds.

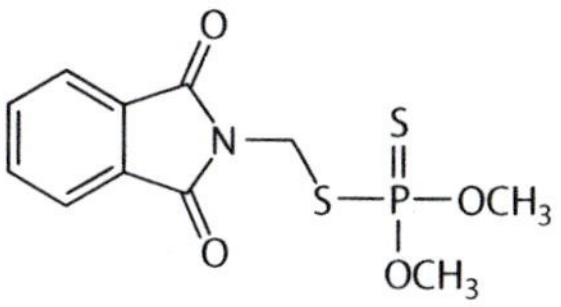

(9.14)

CS AND MUSTARD GAS

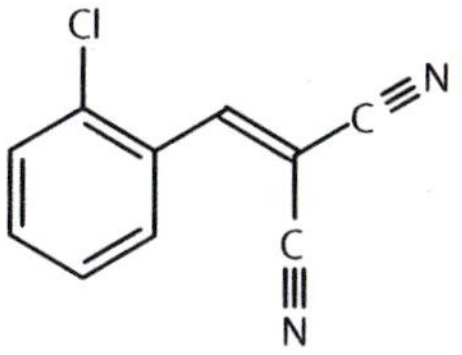

(9.15)

For a start, CS (9.15) is not a gas; 2-chlorobenzylidenemalonitrile is a solid (m.p. 93°C) at room temperature, and the riot control agent is usually a 1% solution of 2-chlorobenzylidenemalonitrile in the solvent methylisobutylketone. It is normally used as an aerosol spray, though it can also be projected in hand grenades or shells. The key molecule was discovered by Ben Corson and Roger Stoughton, two teachers at an American college, who published their findings in 1928 in a paper that reported several compounds with similar structures. They noted that several of these compounds were harmless when wet but that "to handle the dry powder is disastrous". Sneezing due to 2-chlorobenzylidenemalonitrile (CS) caused the face to smart; when 3-nitrobenzylidenemalonitrile was inhaled, sneezing produced a mucus that turned bright yellow on exposure to air.

No one took much notice of CS until the late 1950s, when scientists at the British Chemical Defence Unit at Porton Down learned how to use it. It is regarded as a safe riot control gas. Concentration of around 0.005 mgm^{-3} in air causes discomfort and skin irritation; between 1 and 5 mgm^{-3} in air will incapacitate, but the lethal dose is reckoned to be around 70,000 mgm^{-3}. It is believed to be safe around animals but not near small children. CS molecules are effective because they attach to enzymes in the mucous membrane of the eye, inducing a protective release of tears to dissolve and wash away the offending molecules. It produces a burning sensation; the subject's eyes will jam shut and there will be copious generation of tears as well as mucus. Other symptoms can include retching and vomiting, restricted breathing, and tightness in the chest. Symptoms will persist for a few minutes. If it is not washed from the skin, blistering can result.

In the 1960s Vietnam War, the American army used CS to drive the Vietcong out of underground bunkers, and it was used on the streets during rioting in Northern Ireland in 1969–1972. Its use indoors by police is not recommended because it can bring on panic attacks, and this is believed to have happened during the action terminating the Branch Davidian siege near Waco, Texas, on 19 April 1993.

Phenacyl chloride (CN; better known as "mace"; 9.16) was the standard US riot control gas until it was replaced by CS in 1969. Its symptoms are milder that CS, but longer lived. Pepper spray, an aerosol of capsicum in water, is an alternative to CS as a short-range lachrymatory agent for personal defence. All riot control agents are banned from military use under the Chemical Weapons Convention, but they are used as riot control sprays by police in many countries.

(9.16)

MUSTARD GAS

(9.17)

1,5-Dichloro-3-thiapentane (9.17) is the best known sulfur mustard agent. Widely used in World War I, it leaves large blisters on skin that comes into contact with it and freely penetrates clothing. The pure substance is a colourless, odourless liquid (b.p. 218°C), but the impure material is often a yellow–brown colour with a sharp mustard-like smell—hence, the name. It is usually dispersed in aerosols or by bombs and shells. Mustard gas was first used by the German army at Ypres in 1917 and subsequently used on both sides. Since then it has been used in other conflicts, allegedly most recently by Iraq against Iran in the 1983–1988 conflict and by Saddam Hussein against the Kurds. Its stockpiling and use are prohibited by the 1993 Chemical Weapons Convention. Mustard agents like this are insidious: Symptoms take hours to develop, with itching and irritation giving way to large, fluid-filled blisters and the eyes and lungs burned. Only patients with severe burns are likely to die; in practice, most victims have been incapacitated.

1,5-Dichloro-3-thiapentane and other sulfur mustards are toxic because they readily lose a chloride ion forming a reactive sulfonium ion, which alkylates guanine in strands of DNA, causing severe cell damage. Mustard agents found only limited use in World War II, but in 1943–1944, mustard gas was tested on Australian volunteers, with some unpleasant injuries. German bombing of US ships in Bari Harbour on the evening of 2 December 1943 led to an explosion on board the SS *John Harvey* and release of mustard gas from munitions; 83 died and over 500 people were injured. It was noted at the time that many casualties had significantly reduced white cell counts in their blood, and this led to the discovery of a nitrogen mustard, later known as mustine (mechlorethamine, 9.18), for therapy of Hodgkin's lymphoma.

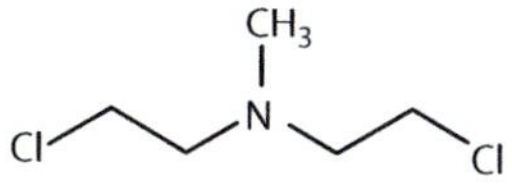

(9.18)

THALIDOMIDE

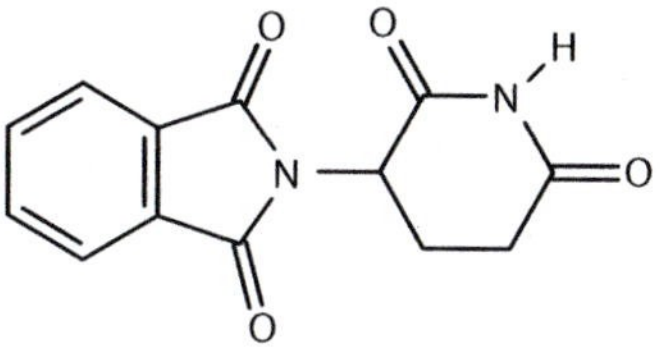

(9.19)

Thalidomide (9.19) is usually cited as the drug that did the most to destroy trust in medicines, and even now the name is notorious. Thousands of women taking it between 1957 and 1961 gave birth to babies with severe birth defects; the most striking and common were undeveloped limbs—hands and feet were attached directly to the body (phocomelia).

In the early 1950s, a West German firm, Chemie Grünenthal, found that thalidomide had a number of useful properties. They patented the drug in 1954, and it was licensed in 1957 to the Distillers Company, who marketed it in the UK as "Distaval". It was also used in 45 other countries (including Europe and Australia). It was seen to be a safe alternative to barbiturates as a sedative and could be taken as a sleeping pill free of side effects. It had no effect upon pregnant rats; it was an antiemetic, and pregnant women were prescribed thalidomide to counter the nausea of morning sickness.

Chemie Grünenthal licensed it to the Richardson-Merrell company for the US and Canadian markets in January 1959. Richardson-Merrell filed a new drug application with the FDA on 12 September 1960, expecting approval to be granted early in 1961.

In August 1960, Dr Frances Oldham Kelsey moved to a post with the Food and Drug Administration in Washington, DC. One month later, she was given her first drug to review: thalidomide. Under FDA law, she had 60 days to review the application; she and colleagues had concerns and returned the application to Richardson-Merrell, asking for more safety information; they had to resubmit the application with the required data. The 60-day period started again. Kelsey's husband, also a pharmacologist working for the FDA, produced devastating criticisms. Meanwhile, she read a letter in the 31 December 1960 issue of the *British Medical Journal,* by Dr A. Leslie Florence of Aberdeen, reporting four cases of peripheral neuritis (irreversible numbness in the limbs) in patients taking thalidomide, and she asked Richardson-Merrell for more evidence from animal tests. She then demanded evidence that thalidomide was safe to take during pregnancy.

By this time, an Australian obstetrician named William McBride had realised that three mothers of deformed babies who died after birth had been taking thalidomide. He wrote a letter to the *Lancet* reporting this; the letter was published on 16 December 1961. Such deformities (phocomelia) had already been reported by two West German doctors, and in spring 1961 children with malformed limbs were born to both the sister and the wife of a young Hamburg lawyer called Karl Schulte-Hillen. He went to see Professor Widukind Lenz, head of the children's clinic at Hamburg University, who started to make enquiries; by November 1961, Lenz knew of 14 such cases and contacted Grünenthal, who withdrew thalidomide from the German market after the story broke in the leading newspaper, *Welt am Sonntag*, and informed licensees. Richardson-Merrell withdrew its FDA application on 8 March 1962, as well as withdrawing the drug from Canada. Cases of phocomelia stopped as soon as the drug was withdrawn.

By her very thoroughness, Frances Oldham Kelsey's actions meant that people became aware of the emerging birth defects due to thalidomide before it could be marketed in the United States. As a result of this case, the US Congress passed legislation requiring tests that showed drugs were safe for pregnant women, as did many other countries. Around 5,000 thalidomide babies survived past childhood; it is believed that at least twice as many died from it. Many survivors have gone on to have successful and fulfilling lives, despite their handicap.

One theory about the teratogenic action of thalidomide is that, because its structure resembles two of the DNA bases (guanine and adenine), thalidomide is believed to intercalate (slide) into the double helix structure between the paired bases, particularly at guanine-rich sites. It interferes with synthesis of integrin proteins; this in turn inhibits angiogenesis, stopping blood vessels forming to supply growing embryonic tissue, so certain limbs do not grow.

A team of scientists led by Dr Neil Vargesson of the University of Aberdeen studied the effect of thalidomide itself and some of its metabolites upon developing chick embryos (2009). These metabolites have either anti-inflammatory or antiangiogenic properties. None of the anti-inflammatory compounds had any effect upon limb development; just one molecule that they studied, CPS-49 (9.20), a fluorinated analogue of some thalidomide metabolites with antiangiogenic properties, produces the same limb defects as thalidomide itself. It inhibits the growth of new foetal blood vessels. Limb formation in the human foetus occurs at ca. 5–8 weeks, and at this time the limbs have a rapidly changing vessel pattern. By this time, the vascular system in the rest of the body is complete. Thalidomide causes the death of cells and stops limbs from forming, which is why only women taking thalidomide in the 34- to 50-day window following the last menstruation were at risk of giving birth to deformed babies.

Because there is a chiral carbon atom in the thalidomide molecule, it exists as two enantiomers. Both isomers of thalidomide have sedative properties; only the (S)-isomer is a teratogen (a producer of birth defects). It was used as a racemic mixture of the isomers; however, the isomers quickly interconvert in vivo (the half-life at pH 7.4 and 37°C is about 9 hours) so that even if only the pure (R)-enantiomer were used, both isomers would soon be found in the body and birth defects could still occur.

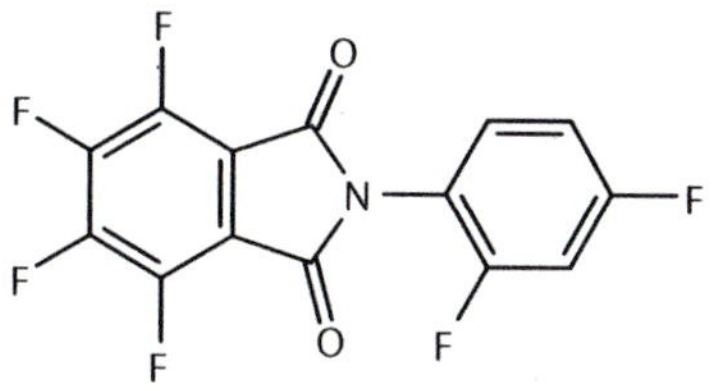

$$(9.20)$$

After thalidomide was withdrawn from sale, research continued. In 1965, an Israeli doctor, Jacob Sheskin, was faced with a leprosy patient who had a very painful complication of erythema nodosum laprosum (ENL; also known as Hansen's disease). This bedridden patient was on the verge of death and totally unable to sleep. No sedative worked. Dr Sheskin recalled that thalidomide was a sedative and gave him two tablets. The patient slept for 20 hours and, remarkably, was able to get out of bed. Two more tablets caused the pain to leave and his leprosy sores began to heal. Further study confirmed the treatment.

Thalidomide is the drug of choice worldwide for this form of leprosy. The FDA has given its approval to this use of thalidomide, marketed as "Thalomid", under very restricted conditions. It has also been used to treat multiple myeloma and is effective in some forms of cancer therapy because it shuts down the blood supply to tumours. It can still be used as a sedative. Some people feel it should still be banned; others, including former thalidomide victims, believe that it can be used under carefully controlled conditions because it is such a good treatment for leprosy and multiple myeloma. But it must never again be administered to pregnant women.

Canadian-born Frances Kathleen Oldham married Dr Fremont Ellis Kelsey in 1943 and is thus usually known as Frances Oldham Kelsey. While she was carrying out her PhD research, her PhD supervisor asked her to carry out animal tests, which established that the toxicity of elixir of sulfanilamide (see p. 193) was due to the solvent, not the sulfanilamide. In tests on antimalarials a few years later, she noted that pregnant rabbits could not metabolise quinine as well as normal rabbits and that rabbit embryos could not break it down at all. She remembered these tests later when it came to considering approval of thalidomide. On 8 August 1962, Frances Oldham Kelsey received the President's Award for Distinguished Federal Civilian Service, the highest honour that could be given to an American civilian, from the hand of President John F. Kennedy. Kelsey became head of the investigational drug branch of the FDA. She continued working for the FDA until she retired as a deputy director in 2005, aged 90, a heroine.

10 Explosives

As mentioned in Chapter 4, ammonia has a vital role in the manufacture of fertilisers, without which current world populations could not be sustained and fed. Ammonia can also be converted into nitric acid, an essential material in the manufacture of many explosives. Explosives are substances that release large quantities of hot gas in a very rapid, violent, usually exothermic reaction. They generally contain an oxidizing agent together with a fuel that is easy to oxidize, producing lots of gas.

Gunpowder is a mixture of potassium nitrate oxidant that, together with sulfur and carbon, produces large amounts of nitrogen and carbon dioxide gas:

$$S(s) + 3C(s) + 2\,KNO_3(s) \rightarrow 3CO_2(g) + N_2(g) + K_2S(s)$$

It is also known as black powder (that is what it looks like). It was first made in China, around the first millennium, whilst in the West the Franciscan monk Roger Bacon had a recipe for making it in the mid-thirteenth century. It is straightforward to make up and handle and can be set off easily. However, it is a weak explosive and also produces smoke when ignited, giving away the user's position on the battlefield. It is also useless when damp, giving rise to the saying "keeping your powder dry". Thus, in warfare, gunpowder was superseded as a propellant by smokeless powders, such as cordite, and supplanted by nitrocompounds as a shell filling.

Ammonium nitrate, $NH_4^+\,NO_3^-$, was first used in explosives in the late nineteenth century. It is a very cheap oxidant, but the moisture absorbed from the air can be a problem because it needs to be sealed in waterproof wrappings. The Texas City disaster of 1947, which killed nearly 600 people, resulted from the explosion of some 3,000 tons of ammonium nitrate aboard the *SS Grandcamp* and *SS High Flyer;* the ammonium nitrate was in paper bags whose carbon content assisted the explosions. Other ammonium nitrate explosions have occurred down to the present day, but it has continued to be used in explosives—not least because it is 25% stronger than TNT.

At 300°C, ammonium nitrate decomposes:

$$NH_4NO_3(s) \rightarrow N_2(g) + 2\,H_2O(g) + \tfrac{1}{2}\,O_2(g)$$

Unusually for an explosive, it contains more oxygen than it needs to burn completely; it is said to be "overoxidized" and this property can be put to good use in mixtures with other explosive materials. In the 1950s, it was discovered that ammonium nitrate fuel oil (ANFO) mixtures made good mining explosives:

$$1/n\,(CH_2)_n(l) + 3\,NH_4NO_3(s) \rightarrow 3\,N_2(g) + 7\,H_2O(g) + CO_2(g)$$

This material has also been used by terrorists, as in the unsuccessful 1993 attempt to destroy the World Trade Centre in New York. A similar mixture, ammonium nitrate–nitromethane (ANNM), was used by Timothy McVeigh and Terry Nichols in the destruction of the Federal Building in Oklahoma City on 19 April 1995 that killed 168 people.

$$2 \, CH_3NO_2(l) + 3 \, NH_4NO_3(s) \rightarrow 4 \, N_2(g) + 9 \, H_2O(g) + 2 \, CO_2(g)$$

Following the discovery of nitration reactions, usually employing mixtures of concentrated nitric and sulfuric acids ("nitrating mixture"), the nineteenth century was to see the discovery of many explosive materials.
In 1845, Christopher F. Schönbein was one of a number of chemists to investigate the nitration of cotton, using "nitrating mixture" to form **guncotton,** trinitrocellulose (10.1). The first guncottons were unstable, until the need for removal of residual acid was appreciated. Nowadays paper or wood shavings are used as the source of cellulose.

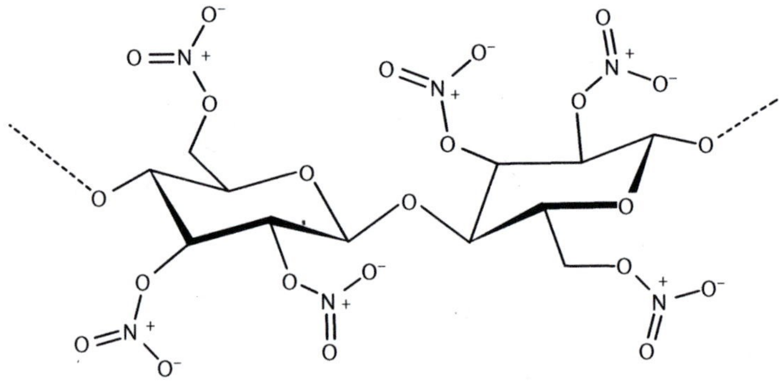

(10.1)

A mixture of guncotton and colloidon (dinitrocellulose) was used to make "Poudre B", a smokeless powder that was also three times stronger than gunpowder (1886). This was rapidly succeeded by much better versions in the form of **ballistite** (Nobel) and **cordite;** the latter is a mixture of nitroglycerine with guncotton and petroleum jelly stabiliser mixed into a paste with acetone solvent and then extruded through dies, like macaroni or spaghetti. This "cord powder" became known as cordite.
Nitroglycerine (10.2) was first made from glycerine (glycerol) and nitric acid in the presence of sulfuric acid by Ascanio Sobrero, in 1846.

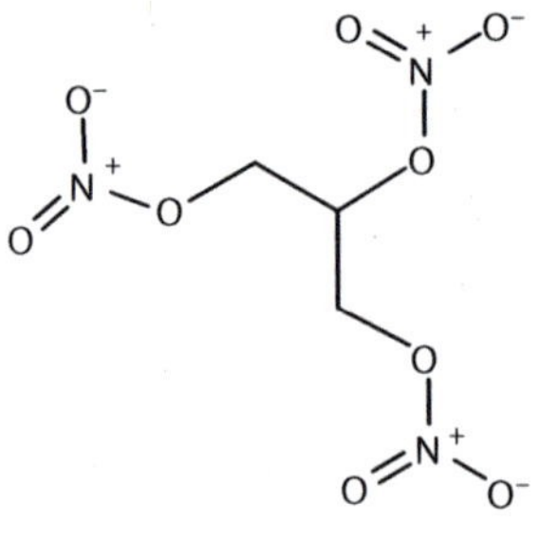

(10.2)

When nitroglycerine is detonated, a lot of gas is formed and, like ammonium nitrate, it is "overoxidised":

$$C_3H_5O_9(l) \rightarrow 1.5\ N_2(g) + 2.5\ H_2O(g) + 3\ CO_2(g) + 0.25\ O_2(g)$$

Nitroglycerine is a colourless, oily liquid at room temperature and has the major drawback of being very shock sensitive (a property exploited in numerous films). Alfred Nobel circumvented this (1867) when he found that nitroglycerine was soaked up by the porous rock known as kieselguhr, making dynamite (and Nobel's fortune). This was a much more stable explosive, with greater power than TNT. A mixture of nitrocellulose and nitroglycerine, known as **blasting gelatine,** is used in road construction to this day.

Nitration of phenol affords **picric acid** (10.3), 2,4,6-trinitrophenol.

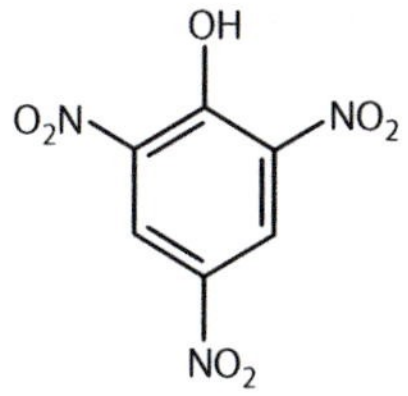

$$\text{(10.3)}$$

Known as lyddite, it was first used by the British Army at the Battle of Omdurman (1898). To their cost, the British carried this into World War I as their major explosive; the Germans had adopted the superior TNT. The lesson was brought home at the Battle of Jutland (1916), when British lyddite shells tended to explode on impact with German armour rather than penetrating. The markedly acidic nature of picric acid also meant that it tended to react with metals in shells, forming dangerously unstable heavy metal salts (e.g., lead picrate).

The explosive yellow **TNT** is 2,4,6-trinitrotoluene (10.4). First isolated pure in 1880, it is the most important of several possible isomers of trinitrotoluene.

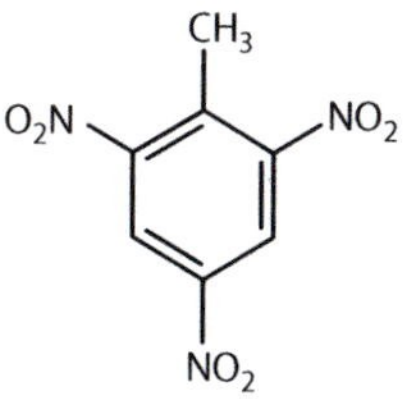

$$\text{(10.4)}$$

Its commercial manufacture involves three steps to insert the three nitrogroups successively into methylbenzene; substitution of the third requires the use of fuming nitric and sulfuric acids. It produces a great deal of gas on explosion:

$$2\ CH_3C_6H_2(NO_2)_3(s) \rightarrow 3\ N_2(g) + 5\ H_2O(g) + 7\ CO(g) + 7\ C(s)$$

Neither such a powerful explosive as picric acid nor as easy to detonate, TNT has several advantages—not least its high-velocity shockwave upon explosion, as the

Germans found, adopting it as their standard military high explosive in 1902. Its low melting point (80°C) means it is easier to fill shells with TNT than with some other explosives. TNT is toxic to humans, and in World War I there were numerous fatalities amongst munitions workers in shell-filling factories. TNT became in short supply; hence, a mixture of TNT with ammonium nitrate, known as **amatol,** came into widespread use. A further advantage of adding oxygen-rich ammonium nitrate is that amatol explodes with a white or grey smoke instead of the black smoke of TNT.

First made in 1898 from hexamine and conc. nitric acid, **RDX** (research department explosive; 10.5), also known as cyclonite, was in widespread use during World War II. It was recently employed in the 2006 Mumbai train bombings.

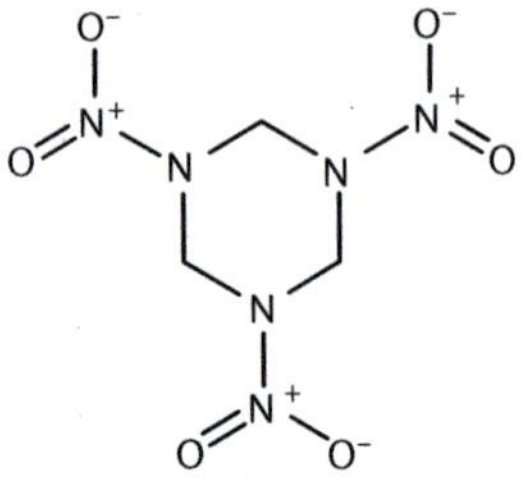

(10.5)

Along with PETN, it is an ingredient in the plastic explosive Semtex; when mixed with TNT and powdered aluminium, it forms **Torpex,** the explosive filling for the Upkeep, Tallboy, and Grand Slam bombs used by the RAF's 617 Squadron in World War II. Aluminium is added to "overoxidised" explosives because its reaction with oxygen will yield extra heat and a greater volume of hot gas. Although not as stable to shock and impact as RDX, **PETN** (pentaerythritol tetranitrate; 10.6) is another very powerful high explosive, similarly discovered half a century before finding an application. Prepared by reaction of pentaerythritol with nitric acid, it is another key ingredient of Semtex.

$$C(CH_2OH)_4 + 4\ HNO_3\ (conc.) \rightarrow C(CH_2ONO_2)_4 + 4\ H_2O$$

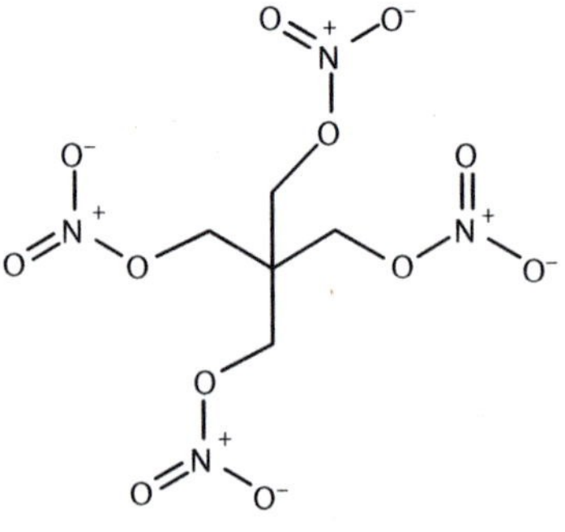

(10.6)

Plastic explosives are what it says on the tin: explosives that are soft and easy to shape by hand pressure. One tends to associate these with the materials used by the French Resistance or parachuted agents to sabotage railways and bridges in German-occupied France in World War II. This material ("Nobel 808") looked like green plasticine and smelled like marzipan (almonds), doubtless due to aromatic nitrocompounds, and was usually detonated by time-pencils. Made by a Czech firm since the 1960s, **Semtex**, a mixture of PETN and RDX in a polymer matrix, is the most infamous plastic explosive. It is very malleable, stays plastic between –40°C and +60°C, and is waterproof and very stable unless detonated. A Semtex-based device was used to bring down PanAm Flight 103 over Lockerbie, Scotland, on 21 December 1988. All 259 people on board died, as well as 11 on the ground. Semtex manufactured since 1991 incorporates a taggant so that it can be detected, usually *p*-nitrotoluene or DMDNB, 2,3-dimethyl-2,3-dinitrobutane. **C-4** is another present-day plastic explosive, largely RDX, with a plasticiser and a taggant. On 12 October 2000, terrorists used C-4 to attack the *USS Cole* while it was refuelling in the port of Aden, Yemen; 17 sailors were killed.

Liquid explosives are one of the terrorist weapons to come into the limelight following the attack on the Twin Towers in New York on 11 September 2001; this has meant restrictions upon liquids that can be carried by airlines as luggage. The main reason for this is **triacetone triperoxide** (**TATP**; 10.7), which can readily be produced by the reaction of acetone (propanone) with hydrogen peroxide in the presence of an acid, usually H_2SO_4.

$$\text{(10.7)}$$

Several forms are known, but the trimer is what is usually produced, rather than the dimer and tetramer. A white solid, it is especially unstable to being set off by shock or friction when dry. It has become of interest to potential terrorists since, unlike most explosives, it does not contain the type of nitrocompounds detected by airport scanners and it can be produced easily with starting materials that are readily obtainable. Its instability is a problem for any terrorist, and at least one person using a TATP device has been seriously injured. It has been linked with several arrested members of terrorist groups. It is believed to have been used by the bombers in London on 7 July 2005, whilst the "shoe bomber", Richard Reid, is reported to have used it in the fuse of his explosive device on 22 December 2001.

$$2\ C_9H_{18}O_6(s) + 21\ O_2(g) \rightarrow 18\ H_2O(g) + 18\ CO_2(g)$$

TATP is an entropy-driven explosive; little heat is released, but several vapour molecules are generated for each molecule of TATP.

Most explosives are organic nitrocompounds for a combination of reasons:

- The products of the reaction contain very strong $N\equiv N$, $C=O$, and $O-H$ bonds, which make the reaction very exothermic.
- The products of the reaction are gases, contributing to a very favourable entropy change.
- The large volume of gas generated in a very short space of time leads to a strong explosive force.
- The molecules are self-oxidising. This means that they contain within themselves oxygen atoms needed for the reaction to take place—atoms within a very short distance of the atoms with which they will combine.
- Aromatic nitrocompounds are sterically congested, creating a strain favouring the decomposition.

Lead azide, $Pb(N_3)_2$, is a very shock-sensitive primary explosive used in detonators to initiate secondary explosives. It has been employed in aircraft ejection seats. The azide that finds the greatest application is **sodium azide,** NaN_3, which is the key substance in the gas generators that inflate automobile air bags. When a sensor in the car detects a sudden deceleration, an electrical impulse heats up the sodium azide to over $300°C$; it decomposes very rapidly into sodium metal and nitrogen gas, inflating the airbag:

$$2\ NaN_3(s) \rightarrow 2\ Na(l) + 3\ N_2(g)$$

The gas generator also contains potassium nitrate and silicon dioxide. Their purpose is to remove the reactive and dangerous sodium metal; the final product is an inert, harmless glass:

$$10\ Na + 2\ KNO_3 \rightarrow K_2O + 5\ Na_2O + N_2$$

$$Na_2O + SiO_2 \rightarrow Na_2SiO_3$$

Disposing of airbags because of the hazards associated with the sodium azide is a serious aspect of car scrappage.

DETECTING EXPLOSIVES

Detecting explosives is big business, whether for airports or for soldiers in the front line, particularly in the post-2001 climate. Dogs are traditionally used for this and even today they are used to detect land mines in Afghanistan. When they search for TNT, what they are actually detecting is 2,4-dinitrotoluene (DNT) impurity from incompletely nitrated toluene. DNT is a lighter molecule than TNT, with a much higher vapour pressure. Increasingly, interest lies in instrumental methods (less fallible and fatigue free), such as artificial noses. One of these has seen field trials in Iraq: It sucks a stream of air over a polymeric fluorescent sensor, and any nitroaromatics bind to the sensor and quench its fluorescence. A colourimetric sensor has also been developed to test for TATP vapour by means of detecting the traces of H_2O_2 impurities at levels of 2 ppb.

11 Pleasure Molecules

PHENYLETHYLAMINE

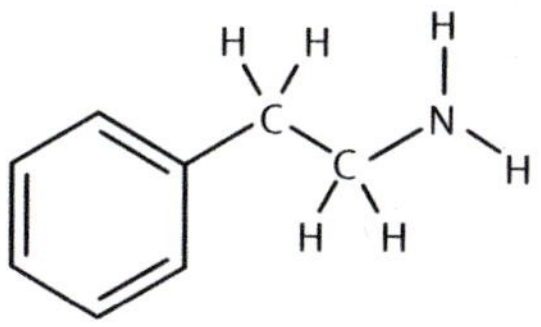

$$(11.1)$$

Phenylethylamine (11.1) or, more strictly, 2-phenylethylamine is the structural parent of the amphetamines, but, unlike them, occurs naturally. The human body makes it by decarboxylation of phenylalanine, albeit in very small amounts; it acts in the brain in a similar way to amphetamines by releasing the neurotransmitters dopamine and norepinephrine. Phenylethylamine levels are low in depressed people; conversely, physical exercise raises phenylethylamine levels, suggesting a cause of "runner's high". Significant amounts of phenylethylamine are found in chocolate, leading to speculation that eating chocolate can make you fall in love, but rapid oxidation of phenylethylamine by monoamine oxidase enzymes in the liver will prevent that.

AMPHETAMINE

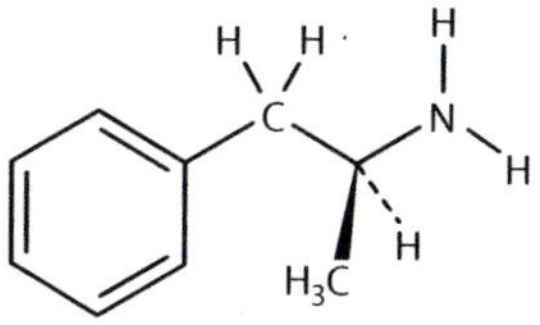

$$(11.2)$$

Replace one hydrogen in phenylethylamine by a methyl group and you have amphetamine (11.2), first synthesised in 1887 by Lazar Edeleanu. It is much longer lived in the body than phenylethylamine because the methyl side chain makes it resistant to the monoamineoxidase enzyme that decomposes molecules like phenylethylamine; the human body has not yet developed a quick way of metabolising amphetamines. Because of its ability to constrict nasal blood vessels, amphetamine was found to be a nasal decongestant and antiasthmatic and began to be marketed as Benzedrine in the late 1920s. In 1932, the first nasal inhalers were marketed; they contained cotton strips soaked in Benzedrine. Like other phenylethylamine-derived molecules, amphetamine promotes release of dopamine and norepiniphrine and also inhibits their reuptake, giving it strong stimulant effects; this led to abuse of the inhalers by people who pried them open and consumed the strips. By 1939, it was a prescription antidepressant drug in the United States.

The stimulant effects of amphetamine were put to a more constructive use by the armed forces of both sides in World War II, who used it to fight fatigue and stay alert, as soldiers, watch keepers on ships, or, especially, pilots (it is reportedly still used in the US Air Force for that purpose). Graham Greene used Benzedrine in 1938 to write 2,000 words every morning and to complete his novel *The Confidential Agent* in 6 weeks, whilst British Prime Minister Sir Anthony Eden relied upon Benzedrine to help get him through the Suez crisis in 1956, though it may not have improved his decision making. A side effect of amphetamines is appetite loss, so after the war they were widely used as prescription drugs to fight obesity (and stress).

Amphetamine has one chiral carbon atom; because only one of its two isomers has stimulant effects, nowadays amphetamine tends to be marked as the single-isomer form (S)-(+)-amphetamine, the dextro-isomer known as Dexedrine. Its wide use as a recreational drug was celebrated in Dexys Midnight Runners, a British New Wave pop and soul group who peaked in the early 1980s ("Come on Eileen", "Geno") and took their name from Dexedrine, which was popular in Mod and Punk culture in the 1960s and 1970s.

METHAMPHETAMINE

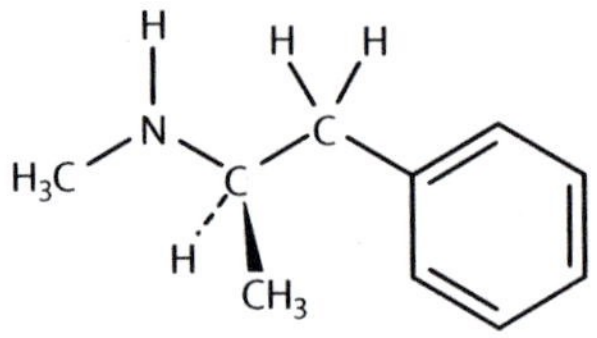

(11.3)

L-methamphetamine

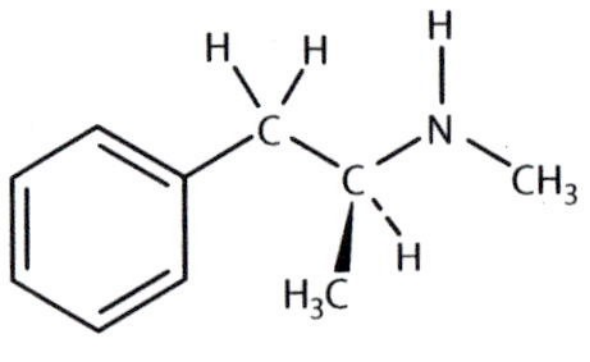

(11.4)

D-methamphetamine

Like amphetamine, methamphetamine has a chiral carbon atom and two isomers. They have identical chemical reactions, solubilities, and melting points, but because they have different 3-D structures, they fit chiral protein receptors differently and therefore have different effects upon the body. L-Methamphetamine ((R)-(–)methamphetamine; 11.3) is simply a decongestant and has no stimulant activity. Its optical isomer, D-methamphetamine ((S)-(+)-methamphetamine; 11.4), is the stimulant commonly known as "speed". L-Methamphetamine is found in American Vicks inhalers (but not the UK version) and was the molecule responsible for the Scottish skier

Alain Baxter losing his bronze medal in the Salt Lake City Winter Olympics in 2002 because, although the two isomers can be separated (and distinguished) by high-pressure liquid chromatography (HPLC) using a chiral column, the International Olympic Commission did not do this.

Like amphetamine, D-methamphetamine raises dopamine levels, producing enhanced amounts of dopamine with concomitant stimulation of the brain. It also stimulates heart action. Like amphetamine, it was widely used by combatants in World War II, and surplus military stock fuelled a big outbreak of methamphetamine abuse in postwar Japan. Low-level abuse continued: for example, long-distance lorry drivers and students as well as celebrities like Johnny Cash. Amphetamine abuse also became a problem in sport, first identified at the 1952 Oslo Winter Olympics, followed by the death of Tom Simpson, 1965 World Champion and the first Briton to wear the famous yellow jersey, on Mont Ventoux during the 1967 Tour De France.

As a "go-faster" drug, methamphetamine started to be known as "speed" in the early 1960s. It was associated with much youth culture, as in the British Mod group The Small Faces, with their 1967 speed anthem "Here Comes the Nice". That was the year of the "Summer of Love" in the Haight-Ashbury region of San Francisco, fuelled by marijuana and psychedelics like LSD; however, it turned sour the following year with an epidemic of injecting methamphetamine intravenously, and its motto changed from "peace and love" to "speed kills".

Methamphetamine abuse next flared up in the 1980s, with the discovery in Far Eastern laboratories that large crystals of methamphetamine hydrochloride could be produced by careful recrystallisation; the material was known as "crystal meth" or "ice". Unlike amphetamine sulfate, methamphetamine hydrochloride can be smoked. Its use spread from Hawaii across America, as well as Japan and Southeast Asia, including countries like Thailand and Cambodia, and South Africa. Its abuse has become a major problem, not least in the United States. It is known as *ya ba* (literally "madness drug") in Thailand and *tik* in South Africa. Crystal meth abusers include the singer Stacy Ann Ferguson (Fergie) from the hip hop/pop group the Black Eyed Peas, who has admitted to her past addiction.

One reason that amphetamine abuse is widespread in countries like the United States is that, unlike cocaine and morphine, methamphetamine can easily be synthesised in the laboratory (not to mention trailer parks and motel rooms). Two methods that have been described both involve the reduction of ephedrine (itself readily available from over-the-counter cough remedies) using either lithium in liquid ammonia as the reducing agent or red phosphorus and iodine to generate HI for the reduction (Figure 11.1).

Another method involves the reductive amination of phenylacetone, first converting it into the formyl amide of methamphetamine and then hydrolysing the amide with HCl(aq) (Figure 11.2).

Methamphetamine promotes release of abnormal amounts of the important neurotransmitter dopamine, which has the effect of stimulating regions of the brain linked with vigilance and the action of the heart. For a short while, the user feels sharper, stronger, and more energetic. It is believed that a significant proportion of the dopamine-producing cells in the brain can be damaged by prolonged exposure to even low levels of methamphetamine, which is responsible for reduced levels of

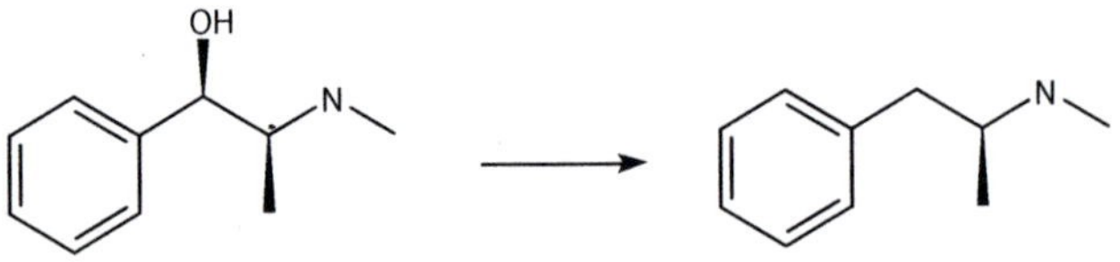

FIGURE 11.1 Methamphetamine can easily be synthesised from ephedrine.

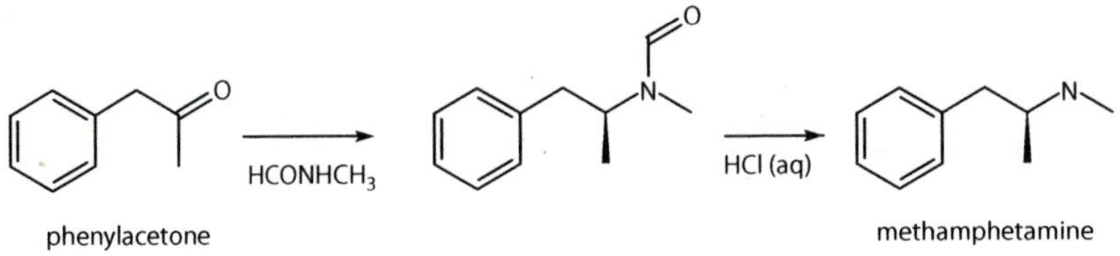

phenylacetone

methamphetamine

FIGURE 11.2 Synthesis of methamphetamine from phenylacetone.

dopamine; this can affect memory, attention, and decision-taking functions. Chronic methamphetamine abuse is reported to lead to significant reduction in grey matter in the brain greater than that in dementia or schizophrenia patients, though this needs further clarification. Associated health risks involve social and family problems, including risky sexual behaviour. Drug-induced psychosis may result.

ECSTASY

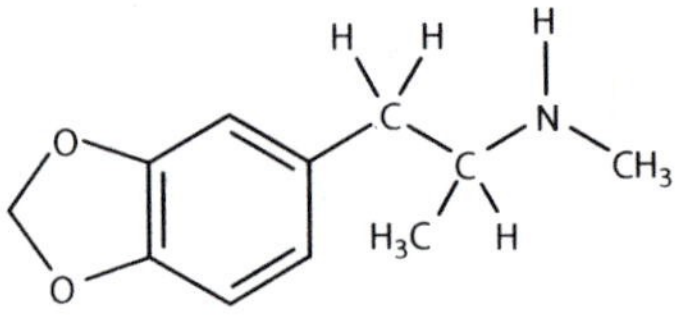

(11.5)

Ecstasy is the common name for 3,4-methylenedioxymethamphetamine (MDMA; 11.5), first synthesised by Anton Köllisch of Merck in 1912 as an intermediate compound in a synthesis and called "methylsafrylamin". Merck immediately applied for and was granted a German patent (274350) on it. After that, MDMA lapsed into obscurity until it emerged as a street drug in 1970. MDMA is usually synthesised starting from safrole, a product of the sassafras tree. As with all chemicals likely to be used by illicit manufacturers of drugs, the supply of safrole is closely controlled. The most obvious synthetic route involves electrophilic addition using HBr(aq), followed by nucleophilic substitution with methylamine (Figure 11.3).

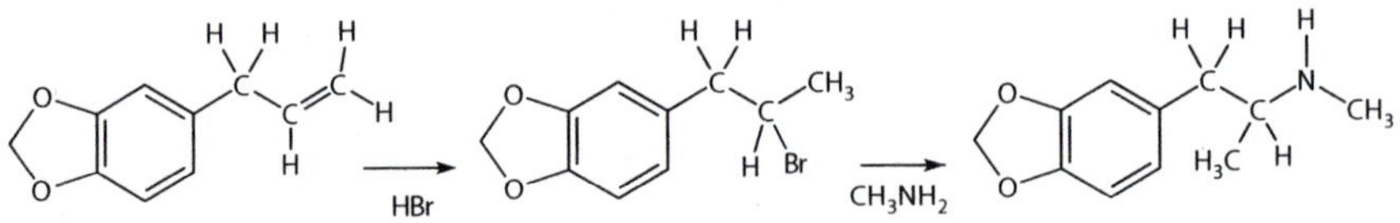

FIGURE 11.3 Electrophilic addition using HBr (aq) followed by nucleophilic substitution with methylamine.

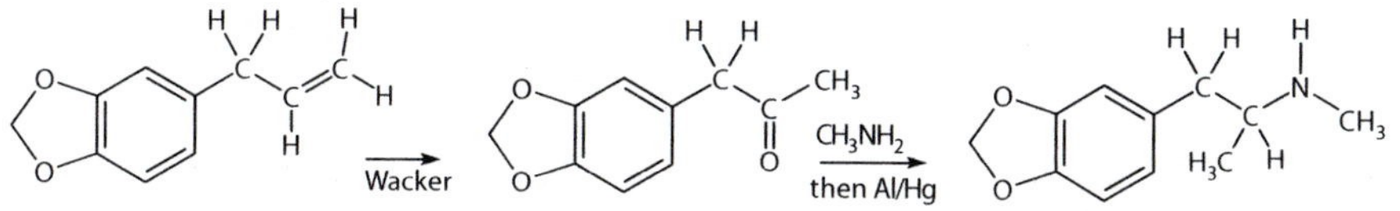

FIGURE 11.4 One alternative method of synthesising MDMA.

One alternative synthesis (Figure 11.4) starts with a Wacker oxidation of safrole to MDP2P (3,4-methylenedioxy-phenyl-2-propanone) with *p*-benzoquinone and palladium chloride catalyst. The MD2P then undergoes a condensation reaction with methylamine, forming an imine that is reduced by aluminium amalgam to MDMA.

Ecstasy came to prominence through Alexander Shulgin (the "godfather of Ecstasy"), who tested it (along with many other substituted 2-phenylethylamines) during the 1960s and commented on its "lightness and warmth". It was seen as an empathy-generating substance and began to be used by psychotherapists in the 1970s, before knowledge of its properties leaked out into the wider community. It was introduced to the American club scene in the early 1980s, when the name "Ecstasy" was coined, and it became a schedule I drug there in 1986 (to be succeeded on the club scene by the much more dangerous methamphetamine). In 1986–1987, it became associated with the dance culture surrounding clubs in Ibiza, leading to the "Balearic Beat", another Summer of Love; its growing use found its way back to the UK, becoming part of the rave culture there and in Europe in the late 1980s.

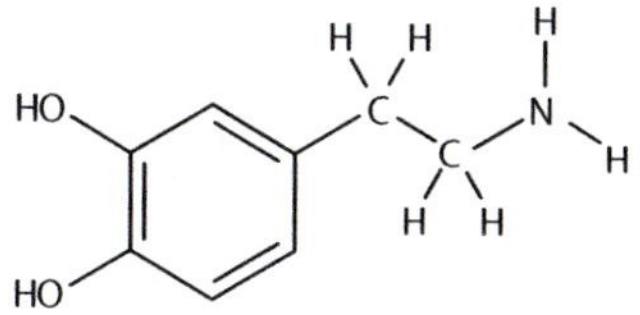

(11.6)

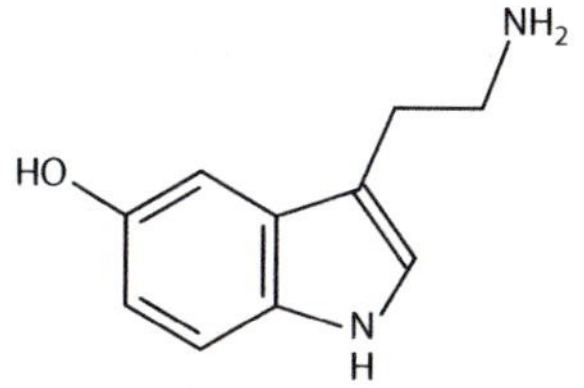

(11.7)

As MDMA use continued at a high level into the early 1990s, the British tabloid press began to take an interest. In the body, MDMA causes release of monoamine neurotransmitters; unlike methamphetamine, which releases dopamine (11.6) in particular, ecstasy especially promotes release of serotonin (11.7). The euphoria produced can last for up to ~5 hours. One effect of MDMA is to cause a rise in body temperature; coupled with the effects of some vigorous dancing, this can cause the user to become both dehydrated and overheated. This was believed to be a contributory factor to the deaths of some MDMA users in the late 1980s and early 1990s.

Clubbers were given some good advice: Chill out, take a break from dancing, and drink plenty of water. Some took this to extremes, which led to deaths caused by water intoxication. On 11 November 1995, an Essex (England) schoolgirl named Leah Betts had her 18th birthday party at home and took a single Ecstasy tablet. Although she was not dancing, she drank around 7 L of water in less than an hour and a half. Within 4 hours, she was in a coma from which she did not emerge; the water intoxication led to her brain swelling, with permanent brain damage. This led to intense tabloid press coverage and condemnation of Ecstasy.

In low doses, Ecstasy *may* be a safer recreational drug than some alternatives. There is currently not enough evidence of occasional low-dose use being harmful, but results of animal testing, in particular, suggest that serotonergic pathways in the brain are damaged. As a substituted phenylethylamine, it is classed as a class A drug under the Misuse of Drugs Act (1971) and attempts to downgrade it to class B have not been successful. From 1997 to 2000, 81 deaths in England and Wales were reportedly due to Ecstasy, but in more than half these cases the victim had been taking heroin or some related opiate. The UK use appears to be slowing down (2010) because of seizures and availability of alternative drugs.

RITALIN (METHYLPHENIDATE)

(11.8)

Ritalin (methylphenyl(piperidin-2-yl)acetate or methylphenidate; 11.8) shares with amphetamines the 2-phenylethylamine backbone. First made in 1944, it has become the most frequently prescribed drug to treat ADHD (attention-deficit hyperactivity disorder) in hyperactive children. ADHD symptoms typically include extreme inattention and impulsiveness, as well as disruptive behaviour in general. The cause of ADHD is not understood; it is three times more common in boys than in girls. ADHD has been recognised for a century and amphetamines have been applied to the problem since the 1930s.

In 1952, GlaxoSmithKline introduced Dexedrine to treat ADHD, and methylphenidate was first applied to the problem in the 1970s. Ritalin is believed to work by affecting the uptake and release of dopamine and norepinephrine in the brain, giving rise to higher levels of dopamine in the synapse. Ritalin is taken orally in low doses, so there is only a slow buildup of dopamine, and it is therefore a safe drug (though, because it is a stimulant, Ritalin abuse in nontherapeutic doses is not unknown). These low therapeutic doses of Ritalin help ADHD patients focus on their tasks and calm down. Methylphenidate is believed to affect particularly the part of the cerebral cortex linked with concentration, and the effects of dopamine enhancement have showed up most clearly when subjects were faced with a challenging mathematical task. Other medications are starting to be used for ADHD, such as Adderall, a mixture of four different amphetamine salts.

In 1996, it was estimated that 5% of boys in the United States were taking Ritalin and it has been suggested that only those exhibiting all three classic ADHD symptoms—inattentiveness, impulsiveness, and hyperactivity—should receive this medication, given concerns in both the United States and the UK that Ritalin is being used as a "chemical cosh" in inappropriate cases. At present, something like 80% of Ritalin use occurs in the United States. University students have used it as a drug to improve cognitive enhancement, especially when revising for exams.

CAFFEINE

(11.9)

Caffeine (1,3,7–trimethylxanthine; 11.9) is found particularly in coffee (especially *Coffea arabica*), tea, and cola nut and, to a lesser degree, in chocolate. Tea has been consumed for up to 5,000 years and coffee for a thousand years or more. First isolated in 1820 by the German chemist Friedrich Runge, caffeine is made by plants to protect them against insects because methylxanthines inhibit insect feeding and are also pesticides, owing to their ability to inhibit phosphodiesterase activity. Biosynthesis of caffeine begins with xanthosine, which is successively transformed into 7-methylxanthosine, 7-methylxanthine, theobromine, and caffeine. The genes encoding three distinct N-methyltransferase enzymes used in the synthesis of caffeine from xanthosine have been isolated and expressed in tobacco plants, leading to caffeine-containing tobacco plants found to be unpalatable to tobacco cutworms (*Spodoptera litura*). Use of these genes has also enabled the engineering of low-caffeine coffee plants.

Caffeine is a stimulant through acting as an antagonist of adenosine, an inhibitor of the central nervous system. Caffeine has a similar structure to adenosine and therefore binds to the adenosine receptor (competitive inhibition), leading to increased dopamine and glutamate activity and to increased neuron firing in the brain and more adrenalin production. Caffeine is thus a slightly addictive drug because it is a cardiac stimulant with mild diuretic effects. A study of genetic variations (2011) found two genes linked with high caffeine consumption; one was next to CYP1A2, responsible for most caffeine metabolism in the human liver. It has been reported that exposure to caffeine is related to poor neuromuscular development in the foetus. Caffeine can penetrate the placental barrier quite easily and thus affect the developing foetus, so it is probably wise for a pregnant woman to restrict caffeine intake severely (two cups a day is the current maximum recommended). It is not just humans who are affected by caffeine. NASA scientists tested the effects of a range of drugs on house spiders, finding that caffeine had the most effect, with webs lacking patterns and consisting of relatively few, random threads.

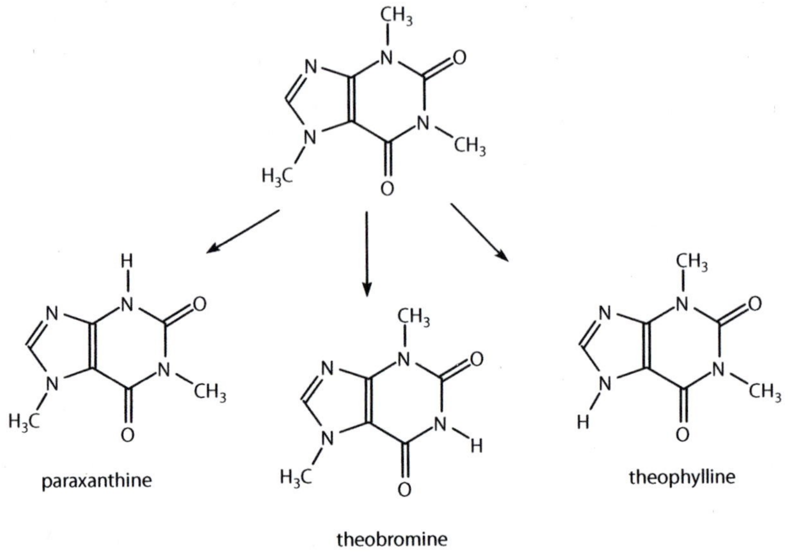

FIGURE 11.5 Metabolism of caffeine in the liver.

Caffeine is metabolised in the human liver by cytochrome P450 oxidase enzymes, initially forming (Figure 11.5) three dimethylxanthines—paraxanthine (84%), theobromine (12%), and theophylline (4%)—followed by further demethylation and oxidation leading to urates and uracil derivatives.

"Normal" doses of caffeine in a cup of coffee or tea are around 50 mg per cup. The lethal dose of caffeine is upward of 5 g, which corresponds to a lot of cups, so normal coffee and tea drinking is absolutely safe. The amount of caffeine in a 250 mL can of Pepsi or Coke is 25–35 mg—less than a cup of tea; a can of Red Bull is said to contain about 80 mg of caffeine, so this is safe too. Real problems come with people taking solid caffeine: There have been fatalities with people consuming several whole packs of caffeine "energy tablets", such as 100 × 100 mg tablets and 400 × 50 mg tablets. In 2010, a 23-year-old Mansfield (England) man died after consuming two spoonfuls of caffeine dissolved in an energy drink. His blood caffeine level was 251 mg/L, and the coroner said that it was 70 times more than that usually obtained from a high-energy caffeine drink. Decaffeinated coffee is produced by using a solvent like supercritical CO_2 to dissolve the caffeine out of the beans (which can be used to make "energy drinks").

There is much more to coffee and tea than caffeine. They contain polyphenolic antioxidants with possible health benefits, and the smell of your freshly brewed tea and coffee has nothing to do with the involatile caffeine. Over 1,000 volatile organic compounds (VOCs) are found in roasted coffee (far more than in wine), but only around 50 of them are important contributors to the overall aroma. When coffee beans are roasted at around 200°C, complex reactions occurring include the Maillard reaction; this involves sugars, amino acids, and peptides, giving a wide range of light, volatile, and smelly molecules responsible for the fragrance of the beans. Some of the important odorants in coffee are substituted pyrazines (p. 61), such as 3-isobutyl-2-methoxypyrazine, 2-ethyl-3, 5-dimethylpyrazine (see Structure 6.8 in Chapter 6), and 2,3-diethyl-5-methylpyrazine, as well as furfuryl mercaptan (11.10), which contributes the "burnt, roasted" note.

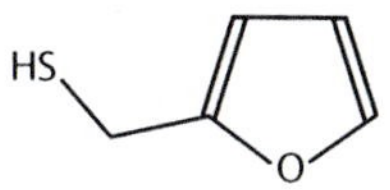

(11.10)

COCAINE

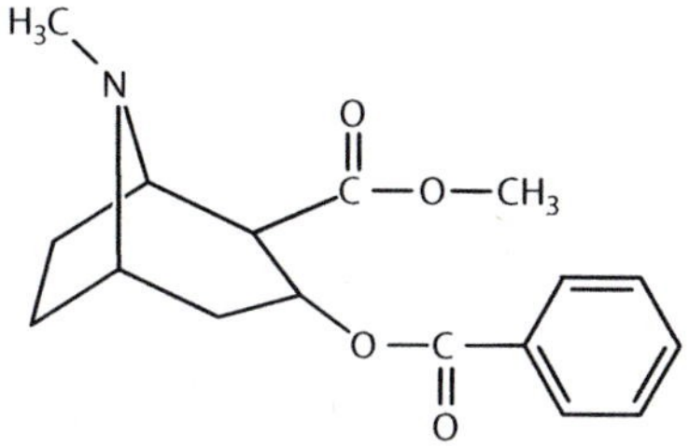

(11.11)

For a molecule famously described by the actor Robin Williams as "God's way of telling you you've got too much money", **cocaine** (benzoylmethylecgonine) (11.11) has humble origins in the leaf of the coca plant, *Erythroxylum coca,* which contains ca. 0.3%–0.7% cocaine. Growing wild in the Andes, the plant makes cocaine from L-orthinine and L-glutamine as an insecticide to deter predators. The molecule has four chiral centres, leading to 16 possible isomers, of which the plant synthesises just the one. People in Bolivia, Peru, Ecuador, and Chile have chewed coca leaves for some 5,000 years. The Incas made pellets of the leaves with a little alkali (lime is used today) and held them in their mouths, giving a gradual release of cocaine that was slowly absorbed from their saliva. Chewing coca leaves suppresses the appetite and increases strength and endurance; it enabled relays of messengers (*chasqui*) to cover up to 300 km a day with messages to and from the Inca Emperor. After the Spanish conquest, the conquistadors supplied Inca slave workers in the silver mines with coca leaves as an endurance enhancing agent; as late as 1909, the Antarctic explorer Sir Ernest Shackleton used "Forced March" cocaine tablets for the same reason.

Cocaine was first isolated by Friedrich Gaedcke in 1855; in 1859, Friedrich Wöhler's student, Albert Niemann, obtained pure crystals. In the late nineteenth century, cocaine was seen as a mild stimulant and added to tonic wines such as *Vin Mariani.* The Georgian pharmacist John Styth Pemberton added coca leaves in his original recipe for "French Wine Cola"; after prohibition came to Atlanta, he used coca leaf and kola nut in what became Coca-Cola, though "decocainised" leaves have been used since the beginning of the twentieth century.

Cocaine was used as a topical anaesthetic in dentistry and ophthalmology, was recommended by Sigmund Freud (nowadays, benzocaine or lidocaine is seen as safer), and, at one stage, was available in over-the-counter medications. Around this time, Sherlock Holmes became the first celebrity with a cocaine problem in *A Scandal in Bohemia* and *The Sign of Four* (1890). Cocaine first became fashionable

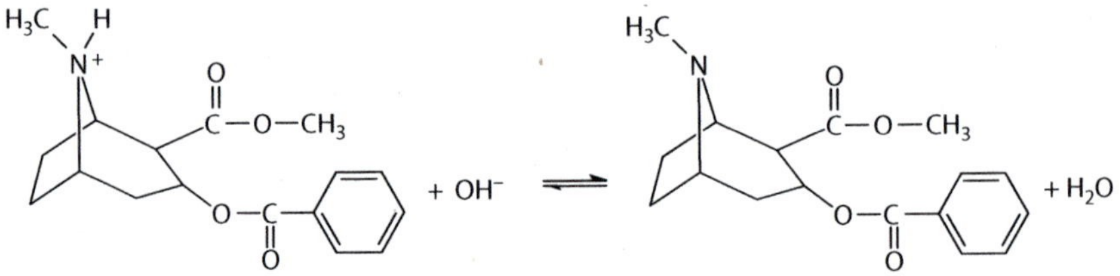

FIGURE 11.6   Equilibrium involved in extracting and processing cocaine.

in 1920s Hollywood film circles; though superseded by amphetamines, it got a jet-set image after the movie *Easy Rider* (1969) was released and from 1970s glam-rockers. By this time, organised South American criminal gangs were controlling the trade—most famously, in the Medellín and Cali cartels in Colombia; the most celebrated *narco* chief was Pablo Emilio Escobar Gaviria (1949–1992).

Over 700 tons of cocaine is extracted each year; most is intercepted before it can come to market, but enough reaches the "consumer" to make it a very profitable industry. Cocaine is traditionally extracted from coca leaf in three stages—usually in three different labs that may be anything up to a thousand miles or more apart. Extraction and purification by solvent extraction rely on the neutral base having low water solubility but high solubility in nonpolar organic solvents, whereas protonated cocaine ions are water soluble but have very low solubility in the organic layer (Figure 11.6):

- Leaves are mashed up with base (e.g., lime) to ensure that neutral cocaine molecules are present; kerosene then extracts them from the leaves. Dilute sulfuric acid is added to the kerosene, so protonated cocaine moves to the aqueous layer. This is separated and treated with an alkali (lime), precipitating neutral cocaine, the crude "coca paste", which is between 30% and 80% pure.
- In another lab, the crude paste is dissolved in dilute H_2SO_4, and aqueous $KMnO_4$ is added until the purple solution turns colourless; this oxidises alkaloid impurities, leaving brown, insoluble, MnO_2, which is filtered off. The aqueous layer, containing protonated cocaine, is neutralised, usually with ammonia solution; precipitated cocaine is filtered off and dried. This is "coke base".
- At a "crystal lab", the coke base is dissolved in an organic solvent such as ether and filtered to remove insoluble impurities. A mixture of conc. HCl and propanone is added to the filtrate, and cocaine hydrochloride crystallises out.

The hydrochloride salt is the form traditionally injected or snorted; heating it destroys much of the cocaine. If this salt is mixed with baking soda and smoked, it generates free, volatile "crack" cocaine (the name arising from the crackling noise of the carbon dioxide gas given off).

Cocaine is a central nervous system stimulant that prevents reuptake of neurotransmitters such as dopamine, serotonin, and noradrenaline and boosts their levels in the synapse. In small amounts, it makes the user sociable and more alert and raises self-esteem and sex drive; pupils dilate and heart rate and blood pressure will rise. Large amounts cause more erratic behaviour, including paranoia, twitching, and delusions ("bugs under the skin")—even possibly seizures or respiratory arrest. Users can find that the "rush" is succeeded by a nasty downswing (crash), thus leading to taking heroin with the cocaine—the lethal "speedball" that killed John Belushi and River Phoenix. There is the possibility of damage to nose and tissues, famously the nasal septum.

Snorting cocaine increases the risk of heart attacks, especially in those over 40; cocaine can cross the placenta and pregnant mothers can abort. Peak levels are reached within 3 minutes if it is injected, 15 minutes by snorting (insufflation), and 60 minutes by oral absorption. If it is swallowed, most is destroyed in the liver before it gets to the brain. Most cocaine entering the human body is metabolised in the liver using the carboxylesterase enzymes hCE1 and hCE2, mainly by hydrolysis of one ester linkage, forming benzoylecgonine (Figure 11.7). Only about 1% is excreted unchanged. Metabolites can be detected in the urine within 4 hours of ingestion, and the presence of benzoylecgonine has implicated sportsmen such as tennis player Martina Hingis and jockey Kieron Fallon in cocaine use. Mixing cocaine and ethanol is reckoned to be more toxic than taking either on its own because hCE1 can combine them (Figure 11.7) into cocaethylene (the ethyl ester).

Benzoylecgonine, which is contained in sweat, can be detected in the fingerprints of cocaine users. The presence of cocaine in a fingerprint could be argued to result from accidental contact with the drug, but the presence of a *metabolite* of cocaine shows that the drug has passed through that person's body, and the

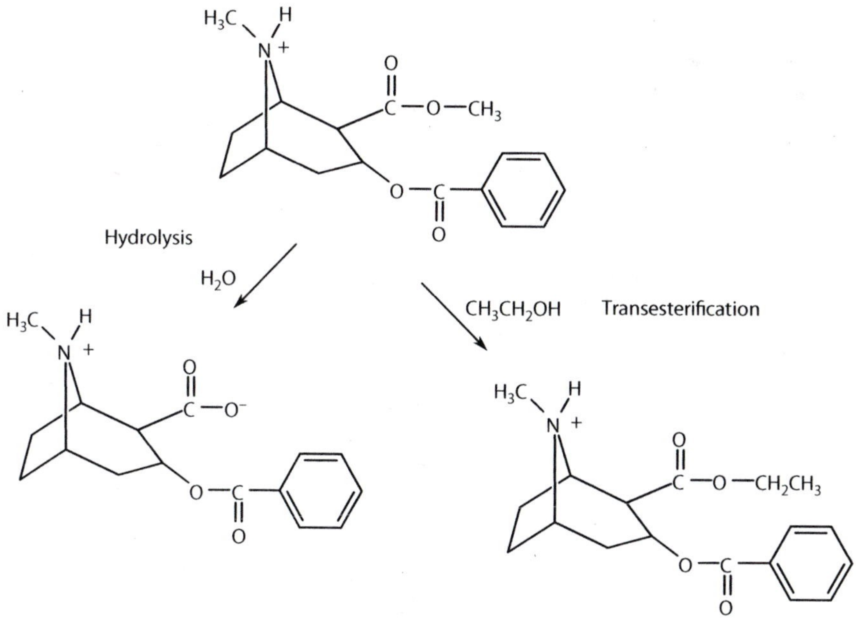

FIGURE 11.7 Metabolism and transesterification of cocaine.

fingerprint is a unique identifier. A very high percentage of banknotes is contaminated with cocaine—partly due to cocaine users rolling up a note to make a tube to "snort" cocaine, but mainly due to transfer of the drug to other notes (e.g., in a counting machine at a bank). The level of cocaine contamination indicates if it was accidental or not. Forensic scientists can detect cocaine on notes, using tandem mass spectrometry and studying the fragments of molecules vapourised from the surface of heated notes. For bulk samples of notes, this is fast and sensitive, scanning 50 notes in about 4 minutes. If more accurate results are needed, solvent extraction gives 90%–100% recovery of the drug from the note. For the spectrometer to look for cocaine, it is programmed to select the peaks with mass-to-charge ratio values of 182 and 105, the two main fragments of the molecular ion; it measures their intensities.

Law enforcement officers are obviously interested in the source of seized narcotics. Research has shown that cocaine from different coca-growing regions has different isotopic ratios of $^{13}C{:}^{12}C$ and $^{15}N{:}^{14}N$ (which appear to depend upon factors like soil type and local weather conditions). Combined with the variable levels of the trace alkaloids truxilline and trimethoxycocaine, this can indicate which region of Bolivia, Colombia, or Peru is the origin of the drug.

DESIGNER DRUGS

The phrase "designer drug" conjures up a vision of a white tablet covered in stubble and wearing miniature Ray-Bans. The truth is more prosaic. Designer drugs are synthetic chemicals with structures just a bit different from known drugs of abuse and thus not prohibited by existing legislation—in other words, a legal version of an illegal substance. The American Controlled Substances Analogue Enforcement Act (1986) makes it illegal to manufacture (and to sell or possess) "analogues"— molecules broadly similar in their chemistry and action to existing schedule I and schedule II drugs. Australian legislation is even more stringent, based on similarity of chemical structure. In the United Kingdom at present, a drug is simply banned when it becomes a problem; thus, whenever a drug is banned, underground chemists can simply tweak the molecule (e.g., by changing an R group) and then rapidly market an analogue that is not banned.

Although the phrase "designer drug" was coined by the American chemist Gary Henderson less than 30 years ago, the concept goes back further. Shortly after World War I, chemists made analogues of diacetylmorphine (heroin) with different acyl groups that were not covered by the existing bans. In the 1960s and 1970s, alternative amphetamines, often hallucinogens like 2,5-dimethoxy-4-methylamphetamine (known as DOM or STP), were made. **Fentanyl** (11.12) was synthesised as an alternative to heroin and other opiates and became popular when heroin was in short supply on the streets. It is about 200 times more potent than morphine. **Alphamethylfentanyl** (11.13), with a related structure, is about 5,000 times stronger than morphine and has a "high" lasting as long as that of heroin. It came on the streets in 1979 as "China White" (a term also used to describe smokeable heroin) and has been associated with around 100 deaths that occurred when people underestimated its potency.

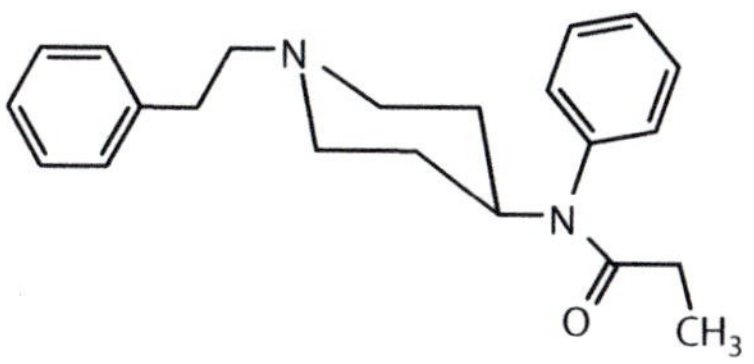

(11.12)

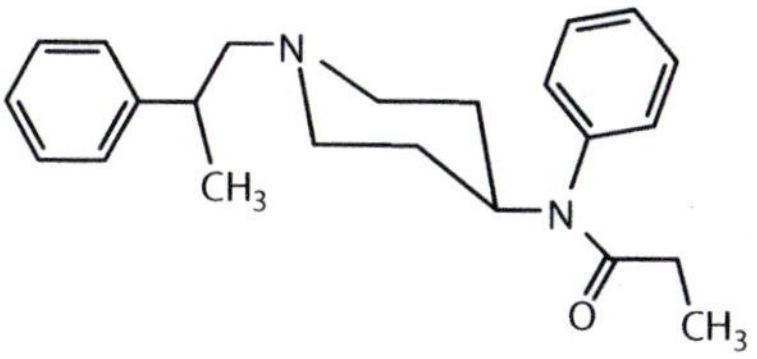

(11.13)

When people buy an illegal drug, they have no way of knowing if it is what it says on the label, let alone how pure the substance is. In 1976, a 23-year-old American called Barry Kidston started to make and use 1-methyl-4-phenyl-4-propionoxypiperidine (MPPP). It had about three times the potency of morphine; people called it "new heroin". Within 6 months, he had developed Parkinson's disease, a disease normally affecting people over 60. In 1982, the same thing happened to seven heroin abusers, aged 20–40, who were using MPPP. Unfortunately, if the reaction mixture used to make MPPP is too acidic or the mixture is heated too much, another product is obtained: MPTP (Figure 11.8). In the human brain, monoamine oxidase (MAO) enzymes convert MPTP into MPP^+, and this destroys the dopamine neurons (i.e., some of the nerve cells) in the *substantia nigra* in the brain. It is thought that MPTP impurities may have caused over 100 cases of Parkinson's disease in people in their 20s and 30s who have used MPPP.

Modified amphetamines have been marketed for over 40 years. The psychoactive effects sometimes credited to "designer" amphetamines are often based on anecdotal evidence, and the toxicity of any particular amphetamine can be hard to assess, given the fact that drugs are often taken in combination. Thus, **2,5-dimethoxy-4-propylthiophenylethylamine** (aka 2C-T-7, "Tripstacy" or "Blue Mystic"; 11.14) has been suggested to produce psychedelic and Ecstasy-like effects. Several fatalities have been associated with 2C-T-7, some involving it in combination with other drugs.

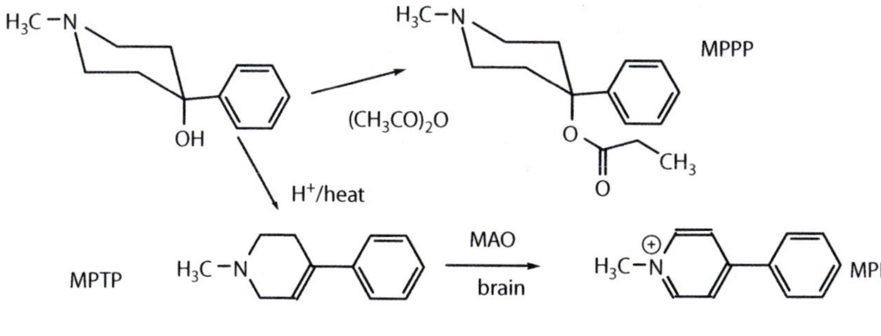

FIGURE 11.8 Formation of MPPP and MPTP.

4-Methylthioamphetamine (4-MTA, aka "Flatliners"; 11.15) is said to be an entac-togen. Its slow onset of action may encourage users to overdose, and it has been associated with half a dozen deaths.

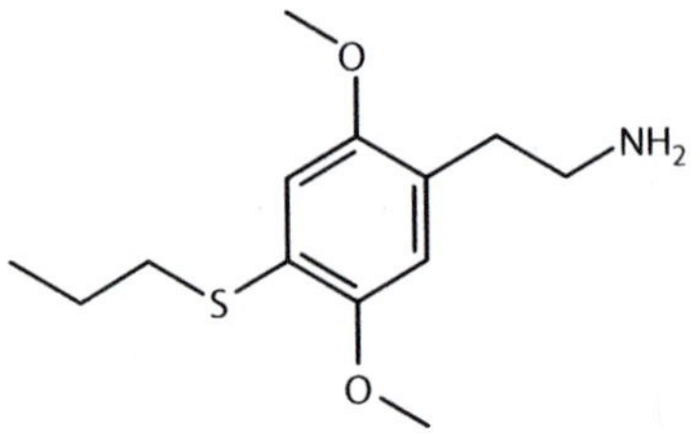

(11.14)

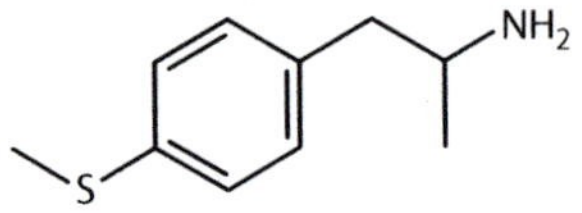

(11.15)

2,5-Dimethoxy-4-methylamphetamine (DOM or STP, allegedly "serenity, tran-quility and peace"; 11.16) has been around since the days of Flower Power in the mid-1960s; Owsley Stanley, the sound engineer for the Grateful Dead, is credited with distributing STP in 1967—in addition to vast quantities of high-purity LSD—in 1960s San Francisco. A psychedelic, slow-acting amphetamine that is a serotonin receptor agonist, it is said to bring on visual hallucinations. The related **2,4,5-trimethoxyam-phetamine** (TMA-2; 11.17) is also reported to be a psychedelic stimulant.

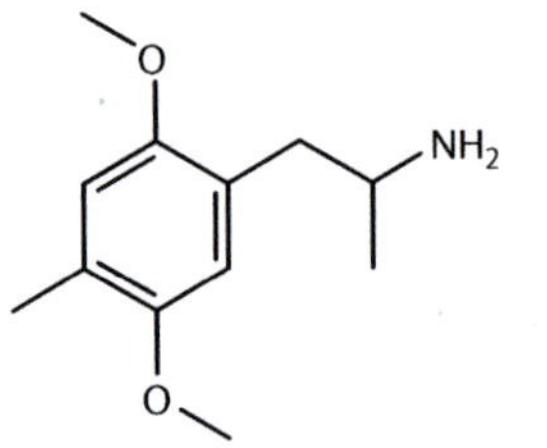

(11.16)

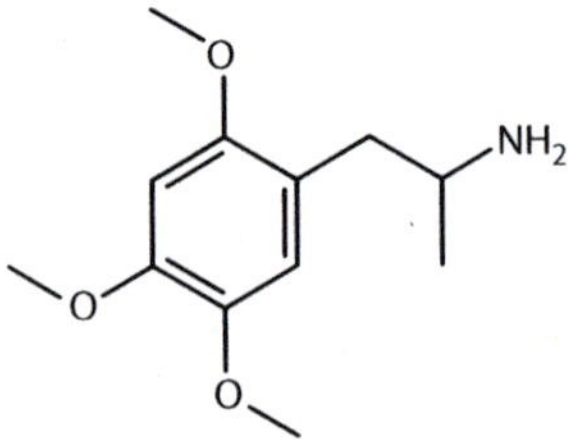

(11.17)

2,5-Dimethoxy-4-bromophenethylamine (2C-B or Nexus; 11.18) is an erotic empathogen that is claimed to be an aphrodisiac. At low doses, it gives a fast acid high but, at higher doses, mescalin-like visual effects are reported. At least two fatalities have been associated with its use. Chloro- and iodo-derivatives have also been studied; the presence of hydrophobic substituents in the 4-position of the ring has been credited with stronger hallucinogenic activity. **2,5-Dimethoxy-4-bromoamphetamine** (DOB; 11.19) is described as a very potent, slow-onset amphetamine—in some ways like LSD but giving much longer trips. Fatalities have been described, very likely due to underestimating the potency of the drug.

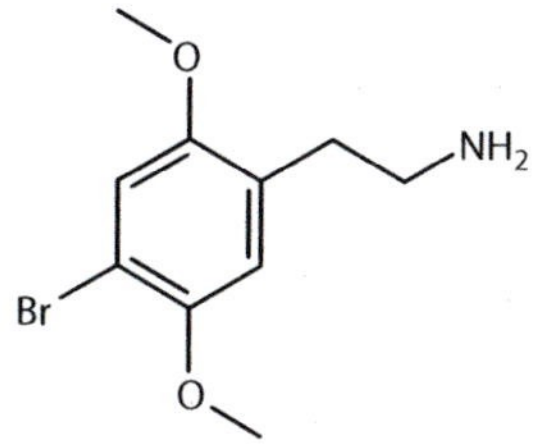

(11.18)

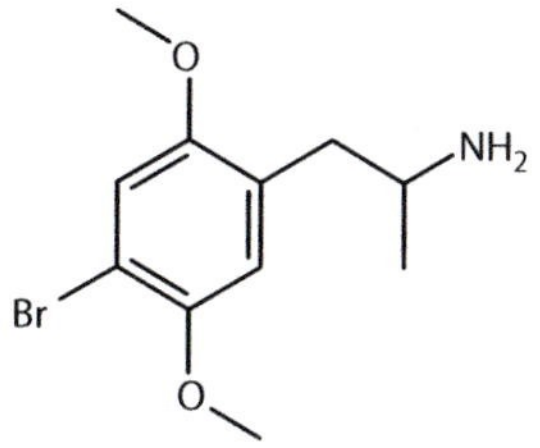

(11.19)

These molecules are nearly all 4-substituted 2,5-dimethoxyamphetamines. This seems to reflect a genuine preference of the receptor sites.

Within the last few years, the focus seems to have moved to different families of analogues, reflecting a weakness in UK drug legislation. It has seemed that no sooner has one drug been banned than an analogue is being marketed. Allegedly, the synthetic chemistry behind this is largely based in China. Thus, in 2008–2010, interest in the UK turned to 4-methylcathinone (**mephedrone**) and its analogues. 4-Methylcathinone (11.20) is a synthetic molecule derived from cathinone, the main psychoactive molecule found in khat, the leaves of an East African evergreen shrub (*Catha edulis*) that are widely chewed as a stimulant by Yemenis and others (e.g., Somalis, Ethiopians, and Kenyans). Cathinone is not an illegal substance in England or the Netherlands at the time of writing, though it is in France and most other countries.

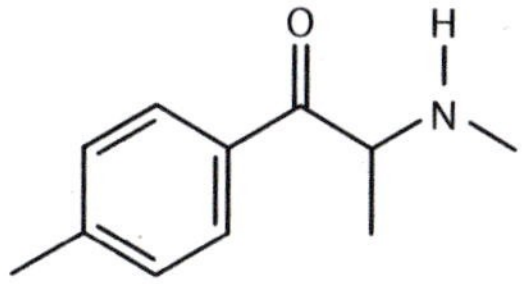

(11.20)

Cathinone has a structure closely related to methamphetamine and is generally regarded as a schedule I drug. Intense media speculation in the UK linked mephedrone, described as an Ecstasy-like stimulant, with the deaths of as many as 20 teenagers, though, as of June 2010, it has been established as a cause of death of just one person. Mephedrone was described by sellers as "plant food"—a convenient fiction for them because these molecules are nothing of the kind. In the UK it was classified as a class B drug with effect from 16 April 2010, and it is illegal in most countries in Western Europe; in the United States and Australasia, it is covered by analogues legislation. Soon after mephedrone was banned in the UK, it was succeeded by stimulants with related structures: **naphyrone** (naphthylpyrovalerone; 11.21), sold as "pond cleaner", and **MDPV** (methylenedioxypyrovalerone, "Ivory Wave"; 11.22), sold as "bath salts".

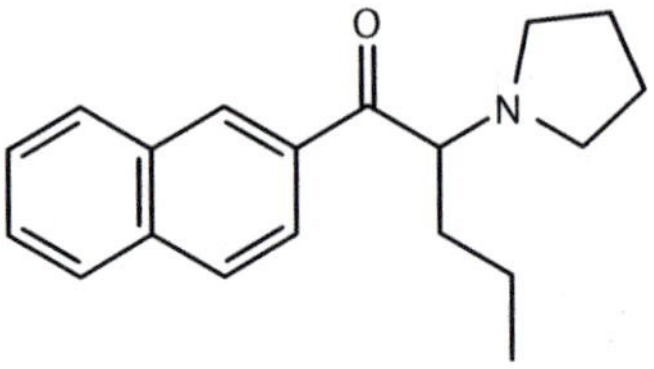

(11.21)

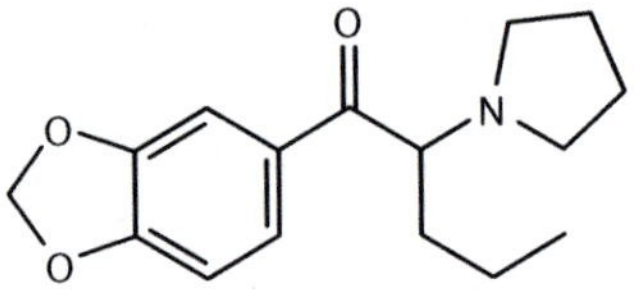

(11.22)

Methylone (3,4-methylenedioxy-*N*-methylcathinone, bk-MDMA; 11.23), sold as a "room odouriser", and **butylone** (β-keto-*N*-methylbenzodioxolylpropylamine; bk-MBDB; 11.24) are examples of other drugs of this type, reportedly entactogens. In general, these cathinones (β-keto amphetamines) seem to act as lower potency versions of the corresponding amphetamines.

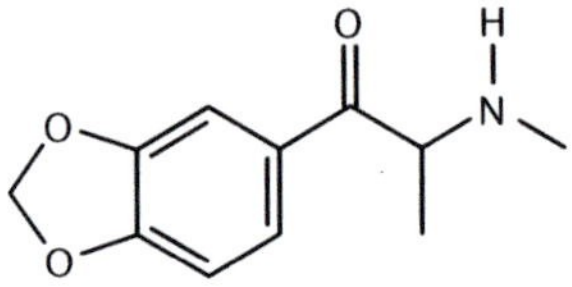

(11.23)

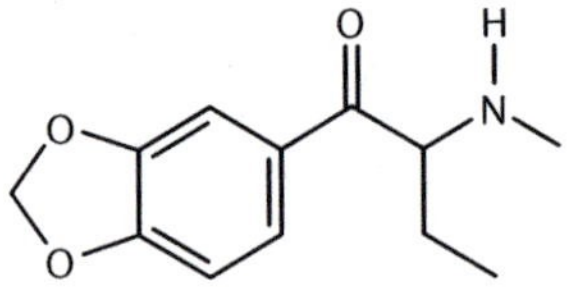

(11.24)

A group of napthaloylindole compounds, originally synthesised as cannabinoid receptor agonists to study the structure of the receptor site, has also attracted attention from abusers in recent years. **JWH-018** (naphthalen-1-yl-(1-pentylindol-3-yl) methanone; 11.25) has been an ingredient in products with names like "herbal spice" and "herbal gold". It is reported to be a very potent agonist for the CB1 receptor and to have four times the potency of THC; it is said to give a marijuana-like "buzz" similar to cannabis when smoked. JWH-018 has been banned across much of Europe, but the synthetic chemists have turned their attention to other molecules of this family.

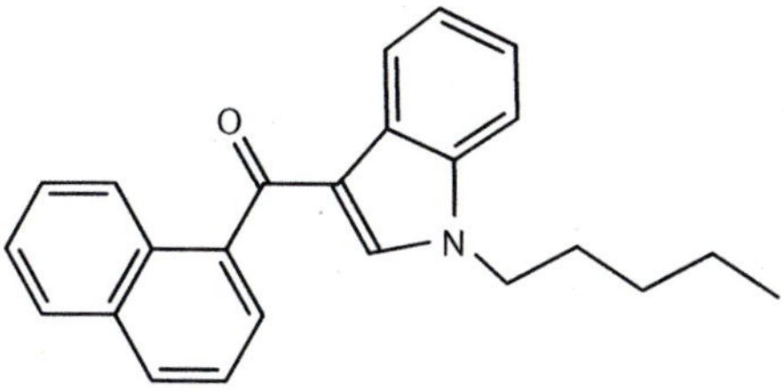

(11.25)

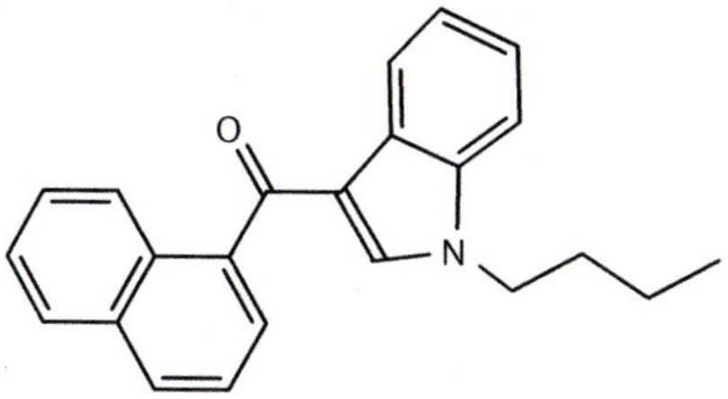

(11.26)

JWH-073 (naphthalen-1-yl-(1-butylindol-3-yl)methanone; 11.26) was sold as a "fertiliser" product called "forest humus" as well as **JWH-250** (2-(2-methoxy-phenyl)-1-(1-pentylindol-3-yl)ethanone; 11.27) and **JWH-200** ((1-(2-morpholin-4-ylethyl)indol-3-yl)-naphthalen-1-ylmethanone; 11.28) in "herbal" smoking blends.

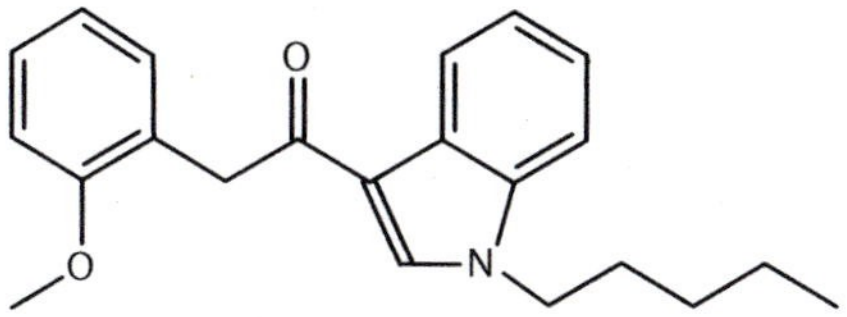

(11.27)

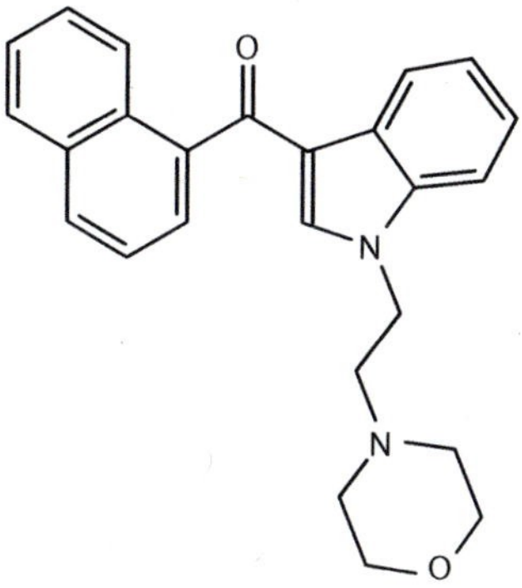

(11.28)

ETHANOL

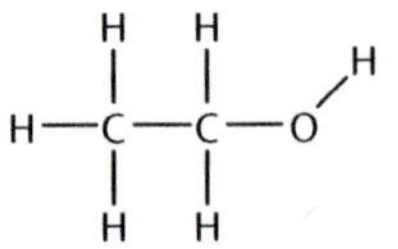

$$H-\overset{\displaystyle H}{\underset{\displaystyle H}{C}}-\overset{\displaystyle H}{\underset{\displaystyle H}{C}}-O^{\diagup H} \tag{11.29}$$

Probably the most socially acceptable drug in most of the world, **ethanol** (11.29) is usually made following one of two routes. Traditionally, this is used to make alcoholic drinks, using a renewable source in the form of carbohydrates. Enzymatic breakdown of starch yields glucose, which upon anaerobic fermentation using the enzyme zymase in yeast, produces ethanol. This process can be carried out at room temperature or just above (the optimum is 37°C–38°C) in batches, using carbohydrate sources like potatoes, fruit (especially grapes), sugarcane, or corn. This is a relatively slow process and only gives up to about 15% yield before the ethanol inhibits the zymase and stops the reaction; fractional distillation is required to produce more concentrated solutions like spirits:

$$C_6H_{12}O_6 \rightarrow 2\ CH_3CH_2OH + 2\ CO_2$$

The more modern, high-tech version employs catalytic hydration of ethene using a phosphoric acid at around 300°C and a pressure of about 60 times atmospheric:

$$C_2H_4 + H_2O \rightarrow CH_3CH_2OH$$

Although this is a very rapid reaction that gives high yields of ethanol of very high purity, there are disadvantages, such as the high-tech plant and high energy input needed, as well as the source of ethene (cracking oil fraction) being a finite resource.

People have consumed alcoholic drinks since at least the fourth millennium BC and possibly back in Neolithic times. It has been suggested that humans like alcoholic drinks because they and their primate ancestors ate a lot of fruit. Ripe fruit emits ethanol as a signal of its ripeness, and the ethanol would have been a useful calorific source (the "drunken monkey hypothesis").

The body gets rid of ethanol by oxidising it in the liver to the aldehyde ethanal (acetaldehyde), a reaction catalysed by an enzyme alcohol dehydrogenase (Figure 11.9). The liver processes, on average, 12 mL of ethanol an hour in an adult (less in a child). The ethanal is in turn converted to ethanoic acid (using the enzyme acetaldehyde dehydrogenase), which the body then breaks down to form CO_2 and

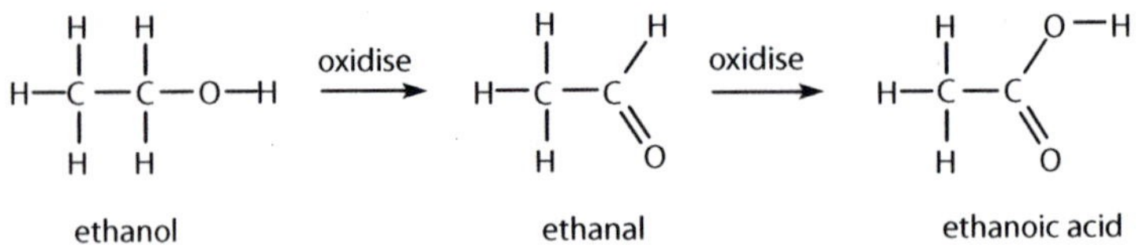

| ethanol | | ethanal | | ethanoic acid |
| oxidise | | oxidise | | |

FIGURE 11.9   Metabolism of ethanol in the body.

H_2O (using the Krebs cycle). Ethanal is a lot more toxic than ethanol itself, and part of the hangover is the body's reaction to ethanal; of course, the ethanol also affects the brain and the central nervous system, and there are other effects, such as dehydration, caused by enhancement of urination.

About 10% of Asian individuals are believed to have genes that code for an inactive form of aldehyde dehydrogenase; even small amounts of ethanol consumed can produce marked flushing and other unpleasant symptoms, so they will largely abstain from alcoholic drinks. About 40% of Asians have genes coding for both active and inactive forms of the enzymes and thus still have lower than normal alcohol tolerance. The drug disulfiram ($Et_2NC(=S)\text{-}S\text{-}S\text{-}C(=S)NEt_2$), which is used to treat alcoholism, works by inhibiting aldehyde dehydrogenase. High levels of ethanal are produced when even a small amount of ethanol is consumed; the patient immediately experiences extreme hangover and other unpleasant symptoms.

Ethanol acts as a depressant by binding to the γ-aminobutanoic acid (GABA) receptor, which controls a chloride ion channel. Ethanol binds to the receptor next to the GABA site and distorts the receptor slightly, making it easier for GABA to bind and thereby opening up the chloride channel. Benzodiazepine drugs like Valium bind to a different site; alcohol and benzodiazepines act synergistically, greatly enhancing the depressant effect on the central nervous system and can prove fatal.

Many other alcohols are more toxic than ethanol. Methanol gets oxidised by alcohol dehydrogenase to methanal, which is in turn converted into methanoic acid, analogously to ethanol's reactions. Both of these products attack the human optic nerve, leading to blindness. A lethal dose of methanol can be as little as 30 mL or so. Some alcoholics drink methylated ethanol because it is a cheap way of getting intoxicated; it is also a cheap way to oblivion and to an early grave. Ethane-1,2-diol (ethylene glycol) is used in automobile antifreeze; it is extremely toxic but tastes sweet, and thousands of children and animals are poisoned every year by it. In the hands of unscrupulous people, this has been used as a murder weapon when it has been added to food that masks its taste. In the liver, alcohol dehydrogenase converts ethylene glycol first into glycoaldehyde and then into other toxic molecules (Figure 11.10), including ethane-1,2-dioic acid (oxalic acid). Ultimately, oxalic acid removes calcium from the body as insoluble calcium oxalate, which is deposited in the kidneys, and it is kidney damage that is usually the cause of fatalities.

If doctors diagnose ethylene glycol poisoning early enough, they can administer an antidote, usually ethanol. This binds to the alcohol dehydrogenase enzyme more strongly than ethylene glycol does; thus, by preventing the ethylene glycol from attaching itself to the liver, the kidneys have a chance to excrete the ethylene glycol and eliminate it harmlessly from the body. The Austrian wine scandal of 1985 was

| ethylene glycol | glycoaldehyde | glycolic acid | oxalic acid |

FIGURE 11.10 Metabolism of ethylene glycol in the liver.

caused by a few wineries adding the sweet but much less toxic diethylene glycol to make cheap white wines appear sweeter and more valuable; as far as is known, no one was injured by this, but it was years before the Austrian wine industry recovered its reputation.

Ethanol burns exothermically, forming carbon dioxide and water, and has been used as a fuel for years. "Green" considerations, plus diminishing oil reserves, have led to its increasing adoption in automobile fuels, though there are disadvantages. The ability of ethanol to hydrogen-bond makes it miscible with water in all proportions; this is invaluable for its use in alcoholic drinks, but it can ruin its value as a fuel. Although Brazil has used ethanol in petrol–ethanol blends since 1976 (as well as the many vehicles running on neat ethanol), it is only recently that other countries, especially the United States, have taken major steps in that direction.

Bioethanol has become big business, with ethanol produced from crops like corn (starch) and sugarcane (sucrose). Some analysts say that, when all the steps are considered, the scientific and economic value of getting ethanol from corn is marginal, but getting it from cellulose could be much more valuable. At present, though, industry is still developing the enzymes to convert cellulose into sugars and then into ethanol, and research continues into ways of doing this. On the other side of the debate ("food vs. fuel") are factors such as taking land out of food production, with concomitant price rises of food particularly affecting the poor.

LSD

In wet summers, a fungus, *Claviceps purpurea,* develops on crops like rye, causing epidemics if bread made from this grain is eaten; people can die after developing symptoms like convulsions and gangrene. This is caused by ergot alkaloids, which are mainly derivatives of lysergic acid. It is possible that ergotism was a trigger factor in the events leading up to the Salem Witch Trials of 1692 and to the French Revolution (1789).

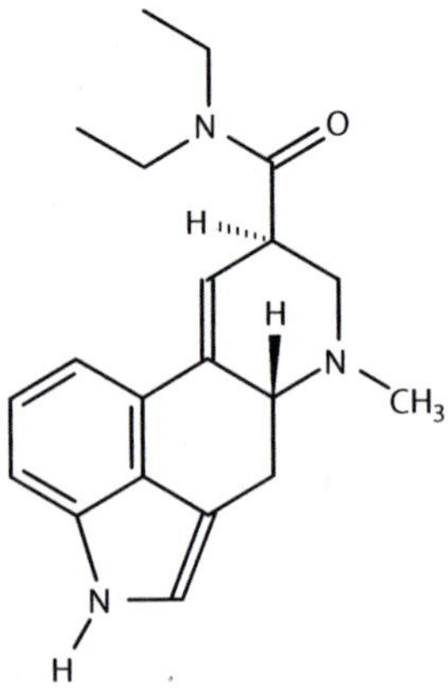

(11.30)

LSD (11.30)—(6a*R*,9*R*)-*N*,*N*-diethyl-7-methyl-4,6,6a,7,8,9-hexahydroindolo[4,3-fg]quinoline-9-carboxamide—was first synthesised on 16 November 1938 by Albert

Hofmann, working at the Sandoz Laboratories in Switzerland. He was trying to make ergot alkaloid derivatives that would be useful medicines; the molecule was given the code name LSD-25 because it was the 25th lysergic acid derivative synthesised by Hofmann. It was not until over 4 years later that he accidentally took in some of the chemical (probably via his hands) and discovered its psychedelic properties. Three days later, on 19 April 1943, he deliberately ingested 250 µg (0.25 mg) of LSD and discovered that it was much more potent than other ergot alkaloids; he experienced dizziness and visual distortions for several hours. When he took a bicycle ride, he felt that he was stationary and the fields were whizzing past.

Hofmann awoke the next day feeling remarkably clear-headed and survived to the age of 102, dying on 29 April 2008. LSD is, in fact, a very powerful psychoactive drug—even 10 µg can produce symptoms, though a "normal" dose would be in the range of 50–100 µg. It has two chiral centres; though only (+)-LSD is psychoactive, it has the absolute configuration (5R,8R). Because the dose is so low, LSD is usually administered either as microdots or small squares of blotting paper that have been infused with a drop of LSD solution.

In the decade from 1950, LSD was used as a research chemical tested by hundreds of psychologists and psychiatrists for applications in psychoanalysis. It was also being tested by the American Central Intelligence Agency, as a possible mind-control agent, on young servicemen and also on members of the public. One of these psychologists was a young Harvard academic named Timothy Leary, who became consumed with enthusiasm for the possibilities of LSD and acted as a spokesman for the molecule, which entered into the counterculture of the 1960s as a recreational drug and entheogen. The jury's still out on whether the Beatles' "Lucy in the Sky with Diamonds" is a reference to LSD, but many groups of the time did use LSD.

People found that the psychological effects depended upon the person involved, the environment during the "trip", and even the their state of mind. Whilst euphoria could be brought on by the sight of a blue sky, passage of a cloud over the sun could produce a tremendous "down". Distortions could take in the movement of inanimate objects, with buildings coming alive and colours being intensified. Crossover between the senses could occur so that people felt that they could see sounds or hear colours. Flashbacks that occur years later have been linked to LSD. Much of this is still inexplicable. The mechanism of LSD action is still not clear, but it is believed that it interferes with serotonin (5-hydroxytryptamine, 5-HT) levels by competing for some types of serotonin receptors in the brain, particularly due to agonist effects at the 5-HT_{2A} receptors. Serotonin and LSD share an indoleamine core.

LSD use peaked in the 1960s. Suffering from terminal throat cancer, Aldous Huxley took an intravenous dose of 100 µg of LSD on his dying day, 22 November 1963 (John F. Kennedy and C. S. Lewis died on that same day). The Haight-Ashbury subculture during the San Francisco "Summer of Love" in 1967 was greatly influenced by LSD-induced psychedelic experiences and psychedelic performers, as well as groups whose music reflected the philosophy: Think of the Byrds, the Beach Boys, Jefferson Airplane, and the Grateful Dead. The Summer of Love passed into history as less peaceful drugs like methamphetamine intruded. It has been stated that there have been no fatalities from LSD overdoses, but it may well have caused deaths through impairing the judgment and capacity of the person involved. It was probably a good thing that Albert Hoffmann was riding a bike whilst under the influence of LSD in April 1943, rather than driving a car. There is also the possibility of a bad trip turning into a psychotic episode. LSD is an unpredictable drug.

Marijuana is obtained from the Indian hemp plant, most usually *Cannabis sativa,* which probably originated in Central Asia. Six thousand years ago, the Chinese grew the plants for the fibre to make rope. Clothing and paper are more recent uses, but it is thought that the drug may have been used medicinally as long ago as 2000 BC. Hindus and Buddhists used it for its intoxicating properties. When people smoke it, they feel happy and self-confident. It makes time seem longer and affects peripheral vision, reaction time, and coordination, as well as short-term memory.

Marijuana traditionally comes in three types, of increasing strength: bhang (from leaves, seeds, stem), ganja (from the flowering tops), and hashish (pure resin from the flowers). Since the 1960s, the content of the principal psychoactive component in cannabis, Δ^9-THC, has increased with up to 15%–20% in some varieties of "skunk". The aroma of cannabis is not an indication of its potency; this comes from various aromatic terpenes. The cannabis plant grows well in all climates, in urban settings like roofs of tower blocks, inside "unoccupied" houses and warehouses, in remote farming areas, and in "clandestine greenhouses". As the twenty-first century sets in, cannabis farming has spread across the UK, as in many other countries.

Marijuana was widely used during the first millennium in the Middle East; because it is not intoxicating, it is not prohibited by the Koran, which bans alcohol. It probably reached France in the early nineteenth century, with Baudelaire among the users. Queen Victoria's personal physician promoted the medicinal use of cannabis for dysmenorrhoea, but it is unlikely that the Queen took it. Dr J. Collis Browne's *chlorodyne* was first marketed in the UK around 1850 to treat a range of complaints, including dysentery, diarrhoea, cholera, insomnia, neuralgia, bronchitis, and asthma. Apart from alcohol, it contained "tincture of hemp" as well as morphine and, not surprisingly, proved addictive. The current mixture contains no opiates.

Indian workers introduced marijuana to the West Indies, and from there it spread to Central America, Mexico, and the United States. Famously, facing a publisher's deadline, Alice B. Toklas took a recipe from a friend for *The Alice B. Toklas Cookbook,* and "hashish fudge" made the British edition (though not the American one).

Marijuana had been largely confined to immigrant communities and jazz clubs, but in the 1950s and 1960s it moved to student campuses. The opening song of Bob Dylan's 1966 *Blonde on Blonde* album, "Rainy Day Women # 12 & 35", features the lyrics "everyone must get stoned"; he was allegedly high on cannabis at the time of the recordings. Dylan introduced the Beatles to marijuana and it was subsequently spread by the Vietnam War and Summer of Love (1967). Recent research suggests that about 9% of marijuana users become dependent. It is stated that no study has ever shown that cannabis use leads to later use of heroin and cocaine, though it could be argued that suppliers of cannabis are likely to supply harder drugs too.

Marijuana contains over 100 different chemicals. The most important molecule is **delta-9-THC** (Δ-9-tetrahydrocannabinol, Δ^9-THC; 11.31). There is little of this present in the living plant, but it is readily formed by decarboxylation of certain carboxylic acids even just by drying. As usual, nature just makes the one isomer, (–)-Δ^9-THC. The synthetic (+) isomer is inactive and has no psychoactive properties.

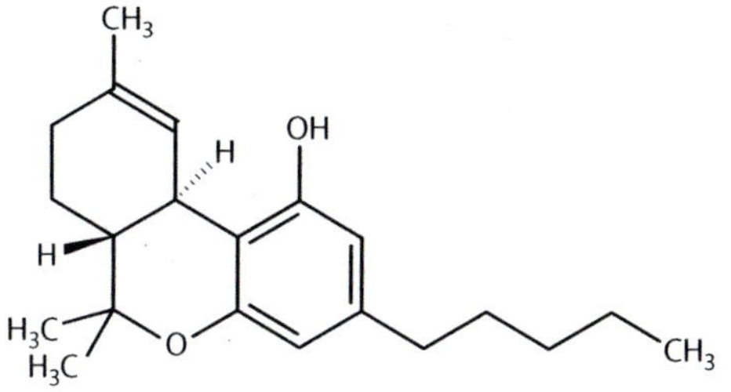

(11.31)

When cannabis is smoked, the blood takes Δ^9-THC straight from the lungs to the brain. The body has two receptors where the drug acts; one of them, called CB1, exists in the brain in particular (it is also found in the lungs, liver, and kidneys), whilst the other (CB2) is only in areas like the immune system. Hallucinogenic effects are produced when molecules like Δ^9-THC bind to CB1. Some natural molecules like anandamide also bind to these so-called "cannabis receptors"; anandamide (11.32), discovered in 1992, appears involved in pain control and the pleasure response. Delta-9-THC is not the only effective chemical in cannabis. By itself, it can produce feelings of panic, but these are alleviated in cannabis by another chemical present called cannabidiol (11.33), which is not psychoactive.

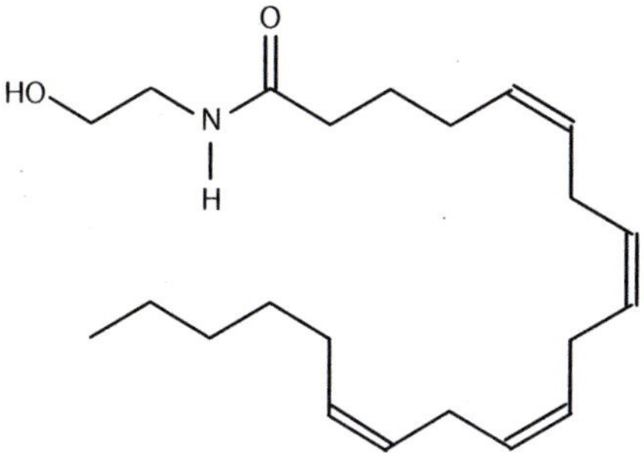

(11.32)

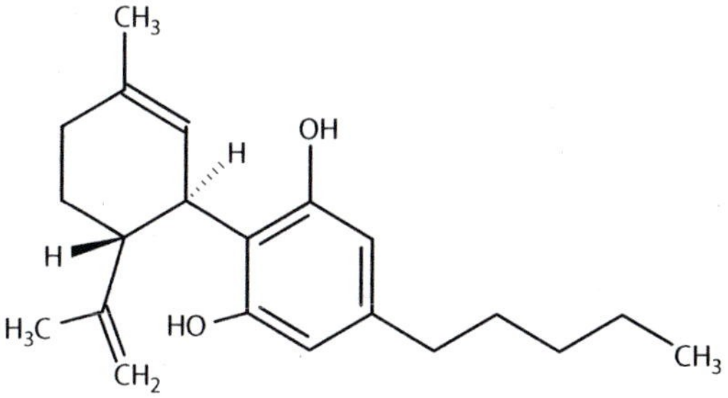

(11.33)

Most people who take marijuana take it for the "high", but there are some who take it for medical reasons. Cancer patients say that it helps reduce nausea and vomiting;

AIDS patients say that it stimulates their appetite and multiple sclerosis patients believe that it reduces muscle cramps. There is evidence from paraplegics and quadriplegics that it helps control involuntary muscle spasms. When the renowned biologist Stephen Jay Gould was under treatment for abdominal mesothelioma cancer, he smoked marijuana to alleviate the severe nausea that the chemotherapy gave him ("marijuana worked like a charm"). Marijuana has been shown to lower intraocular pressure; this is potentially useful in glaucoma patients. In the UK in 2010, a plant-sourced drug, Sativex®, which contains tetrahydrocannbinol (THC) and cannabidiol, was licensed as a prescription-only oromucosal spray administered into the mouth for the treatment of multiple sclerosis patients, for neuropathic pain and other symptoms, and as a painkiller for cancer patients.

Synthetic pure isomer (–)-Δ^9-THC is marketed as dronabinol (trade name Marinol®) in the United States to treat vomiting and enhance appetite, especially in AIDS patients. **Nabilone** (11.34) is a synthetic cannabinoid that is marketed as Cesamet in Canada, the United States, the UK, and Mexico for treatment of chemotherapy-induced nausea and vomiting, as well as treatment of anorexia and weight loss in patients with AIDS. Medical opinion on the benefits of cannabinoids as medications is mixed, though what is not in doubt is that smoking is a very unsafe way of delivering the molecules—not least because of the risk of illnesses associated with smoking, such as lung cancer and bronchitis. Cannabis smoke contains over 150 different compounds; it is not too dissimilar to tobacco smoke because it contains toxic molecules like CO and HCN as well as the known carcinogen benzopyrene. Although no one has yet made any long-term study of marijuana smokers, it is unlikely to be a healthy option.

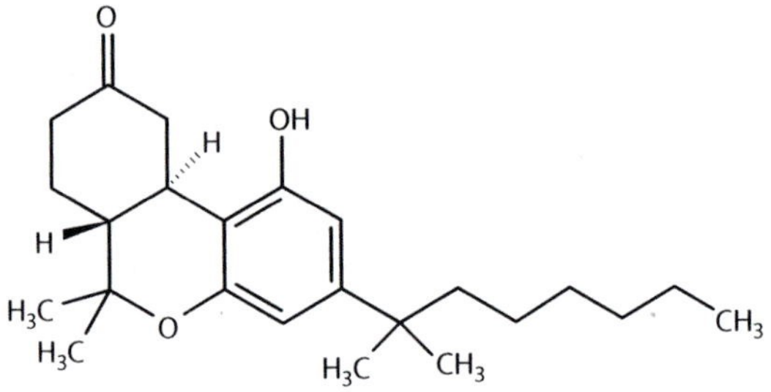

(11.34)

NICOTINE

And surely in my opinion, there cannot be a more base, and yet hurtfull, corruption in a Countrey, then [sic] is the vile use (or other abuse) of taking Tobacco in this Kingdome, which hath mooued me, shortly to discouer the abuses thereof in this following little Pamphlet (from the foreword to *A Counter-Blaste to Tobacco*).

King James I of England clearly hated nicotine; Columbus thought smoking was a heathen practice, and both Urban VIII (in 1642) and Innocent X issued papal bulls against tobacco use. The Jansenists were for it and the Jesuits against. In the end, governments solved the problem by taxing it.

The tobacco plant *Nicotiana tabacum* is the principal source of **nicotine** (11.35) and takes its name from Jean Nicot de Villemain, the French ambassador to Portugal, who sent tobacco and seeds from Brazil to Catherine de Medici in 1560. Nicotine is found in tiny but insignificant amounts in other members of nightshade family (Solanaceae), such as potato, tomato, aubergine, and green pepper. Nicotine is synthesised in the roots of the tobacco plant, starting from the amino acid ornithine, but the majority accumulates in the leaves. Smoking tobacco leads to coronary heart disease, emphysema, and cancers of mouth, throat, and lung, as well as contributing to other cancers. It is responsible for over 4,000,000 smoking-related deaths each year worldwide, with about 10% of those in the United States.

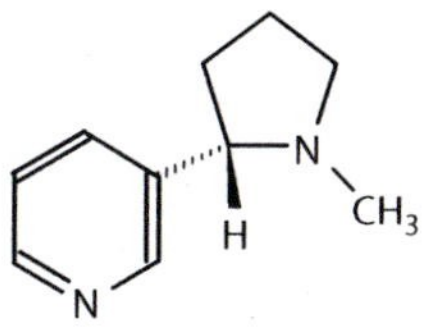

$$(11.35)$$

Nicotine itself is a fast-acting poison that is toxic to animals and insects alike. A lethal dose of nicotine to a human is around 50 mg, so it is much more poisonous than the average alkaloid like cocaine. On average, a cigarette contains about 8 mg of nicotine, but most of the nicotine content of a cigarette is oxidised to other compounds when the cigarette is smoked, so perhaps only 1 mg of nicotine is delivered to the smoker's lungs. Nicotine has a half-life in the body of about an hour, so when nicotine levels drop, the smoker experiences withdrawal symptoms and promptly reaches for the pack of cigarettes again. In the liver, nicotine is metabolised by cytochrome P450 enzymes (largely CYP2A6); the principal metabolite (75%) is cotinine (11.36). This has a lifetime in the body of around 24 hours, so measurements of the cotinine content of blood or urine can be used to see whether someone has been smoking in the last day or two.

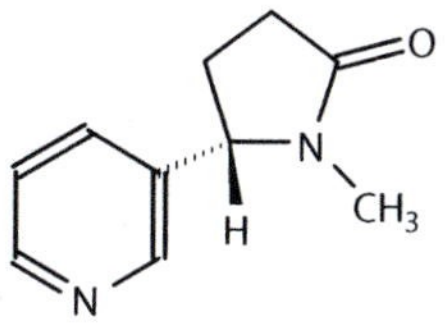

$$(11.36)$$

A cigar contains between 100 and 200 mg of nicotine; if this were extracted from the tobacco and injected directly, it would kill at least two humans. However, a combination of factors—cigars are smoked slowly, giving the body time for metabolism; the fact that most cigar smokers do not inhale, plus the low amount surviving combustion—means that cigar smokers survive. In fact, since nicotine gum usually comes in 2 or 4 mg doses, someone may get more nicotine from these or from snuff or nicotine patches than from smoking a cigarette.

The free nicotine molecule is easily absorbed through human skin and mucous passages, as well as in the lungs. A poisonous dose of liquid nicotine can be absorbed through the skin. When tobacco is smoked, nicotine is transported in the blood from the lungs to the brain in less than 10 seconds. Once in the brain, it stimulates the release of neurotransmitters, and it is the release of dopamine, a "reward chemical", that is responsible for the addictive effects of nicotine. Nicotine's effect on the brain is due to its ability to bind strongly to the brain's $\alpha 4 \beta 2$ acetylcholine (ACh) receptors. In the body, the nicotine molecule is protonated and forms both a strong cation–π interaction with a tryptophan residue and a hydrogen bond to a tryptophan carbonyl group in the brain's acetylcholine receptor.

It had long puzzled scientists that nicotine does not form similar interactions with the ACh receptors in muscle, where the immediate binding site residues, including the tryptophan, are identical. The reason for this was reported in 2009. It transpires that, in the muscular ACh receptors, there is a point mutation, four amino acid residues away from the tryptophan, that affects the shape of the binding pocket that binds acetylcholine. Thus, there is no cation–π interaction between nicotine and this receptor and the hydrogen bond is weaker; therefore, nicotine is much more weakly bound by the muscular ACh receptors. Dennis Dougherty, the chemist at the California Institute of Technology who led the team that made this discovery, commented that "if it [nicotine] activated the receptor in muscles, humans would probably die instantly from smoking".

Cigarette smoke contains around 4,000 different compounds. Its carcinogenic effect is not due to nicotine, but rather to other chemicals such as benzo[a]pyrene. In nature, nicotine occurs as the single isomer (S)-nicotine, but when tobacco is smoked, a small amount of the unnatural (R)-nicotine, which is a very weak agonist, is formed.

There are limited health benefits associated with nicotine: Some studies have suggested that smokers seem to be less likely to get Parkinson's disease, and using a nicotine patch appears to assist the drug haloperidol to suppress Tourette's syndrome. But the benefits of using nicotine are outweighed by the risks of smoking. Dr Jerry Buccafusco and his collaborators in Augusta, Georgia, have found that cotinine may help enhance memory and protect brain cells, so researchers are looking to see if it can help protect against Alzheimer's and Parkinson's diseases.

Because it is toxic to most insects, nicotine has been used as an insecticide since the early eighteenth century; it was whilst looking for alternative insecticides that Gerhard Schrader discovered tabun, the first nerve agent (p. 106). *Manduca sexta,* the tobacco hornworm, mainly feeds on tobacco and tomato; it selectively binds the nicotine so that it is not toxic. The cigarette beetle, *Lasioderma serricorne* (Fabricius), is another exception, tolerating tobacco nicotine levels of up to 8% concentration and apparently excreting nicotine unchanged. The larval stages of the beetle will feed on a varied diet that includes grain and cereal products as well as some fruit and dried fish.

A more subtle use of nicotine has been found in a study of a wild tobacco plant (*Nicotiana attenuata*) from the Mojave Desert in the United States that showed that it uses very low doses of nicotine in its nectar. The plant uses sweet-smelling benzylacetone (11.37) to attract insect pollinators, but the bitter taste of the very small amounts of nicotine present deter visitors such as hawkmoths and humming-birds from taking too much nectar.

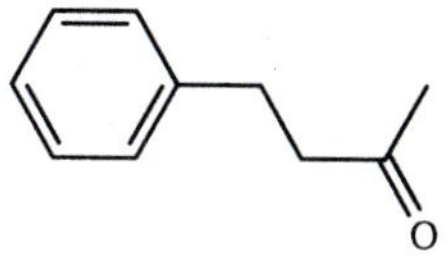

(11.37)

12 Natural Healers

GALANTAMINE

$$(12.1)$$

In the *Odyssey,* written by Homer about 750 BC, the story is told of Hermes giving Odysseus a medicine from a plant that had "a black root but a milklike flower", which the gods called "moly". This medicine enabled him to resist the spells of the enchantress Circe, who had already bewitched half his crew. It is thought that the medicine may have come from a snowdrop. In the 1950s, a Bulgarian pharmacologist noticed that local villagers rubbed snowdrops on their foreheads to ease nerve pain in what seemed to be an old folk remedy.

Further research led scientists to galanthamine (or galantamine; 12.1). It has been extracted from plants that include the Caucasian snowdrop, *Galanthus woronowii;* the common snowdrop, *Galanthus nivalis;* and the summer snowflake, *Leucojum aestivum.* For medical purposes, it has been obtained from daffodils, *Narcissus pseudonarcissus,* but chemists are now making synthetic galantamine. Galantamine, marketed under trade names Reminyl®, Nivalyn®, and Razadyne®, is being used for people suffering from mild to moderately severe Alzheimer's disease. It does not change the course of the dementia process, but it slows decline and appears to improve performance on memory tasks.

Acetylcholine is an important transmitter of messages in the human nervous system. Acetylcholine-producing neurons degenerate in the brains of people with Alzheimer's disease and this is thought to be linked with memory loss. Galantamine crosses the blood–brain barrier and inhibits the hydrolysis of acetylcholine by acetylcholinesterase, thus increasing the concentration of acetylcholine in the brain. It may also help make better use of the acetylcholine receptors. Other alkaloids are found in snowdrops, too, and it is possible that a combination of alkaloids could turn out to be an even more effective treatment for Alzheimer's. Galantamine is not just an Alzheimer's drug. When it was first studied in the 1950s, it was used in Eastern Europe to treat poliomyelitis and muscular pains and, more recently, has been used as a dream enhancer.

J. M. Barrie once said that we are given memories "so that we might have roses in December". Sadly, this is not true for all elderly people, so anything that helps us keep memories is a good thing.

OPIUM AND MORPHINE

Morphine (12.2) is the best molecule available for dulling chronic pain. The opium poppy (*Papaver somniferum*) synthesises morphine starting from the amino acid tyrosine. If the unripe seed capsules are cut, once the poppy has flowered, a milky substance flows out; upon drying, this becomes the yellow-brown paste known as opium. The opium poppy is native to the Eastern Mediterranean and was certainly being used by Sumerians in Mesopotamia ca. 3500 BC. Alexander the Great is said to have taken opium to Persia and India, from where it reached China, where it was used medicinally as a painkiller.

After the Chinese banned tobacco smoking in 1644, many Chinese turned to opium as an alternative and the British East India Company found exporting opium to China a profitable undertaking. Chinese governments were unhappy with this, and the result was the Opium Wars of 1839–1842 and 1856—not the high point of nineteenth century British foreign policy. One consequence of the Opium Wars was that Hong Kong became a British Crown Colony until 1997. Nowadays, opium is a cash crop across a region stretching from the Middle East to the Far East; perhaps 90% of the world's opium and heroin originates in Afghanistan.

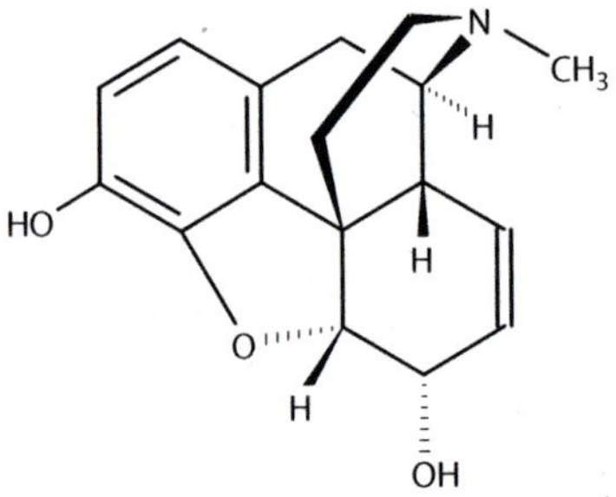

(12.2)

People used opium medicinally, usually as an alcoholic solution known as laudanum, before morphine was isolated (ca. 1804). Laudanum was almost the aspirin of the late eighteenth and early nineteenth centuries; people took it for virtually anything; in an age of dysentery and cholera, it was probably much safer than drinking the water. The great poets and writers of the day, like Byron, Keats, Samuel Taylor Coleridge, and Thomas de Quincy, used it, but so did quite respectable people like Lewis Carroll, Louisa May Alcott, and Florence Nightingale. Wilkie Collins wrote *The Moonstone* whilst under its influence. A certain American president did not inhale, but took laudanum instead: George Washington used it for his dental pains. Poor people used it as a medicine, whether in the city slums or the East Anglian Fenland or as a counter to diarrhoea and pain and, also, one suspects, as a panacea to relieve sad, dull lives.

(12.3)

Opium contains over 20 different alkaloid molecules; the most abundant is morphine (typically 10%). Apart from morphine, **codeine** (12.3) is the only other alkaloid painkiller found in the opium. Morphine is synthesised in the poppy in 17 steps, starting from the amino acid tyrosine ($HO-C_6H_4-CH_2CH(NH_2)COOH$). One molecule of tyrosine is converted to dopamine and a second to 4-hydroxyphenylacetaldehyde; the two undergo a condensation reaction to form (S)-norcoclaurine, catalysed by the enzyme (S)-norcoclaurine synthase. Mutant poppies that can produce thebaine and oripavine but not codeine or morphine have been known for some years; these are valuable because thebaine can be converted into several pharmaceuticals, including semisynthetic opiates.

It was only in 2010 that two enzymes responsible for the conversion of thebaine to morphine were identified. Thebaine 6-O-demethylase (T6ODM) controls the first stage of the conversion of thebaine into codeine, and codeine O-demethylase (CODM) converts codeine into morphine (Figure 12.1). At present, most codeine is manufactured by methylation of morphine, but one possibility is to engineer a poppy that would produce codeine but not morphine, lacking the capability to demethylate codeine. This would both increase the supply of codeine and lower its price, and also reduce the supply of morphine, thus reducing the amount available for heroin synthesis.

Codeine, the second most abundant alkaloid in opium, was isolated some time after morphine. When taken orally, about 10% of codeine is demethylated into morphine in the liver, so you experience painkilling effects; thus, codeine usually acts as a prodrug, though its effectiveness as an antitussive (cough medicine) is believed to be due to codeine, rather than morphine. Codeine is widely used as a treatment for mild or moderate pain and is less addictive than morphine.

In the first decade of the nineteenth century, morphine was isolated more or less simultaneously by Jean-François Derosne, Armand Séguin, and Friedrich Wilhelm Sertürner, but it is the latter who usually gets the credit, after his work was publicised a decade later by Gay-Lussac. It was called morphine after Morpheus, the Roman god of dreams, and is very addictive. The invention of the hypodermic syringe in 1853 facilitated its use (and abuse); addiction first became a problem after the American Civil War, when it was widely used on casualties. Hundreds of thousands of combatants were said to have become addicted to it.

Because morphine can cross the placental barrier, babies born to mothers who are on morphine can go through a long period of withdrawal. However, morphine is the most effective painkiller available for people in severe pain. It can be made in the

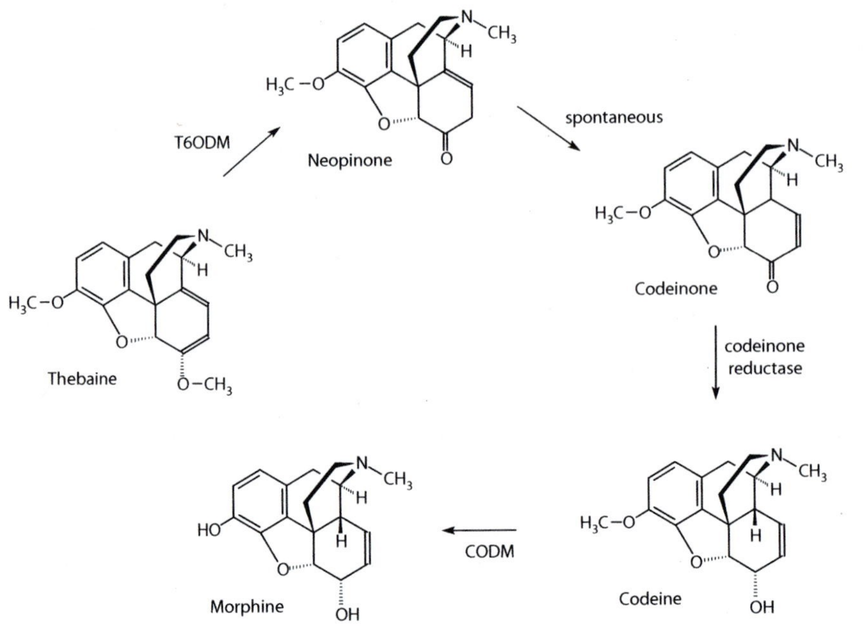

FIGURE 12.1 Thebaine 6-O-demethylase (T6ODM) controls the first stage of the conversion of thebaine into codeine, and codeine O-demethylase (CODM) converts codeine into morphine.

laboratory, but the poppy does it much more efficiently and cheaply. Natural morphine is entirely the (−)-isomer; the unnatural (+)-isomer has been synthesised and binds only very weakly to the opiate receptor. Morphine binds to the same kind of receptor in the brain as endorphins and enkephalins, the molecules produced naturally by the body as responses to pain. The signalling chain includes a G protein; when morphine binds to the receptor, the protein responds in several ways, such as increasing conduction through potassium channels and reducing the release of neurotransmitter molecules so that the pain message gets blocked.

The receptor is believed to have three main areas that bind morphine, endorphins, and related molecules: an anionic site, of dimensions 8 Å × 6.5 Å, which binds the amine nitrogen (A), which can become positively charged; a cavity that accommodates the projecting piperidine ring (B); and a flat surface that binds the benzene ring (C) by van der Waals forces (Figure 12.2).

A number of structural features have to be present for the opioid painkiller to be active, including the tertiary nitrogen with a small alkyl group (A), the presence of a piperidine ring (B), a quaternary carbon (D), a benzene ring (C)—or species with the same shape—attached to it, and a C_2 spacer (E), part of the piperidine ring, between the tertiary nitrogen (A) and the quaternary carbon (D). Several opioid receptors, known as μ (mu), κ (kappa), σ (sigma), and δ (delta), have been identified; the mu receptor has especially high affinity for morphine and opioids.

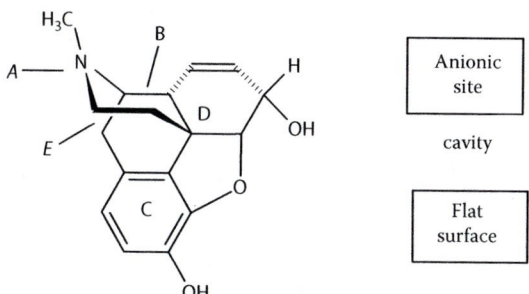

FIGURE 12.2 The receptor is believed to have three main areas that bind morphine, endorphins, and related molecules: an anionic site, of dimensions 8 Å × 6.5 Å, that binds the amine nitrogen (*A*), which can become positively charged; a cavity that accommodates the projecting piperidine ring (*B*); and a flat surface that binds the benzene ring (*C*) by van der Waals forces.

HEROIN

In 1874, a British chemist named Charles Romley Alder Wright, working at St Mary's Hospital in London, heated morphine with ethanoic anhydride and isolated a new substance, diacetylmorphine (used medicinally in the UK under the name diamorphine). This compound was further investigated by the German chemist Felix Hoffman in the Bayer laboratories. Heroin promised to be a nonaddictive, faster acting alternative to morphine and acquired its name from its "heroic" properties; it was marketed in 1898 as a cough treatment.

This optimism was sadly misplaced. Heroin (12.4) is a prodrug that is metabolised in the body to morphine. It is very easy to make from morphine, though it can retain traces of the by-product, ethanoic acid, whose vinegary smell gives trained sniffer dogs something to detect. Replacing the two polar –OH groups of morphine with ester linkages makes heroin more lipophilic than morphine; thus, it will cross the blood–brain barrier, and this is probably responsible for its faster action. Doctors use heroin, under the less sensational name of diamorphine, as a way of administering morphine. It was Dr Harold Shipman's chosen agent of death for the many patients that he murdered.

(12.4)

After World War II, heroin addiction became associated with the world of pop music and entertainment; the deaths of iconic figures like Gram Parsons, Janis Joplin, Jim Morrison, Sid Vicious, Tim Buckley, and Kristen Pfaff are associated with morphine or heroin, as well as Paula Yates and Talitha Getty. River Phoenix and John Belushi were killed by speedballs—a mixture of heroin and cocaine. Around 10% of the American soldiers in the Vietnam War were believed to have been addicted to heroin.

Modern drug testing is so sensitive that heroin metabolites like morphine and codeine can be detected in urine samples taken from people who have eaten foods containing poppy seeds (such as on bagels). This can have serious consequences in the United States: People have lost jobs or had babies taken away from them. It impacted popular consciousness through the 1996 "Shower Head" episode of *Seinfeld* (in which the character Elaine discovered that her urine test results showing the presence of opium were due to her eating poppy seed muffins). There is currently no general way of distinguishing between "innocent" consumers and people who have taken heroin (the "poppy seed defence"). Although it is possible to test for 6-monoacetylmorphine, a heroin metabolite not formed from either morphine or codeine, this is only formed in small, short-lived amounts.

Another molecule structurally derived from morphine is **etorphine** (12.5), a synthetic cousin about 500 times more effective a painkiller. Etorphine was first synthesised in the early 1960s by a research group at McFarlan-Smith and Co. in Edinburgh, led by Professor Kenneth Bentley, who were trying to make new nonsteroidal antiinflammatory drugs. Its effects were allegedly first noticed when someone in the lab stirred the midmorning cups of tea in the lab with an unwashed glass rod that had last been used with etorphine; the chemists were rendered unconscious. Etorphine is so toxic that it is usually dyed red for ease of recognition and is banned from use on humans. Because it is so much stronger than morphine, it is used to knock out wild game such as rhinos and elephants (attached to darts fired from rifles). In the Kruger National Park, they use shotguns, firing darts based on .22 calibre blank cartridges. One 4 mg dose will immobilise a 5000 kg African elephant; 1 mg will knock out a 2000 kg rhino.

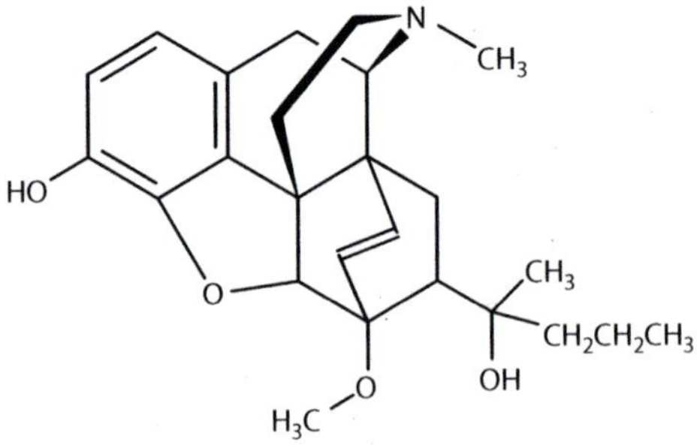

(12.5)

PENICILLIN

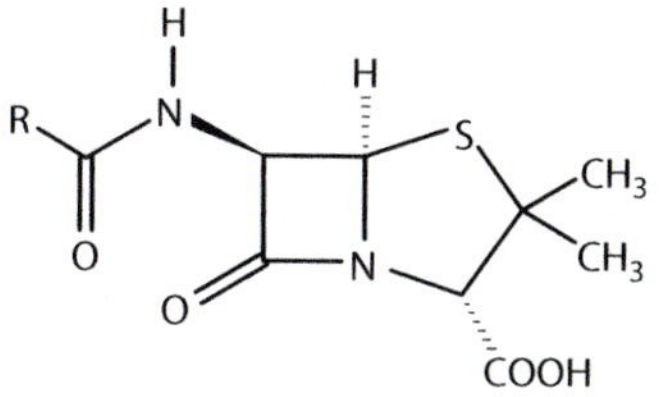

$$(12.6)$$

The name "penicillin" is given to a type of antibiotic featuring the core structure shown in 12.6, where the R group is variable. The antibiotic effects of penicillin were originally discovered in 1896 by a French medical student, Ernest Duchesne; however, an Army career prevented him from investigating further, and it fell to Alexander Fleming to exploit it. On returning to St Mary's Hospital in London (now part of Imperial College) from his late summer break in 1928, Fleming made a fortuitous discovery: a number of petri dishes containing staphylococcal cultures had been left over the holiday, and one had become contaminated by a mould (which proved to be *Penicillium notatum*). There was no staphylococcal growth in the region of the mould. Study of subcultures of the mould showed that they killed many sorts of bacteria.

Fleming made sporadic progress with this research until it was taken up by Howard Florey and Ernst Chain at Oxford University in 1938. In 1940, they found that it cured mice of deadly streptococcal infections and then clinical trials began. Penicillin was in such short supply that it was recovered from patients' urine and recycled. Much of the credit for making penicillin a successful drug was due to the unsung hero of the Oxford team, Norman Heatley (as Sir Henry Harris put it, "... without Heatley no penicillin").

Scaling up production was a problem in World War II Britain, so production was transferred to the United States (which also had a larger industrial base) and was started up in Peoria, Illinois. It was found that higher yields of penicillin could be obtained if the mould were grown on corn-steep liquor, using fructose rather than glucose, whilst the structure of the penicillin could be altered by the choice of chemicals added to the broth. A mouldy cantaloupe melon provided a strain of *Penicillium chrysogenum* that, together with adding phenylacetic acid to the brew, proved an excellent source of what became known as penicillin G, benzylpenicillin (12.7). Mass production ensued, and by the end of the war American factories were making enough penicillin to treat 250,000 patients a month, which saved the lives of many Allied soldiers. Large quantities of penicillin were also used to treat the venereal disease that resulted from fraternisation between women and Allied troops (particularly Americans) in the European nations liberated in 1944–1945.

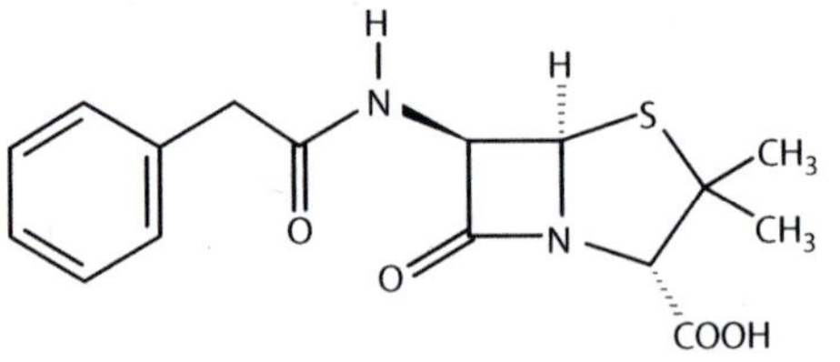

(12.7)

All this time, the structure of penicillin was the subject of conjecture, and it was 1945 before Dorothy Hodgkin determined the structure of penicillin G (12.7) by x-ray diffraction. The presence of the four-membered β-lactam ring was a major surprise. A strained ring with 90° bond angles was something that few chemists would have predicted (as E. P. Abraham did); yet, it is the reactivity that this confers on penicillin that causes its function of preventing cell wall formation by enabling it to attack the cell-wall-forming enzyme.

Many of the earlier penicillins to be made, notably penicillin G, are broken down by acid (as in the human stomach) and have to be administered by injection. Penicillin V (12.8), phenoxymethylpenicillin, which is made by including phenoxyacetic acid (PhOCH$_2$COOH) in the culture broth, was a breakthrough because it was relatively acid-stable and could be taken orally.

Penicillin works by interfering with the ability of bacteria to make their new cell walls. It binds to the enzyme peptoglycan transpeptidase needed for the synthesis of the cell walls and thus stops the polypeptide chains' cross-linking to each other. The bacteria lengthen considerably but cannot divide. Eventually, the high osmotic pressure causes the cell wall to break. Animal cells do not have cell walls and thus do not contain peptoglycan transpeptidase, which is why penicillin is not toxic to humans. But bacteria reproduce and adapt very quickly and, within just a few years, some bacteria had become resistant to penicillin; nowadays all strains of *Staphylococcus aureus* are resistant to penicillin G. Mutations in chromosomes have altered receptor sites so that penicillin is no longer able to bind to them. Others have learned how to make an enzyme called penicillinase, a β-lactamase enzyme that catalyses the hydrolysis of the β-lactam four-membered ring in the penicillin molecule and destroys its activity.

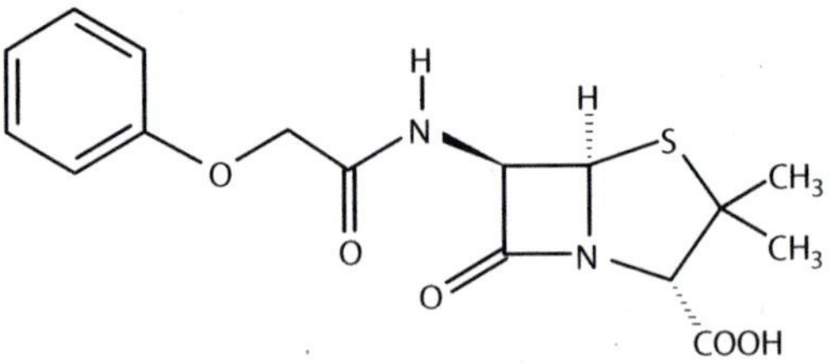

(12.8)

Scientists had to produce new penicillins and the key to this proved to be the discovery that the molecule at the heart of the penicillin structure, 6-aminopenicillanic acid (6-APA; 12.9), was isolable from the culture broth used to make penicillin. It was then a matter of adding the "R group" to make the particular side chain that was desired—semisynthesis. Subsequently, it was found that 6-APA could also be made from penicillin G.

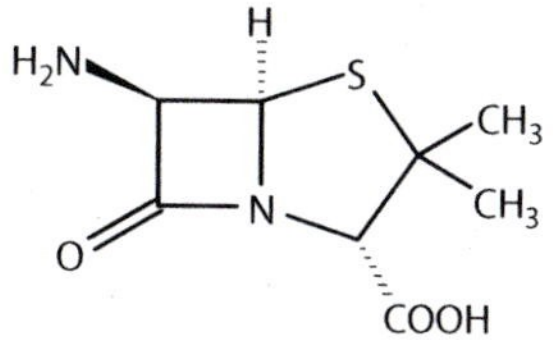

(12.9)

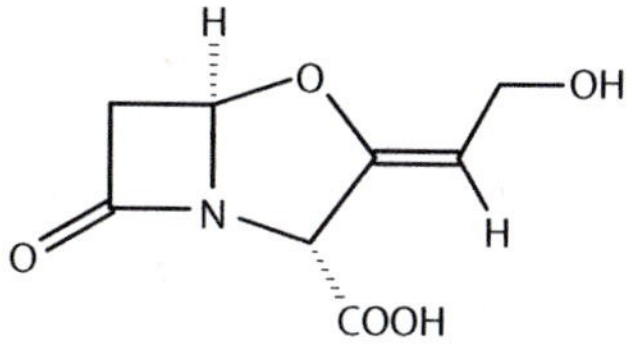

(12.10)

It is possible to add clavulanic acid (12.10)—a molecule with structural similarities to penicillin but no antibacterial activity—which protects penicillin by binding to the β-lactamase through a serine residue and permanently inhibiting it. Combinations of clavulanic acid with amoxycillin are widely used. Amoxycillin (12.11) and ampicillin (12.12) are two of the newer semisynthetic penicillins. Unlike traditional penicillins, they are active against some Gram-negative as well as Gram-positive bacteria. The amine side chain in the R group gives the molecules the ability to penetrate the outer membrane of Gram-negative bacteria.

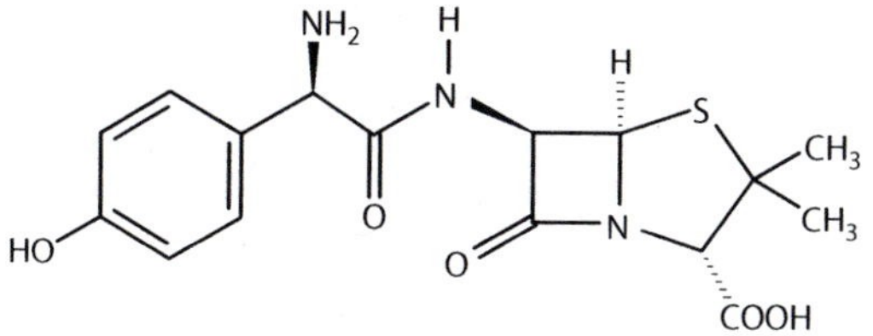

(12.11)

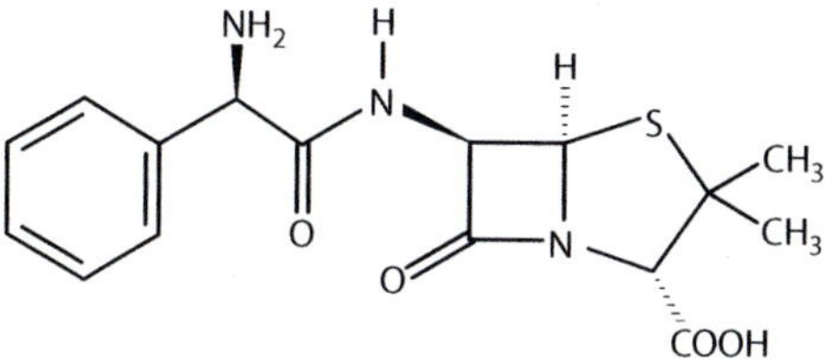

(12.12)

Although it has only a weak antibacterial action, methicillin (12.13) has methoxy groups in the 2,6-positions of the benzene ring that make it resistant to β-lactamase enzymes.

$$(12.13)$$

It is not used now, but its name is familiar in methicillin-resistant *Staphylococcus aureus* (MRSA), the name given to infections resistant to penicillin and other β-lactam antibiotics.

Penicillin biosynthesis is carried out in a series of enzyme-catalysed reactions. A tripeptide, δ-(L-α-aminoadipyl)-L-cysteine-D-valine (ACV), is assembled by the enzyme ACVS from L-α-aminoadipic acid, L-cysteine, and L-valine (the latter having been isomerised into D-valine). The second step in the biosynthesis of penicillin G is to change ACV into the bicyclic core of penicillin, using the enzyme isopenicillin N synthase; the product is isopenicillin N. Isopenicillin N synthase is a nonhaem iron-dependent oxidase that binds ACV by the iron, which then coordinates dioxygen so that iron-oxo species can oxidise the ACV and generate the β-lactam core. Subsequently, isopenicillin N undergoes a side-chain exchange, forming the penicillin. Isopenicillin N is also an intermediate in the biosynthesis of the cephalosporin antibiotics.

QUININE AND ANTIMALARIALS

Deadly disease means more than the Black Death, the 1918 Spanish flu pandemic, AIDS, or viral haemorrhagic fevers. Every year over 300 million people catch malaria, with around 2 million people dying of it. Around 90% of these deaths are in sub-Saharan Africa—mainly young children. It is a continuous pandemic whose famous victims are legion, including, in the distant past, the pharaoh Tutankhamun, Alexander the Great, Alaric the Goth, the Roman emperor Titus, and, later, the poet Dante, possibly the painter Caravaggio, and numerous mediaeval popes and cardinals. (Rome was a very unhealthy place.) Oliver Cromwell, the Lord Protector of England, Wales, Scotland, and Ireland, died of malaria in 1658, whilst the poet Byron fell victim in 1824.

Malaria is caused by a protozoan organism belonging to the genus *Plasmodium; P. falciparum* and *P. vivax* are both the most common and most dangerous. The parasite is transmitted by a bite from an infected female *Anopheles* mosquito. Once inside the human host, the parasite multiplies in the liver and in red blood cells, which periodically release more parasites, concomitantly causing fever. These organisms can infect any mosquito that feeds on the host's blood, thus helping to spread the disease. At a minimum, the victim is subject to fevers and chills, but around 1% of *P. falciparum* infections prove fatal. If the *P. falciparum* parasites accumulate in the brain, the result is lethal cerebral malaria.

Europeans colonising South America found that natives knew that an extract from the bark of Cinchona trees growing there was active against malaria. By 1630–1640,

Jesuits had taken it back to Europe, where it became a recognised treatment. That stern protestant, Oliver Cromwell, refused to take "Jesuit's bark" for his malarial fever as he lay dying. In 1672, an English apothecary called Robert Talbor published a book entitled *Pyretologia: A Rational Account of the Cause and Cures of Agues* stating, "Beware of all palliative cures, and especially of that known as Jesuits' Powder". The cunning Talbor cured King Charles II of malaria with a secret potion based on Jesuits' Powder dissolved in wine and soon afterward cured the Dauphin, son of Louis XIV of France, in the same way. Cinchona cultivation was a Spanish monopoly but, in 1865, cinchona seeds were smuggled out of Peru to start the cultivation of cinchona in the Dutch settlement of Java. Without widely available and cheap quinine (12.14), tropical enterprises like the building of the Panama Canal would not have been possible.

(12.14)

Identifying and extracting the active chemical in the cinchona bark had to wait for more advanced chemical understanding of structure and also organic synthesis. Quinine was first extracted from cinchona bark in 1820 by two French chemists, Pierre Joseph Pelletier and Joseph Bienaimé Caventou. Its chemical formula of $C_{20}H_{24}N_2O_2$ was determined in 1852 by Strecker; whilst the British chemist William Perkin was trying to make quinine by oxidative coupling of *N*-allyl toluidine, he actually synthesised mauve in 1856. It was not until 1907 that Paul Rabe established the correct structure and, with Karl Kindler, in 1918 he reported the synthesis of quinine in three steps from d-quinotoxine. Woodward and Doering (1944) synthesized quinotoxine from 7-hydroxyisoquinoline (15 steps) and relied upon the Rabe–Kindler conversion of d-quinotoxine to quinine to complete their paper, "The Total Synthesis of Quinine".

With its four chiral carbons, stereoselective synthesis of quinine proved a challenge, but it was eventually accomplished by Stork (2001). Despite all this, cinchona trees are still the only economic source of quinine. Quinine remained a frontline weapon against malaria until 2006, when the WHO recommended artemisinins as first choice. If they are not available, quinine is the best alternative, though it has side effects (e.g., dizziness, tinnitus, vertigo, constipation, and nausea). It has other medical uses, such as treating leg cramps, whilst its main nonmedical use is in tonic water (but at levels too low to be a drug). Like chloroquine and other antimalarials, quinine is believed to work by inhibiting haem polymerase, the enzyme that the malaria parasite uses to polymerise and eliminate ferriprotoporphyrin IX (haem). This iron compound is thought to react with O_2 to form reactive species like H_2O_2 and iron-oxo compounds toxic to the malarial parasite.

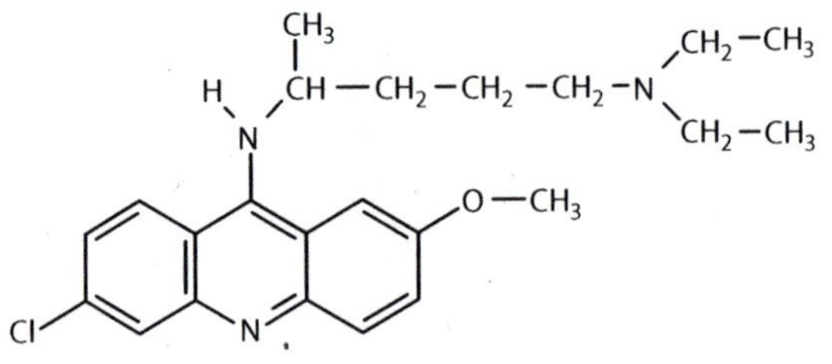

$$(12.15)$$

Synthetic antimalarials were developed because by the 1930s some malarial parasites were becoming resistant to quinine. **Quinacrine** (Mepacrine or Atebrine; 12.15) was issued widely to Allied troops in World War II. At first, persuading troops to take it was a problem because it turned the skin and urine yellow, and there was a widely held belief that it led to impotence. At one stage, General William Slim, the Allied commander on the Burma front, found that for every wounded soldier evacuated, over a hundred troops were evacuated with malaria; upon enforcing the use of Mepacrine, malaria rates dropped below 5%. Mepacrine had side effects such as toxic psychoses in extreme cases; in 1941, Orde Wingate, later to command the Chindits in Burma, attempted suicide while taking higher than recommended doses of Mepacrine for malaria. These side effects led to its eventual replacement by other drugs, but it was a very successful therapeutic and prophylactic antimalarial drug throughout the war.

Chloroquine (12.16), discovered in 1934, was extensively tested in the United States during World War II. Less toxic than Mepacrine and requiring lower doses, it was widely adopted after 1947 as a safe, cheap, and widely available prophylactic. The malarial parasite has now become resistant to it in many parts of the world, particularly in most of Africa, Brazil and the northern parts of South America, India, and much of Southeast Asia. There is evidence that resistance is lost when chloroquine is withdrawn for a few years; chloroquine-resistant parasites have a survival advantage, which they lose when chloroquine is absent. It is possible that chloroquine can be successfully reintroduced in combination therapy when it is administered with another antimalarial drug.

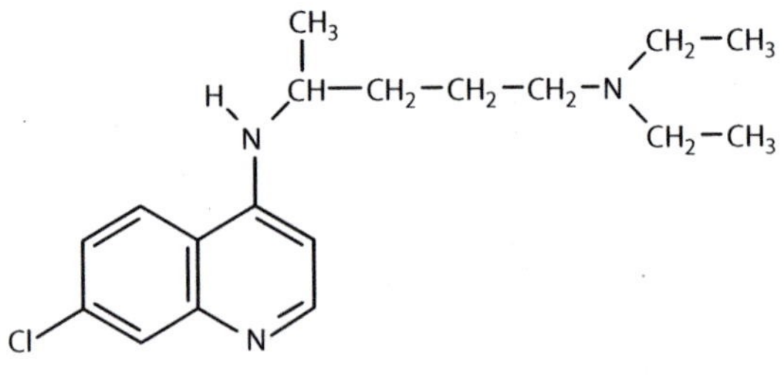

$$(12.16)$$

Mefloquine (also known as **Lariam**; 12.17) was developed in the 1960s and first used by American troops fighting in Vietnam in 1971 to prevent and treat *P. falciparum* malaria, especially in areas of resistance to chloroquine. It came into general use in the early 1980s. It was very successful at first, but the malaria parasite rapidly grew resistant to it. Although effective, Mefloquine is a controversial treatment because side effects such as dizziness, nightmares, headaches, and insomnia are common; serious neuropsychiatric reactions can occur in a very small number of cases, perhaps 1 in 10,000. Mefloquine-resistant parasites are endemic in much of Burma, Cambodia, and Thailand, and alternatives are suggested. Even in Africa, combinatorial therapy is recommended.

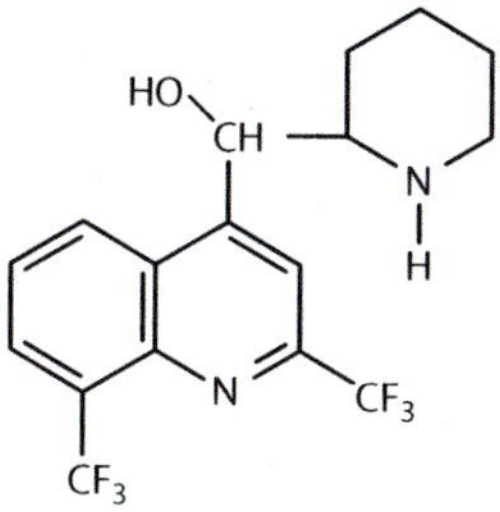

(12.17)

Atovaquone (12.18) was discovered in the 1980s and since the late 1990s has formed part of the popular antimalarial drug Malarone, taken in conjunction with Proguanil. It appears to be tolerated better than alternatives and is a shorter treatment, but is very expensive compared to other medications and is thus currently an option for the more well-to-do tourists.

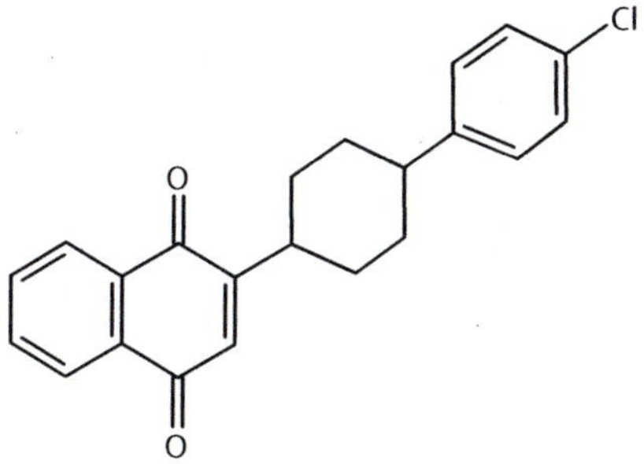

(12.18)

The next breakthrough happened as a result of the Vietnam War; as the North Vietnamese forces were suffering heavily from malaria, they appealed to their ally, China, for help. Many Chinese herbs were tested and attention was drawn to an ancient herbal remedy based on the plant *Artemisia annua*, which had been used by Chinese herbalists for over a thousand years. Peasants in rural areas had made a tea from it (using hot, not boiling, water) or simply chewed fresh plants mixed with brown sugar, as a way of treating fevers. A team led by Professor Tu Youyou obtained artemisinin, the active ingredient, by low-temperature ether extraction; they found the formula ($C_{15}H_{22}O_5$) of the white crystals in 1972 and its structure was determined in 1976. **Artemisinin** (12.19) has relatively short-term potency; because of this, it is not a suitable prophylactic. It also suffers from a lack of water solubility. Semisynthetic derivatives developed to improve its potency were soon in use as antimalarial drugs like artesunate (12.20) and artemether (12.21).

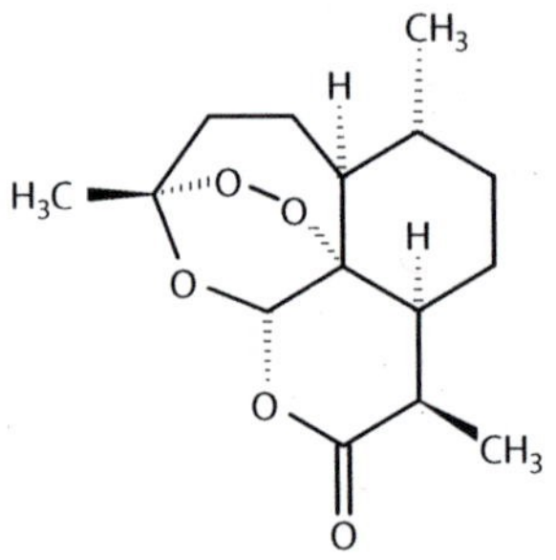

(12.19)

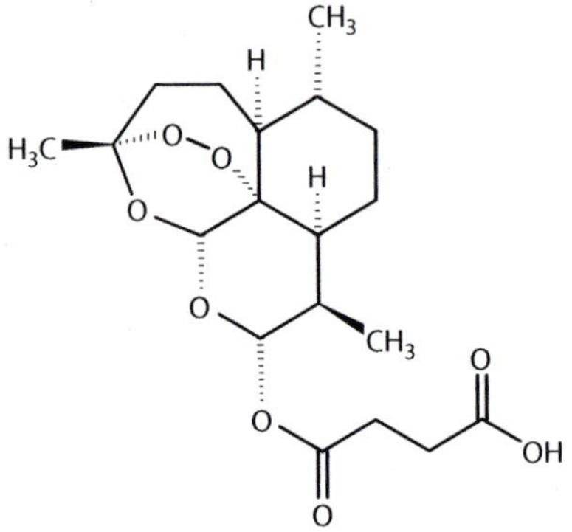

(12.20)

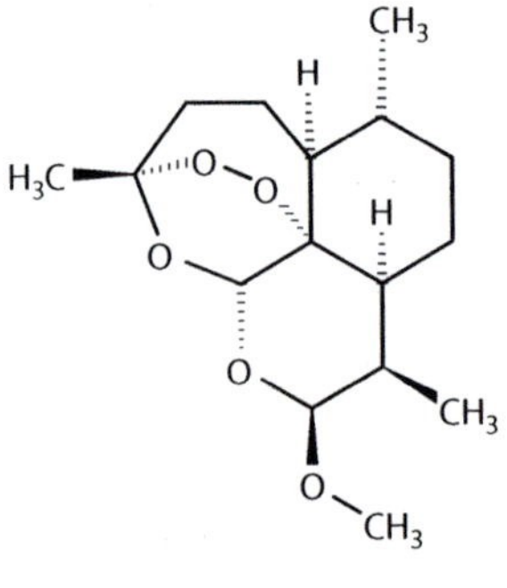

(12.21)

When *Plasmodium* parasites ingest haemoglobin to extract protein for growth, haem groups are freed. Either haem groups or free Fe^{2+} ions attack the peroxide group in artemisinin, resulting in the generation of free radicals that attack and kill the DNA of the parasite. Artemisinin has also been suggested to inhibit an enzyme called PfATP6 that is involved in pumping Ca^{2+} ions into membrane organelles, and this may be a means of its action. There are concerns about the possible emergence of artemisinin-resistant parasites; this is compounded by the widespread availability of fake artemisinin medications, especially in Southeast Asia, and by people not completing their course of treatment. ACT (artemisinin combination therapy using artemether–lumefantrine, artesunate–mefloquine, artesunate–amodiaquine, and artesunate–sulfadoxine/pyrimethamine) has been endorsed by the World Health Organisation as a frontline treatment for severe *P. falciparum* malaria.

The most promising totally synthetic drug to be investigated lately is OZ277 (Arterolane; 12.22). The starting point in its design was the "endoperoxide" bridge that is key to the activity of artemisinin. The first compounds synthesised were too unstable and also not active as a drug, so it was decided to protect the bridge by incorporating a bulky adamantyl group on one side of the bridge. This made the molecule not just more stable but also better at killing malarial parasites than the best artemisinin derivatives. The first compounds of this type were not very soluble, so the next step was to incorporate some polar groups to enhance this. OZ277 is not just more soluble, but also is absorbed well by the body when administered orally and is very effective against malaria in both in vitro and in vivo testing. It is now in phase III human trials and is believed to act in a similar way to artemisinin.

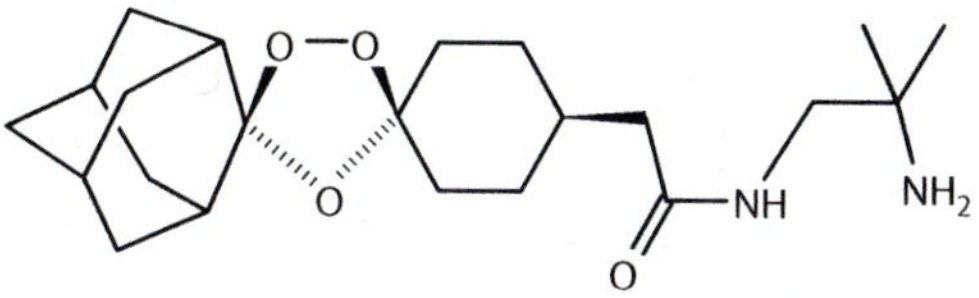

(12.22)

Antimalarial drugs should not, of course, be seen in isolation. They do not remove the need for other countermeasures, such as mosquito nets, covering skin with clothing, and the use of mosquito repellent.

TAXOL (PACLITAXEL)

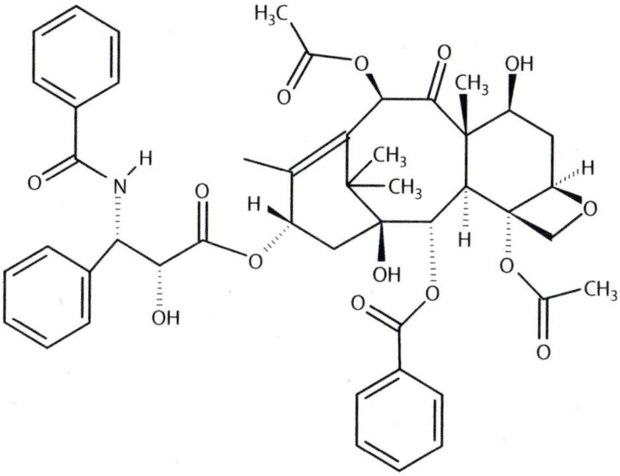

(12.23)

Some drugs have been discovered by accident. Paclitaxel (Taxol; 12.23) was discovered as the result of a systematic search, but between its discovery and its coming into use, a good deal of time elapsed before it became, for a while, the all-time best-selling cancer treatment drug.

In 1958, the American National Cancer Institute launched a programme to discover new pharmaceuticals. Some of these would be synthetic molecules, but the institute also asked researchers to investigate natural sources like plants and marine life. In 1962, Arthur Barclay, a botanist working for the US Department of Agriculture, collected bark samples from Pacific yew trees in Washington state. In 1967, Munroe E. Wall and his colleague, Mansukh Wani, who were working at the Research Triangle Institute in North Carolina, found that a chemical in the samples killed leukaemia cells and then went on to isolate white crystals of the molecule responsible. They called it taxol, after the Pacific yew (*Taxus brevifolia*) and determined its structure in 1971.

It was some years after 1971 before research into taxol really took off. There were various reasons for this; the optimum yield of taxol from the yew bark is 0.014% and, moreover, the drug had not yet been tested against solid tumours. It is also possible that some recalled that much of the (European) yew tree was toxic; in 53 BC a British chieftain named Catuvolcus (Cativolcus) committed suicide by drinking a yew extract. However, in 1978 Susan B. Horwitz, of New York's Albert Einstein College of Medicine, found that taxol stabilised microtubules, the subunits of protein that are intimately involved in cell division. This meant that mitosis (cell division forming two daughter cells) was inhibited, so the cells were unable to multiply and tumours were thus unable to grow. Vincristine and vinblastine (pp. 173–174) act as anticancer drugs by preventing microtubule assembly, in contrast to taxol, which promotes it. This was a completely new mechanism, giving hope that taxol would work on cancers where other drugs were unsuccessful. That same year, results of tests of taxol against human tumour cell lines (breast, colon, lung) transplanted to mice were reported; in particular, they showed considerable regression in the mammary xenograft.

Taxol's structure, which contained fused four-, six-, and eight-membered rings, proved a challenge to the synthetic chemist. Two parallel total syntheses of taxol were reported almost simultaneously in early 1994 by the Holton group at Florida State University and the Nicolaou group at Scripps Research Institute and the University of California, San Diego. Holton's paper was submitted to the *Journal of the American Chemical Society* on 21 December 1993 and appeared on 23 February 1994; Nicolaou's paper was submitted to *Nature* on 24 January 1994 and was published on 17 February 1994.

Several problems had to be faced before taxol could be tested on humans. For one thing, it is insoluble in many solvents, but eventually a mixture of ethanol and a surfactant (Cremophor EL) was found to be suitable. Getting enough taxol was a real problem because a course of treatment required around 3 g. At that time, taxol was obtained from the bark of the slow-growing Pacific yew; the removal of the bark kills the tree. After about 200 years' growth, the bark of a 40-foot tree will yield about 0.5 g of taxol, so treating just one patient would sacrifice several trees. Furthermore, the Pacific yew tree is also the home of the endangered Northern spotted owl and, like Harry Potter, everyone loves owls. Fortunately, a solution appeared. Pierre Potier, of the French National Centre of Scientific Research (CNRS) at Gif-sur-Yvette, was celebrated for looking for natural medicines (*le magasin du bon dieu*, as he put it). He noted in 1981 that the taxane core could be found in a compound called 10-deacetylbaccatin III (10-DAB; 12.24) in the needles of the European yew (*T. baccata*), which grew freely in the CNRS park.

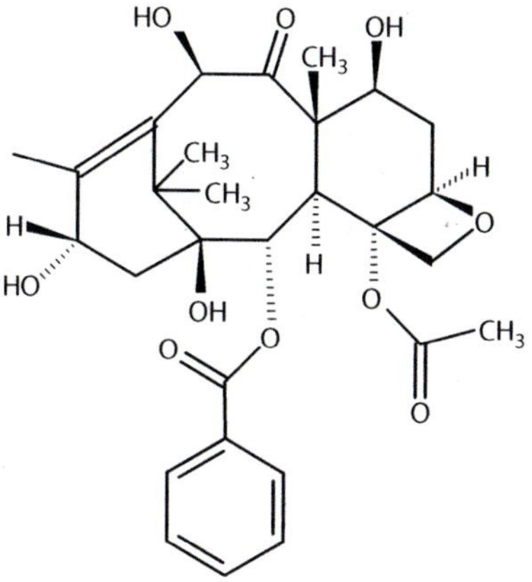

(12.24)

Unlike the bark of the Pacific yew, this was a fast-growing renewable resource and also available on a much larger scale. It was then a matter of semisynthesis to obtain taxol, using the existing 10-deacetylbaccatin and attaching the correct side chain to it. This was achieved by the Potier group, in 1988, and later by other workers, particularly Robert Holton's group at Florida State University (Tallahassee). The Holton route formed the basis of a commercial process adopted by Bristol-Myers Squibb, which brought taxol to market in 1993. (Bristol-Myers Squibb also managed to patent the name Taxol, with paclitaxel as the generic name for the molecule.)

Phase I trials of taxol began in 1984 and phase II in 1987. There were good results against breast cancer and small-cell lung cancer, and outstanding results against ovarian cancer, with several long-term remissions. The FDA approved taxol for treatment of drug-resistant ovarian cancer in 1992, for treatment of breast cancer in 1994, and, subsequently, for non-small-cell lung cancer and Kaposi's sarcoma. In 1986 Potier and his team also produced by semisynthesis a structurally similar but more soluble molecule, docetaxel (12.25), commercialized as Taxotere® and marketed across the world as an anticancer drug—most often for treatment of breast, prostate, and small-cell lung cancers.

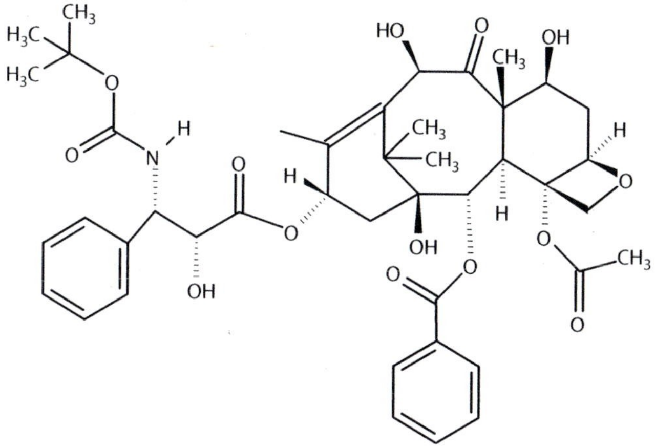

(12.25)

Taxol has a complicated structure: The "taxane" skeleton has three fused rings with four, six, and eight carbon atoms and thus synthetic routes to taxol are correspondingly complicated, with between 35 and 50 steps. Commercially, taxol is now produced from cultured plant cells but scientists are continually on the outlook for more efficient methods that may also present the possibility of making variant molecules that could be better drugs.

Possibly, the key step to taxol biosynthesis in the Pacific yew is the cyclisation process that converts geranylgeranyldiphosphate into taxa-4,11-diene, which subsequently undergoes processes involving oxidation, acylation, and addition of a side chain to form taxol. This step is carried out by the enzyme taxadiene synthase, and its structure was reported in January 2011, giving rise to the possibility of engineering more efficient syntheses. In 2010, scientists at MIT and Tufts University reported that they had engineered *Escherichia coli* bacteria to produce taxadiene, as well as the next molecule in the biosynthetic pathway, taxadien-5α-ol. An obvious next step would be to find ways of making baccatin III, the useful starting material for taxol synthesis.

TETRACYCLINE

The tetracyclines are a family of antibiotics, some of which are produced naturally by *Streptomyces* bacteria, such as chlorotetracycline (12.26), which was the first one to be discovered; it was isolated from *Streptomyces aurofaciens* obtained from samples of Missouri soil in 1945 and marketed as a medicine in 1948. Chlorotetracycline had advantages over the other antibiotics available at the time (including penicillin) because it could be taken orally and was also effective against both Gram-positive and Gram-negative bacteria. Within a few years, tetracycline (12.27) and oxytetracycline were also on the market; the latter was isolated by Pfizer from *Streptomyces rimosus* in the soil near its Indiana plant. Many members of the tetracycline family have been made by semisynthesis, using a naturally produced molecule to provide the core and then modifying its structure.

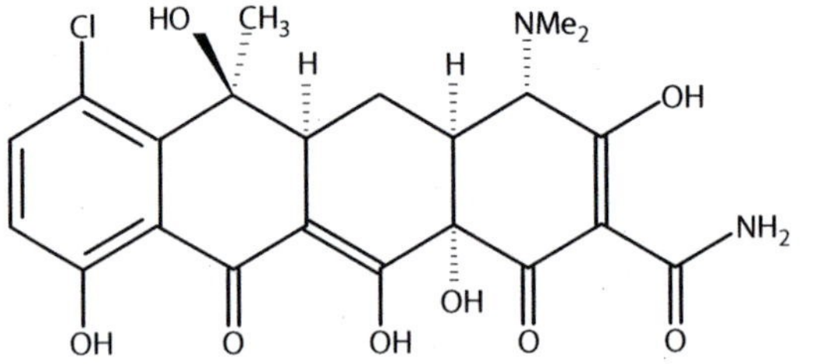

(12.26)

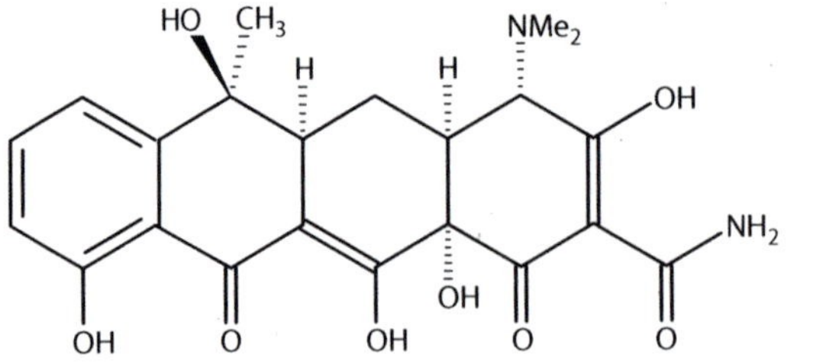

(12.27)

Tetracyclines inhibit bacterial protein synthesis, stopping the transfer RNA-amino acid from attaching to the ribosome and interfering with the messenger RNA codon reading the t-RNA anticodon. This is a key step of protein synthesis, which explains why tetracyclines are effective against a wide range of both Gram-positive and Gram-negative bacteria. But the tetracyclines are not used as widely as they once were because of the emergence of resistant strains of bacteria. In part, this is due to the use of small amounts of antibiotics in animal feeds of animals such as pigs, chicken, and cattle in order to control disease and promote growth.

However, some tetracyclines still find wide application, notably **doxycycline** (vibramycin; 12.28), whose synthesis was reported in 1962; it received FDA approval 5 years later. As a much cheaper and equally effective alternative to Cipro, it was recommended for anyone who might have come in contact with a spore during the anthrax alerts in the autumn of 2001, which followed the 11 September terrorist attacks. Doxycycline is effective against various diseases, including *Yersinia pestis* (bubonic plague), Lyme disease, and Rocky Mountain spotted fever, as well as anthrax.

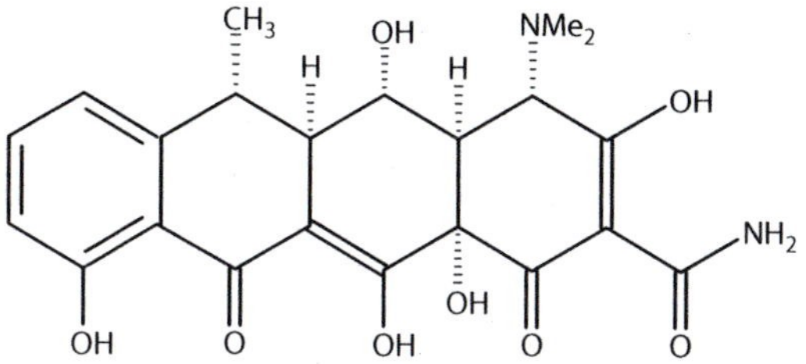

(12.28)

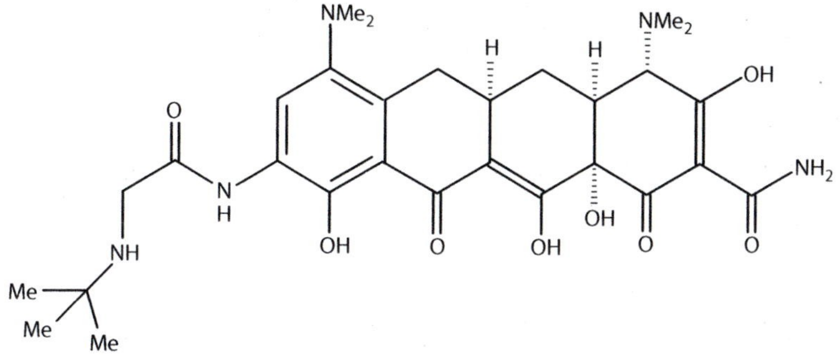

(12.29)

Continuing research means that other tetracyclines have come into use, such as tigecycline (Tygacil; 12.29) in 2005; this is active against resistant bacteria such as *Staphylococcus aureus*. In a surprising twist, it appears that tetracycline itself was in use over 1,500 years ago. Thirty years ago, George Armelagos of Emory University in the United States found tetracycline in human bones from Nubia (in the present-day Sudan) that was dateable to the period between AD 350 and 550. Subsequently, he found that the tetracycline came from the Nubian beer (a syrupy gruel very unlike present-day beers); the grain used to make the beer contained *Streptomyces* bacteria. Further research indicates there are high levels of tetracycline in the bones of even small ancient Nubian children. The ancient Nubians recognised the health effects of the beer and deliberately produced it, even if they did not know about tetracycline.

VANCOMYCIN

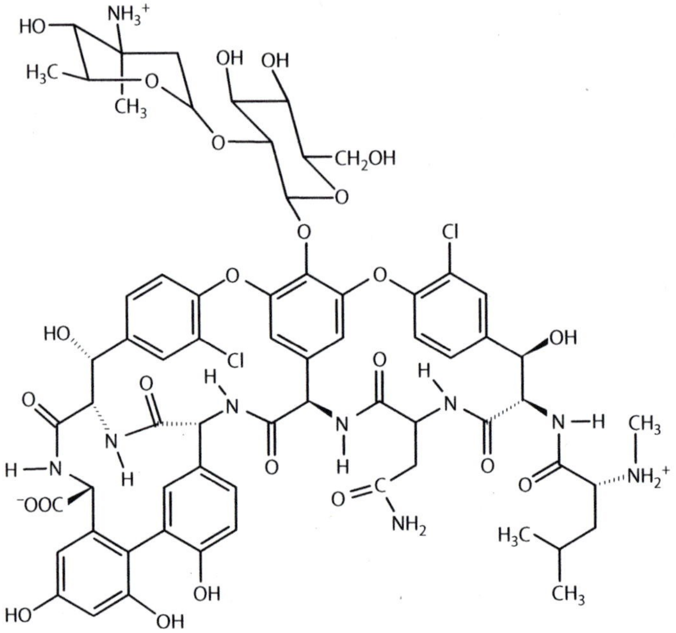

(12.30)

The rise in the number of penicillin-resistant bacteria has presented a serious challenge to pharmaceutical chemists. Vancomycin (Vancocin®; 12.30) was first isolated in 1956 from the microbe *Amycolatopsis orientalis* found in soil from the jungles of Borneo. As a glycopeptide, it represented the first of a new type antibiotic, though it is only effective against Gram-positive bacteria. It is reserved as a "last resort" antibiotic for cases where the bacteria are resistant to all other antibiotics, though there are now a few indications of resistance to vancomycin. It is administered by injection.

Teicoplanin (Targocid®) is a similar glycopeptide that also finds clinical use. Like penicillin, vancomycin works by inhibiting bacterial cell wall synthesis. However, whilst penicillin inhibits the formation of peptidoglycan cross-links by binding to the enzyme controlling this process in the bacterium, vancomycin binds to terminal D-alanyl-D-alanine groups and interferes with peptidoglycan biosynthesis—notably the formation of glycosidic bonds between the sugars and peptide cross-links, leading to a weakened cell wall. Vancomycin forms five strong hydrogen bonds with the terminal D-alanyl-D-alanine units ("five-point interaction"); vancomycin-resistant bacteria have different terminal amino acid terminal units. The presence of D-alanyl-D-lactate and D-alanyl-D-serine units means that only four hydrogen bonds can form, making the interaction only 1/1000 the strength of the "five-point interaction", so the vancomycin has correspondingly less effect on the bacteria.

VINCA ALKALOIDS

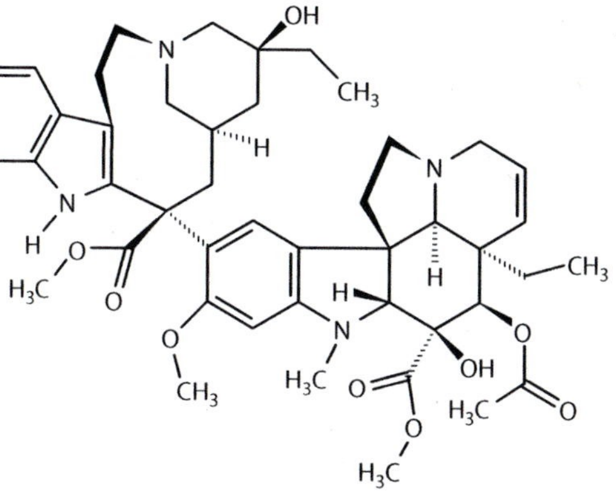

(12.31)

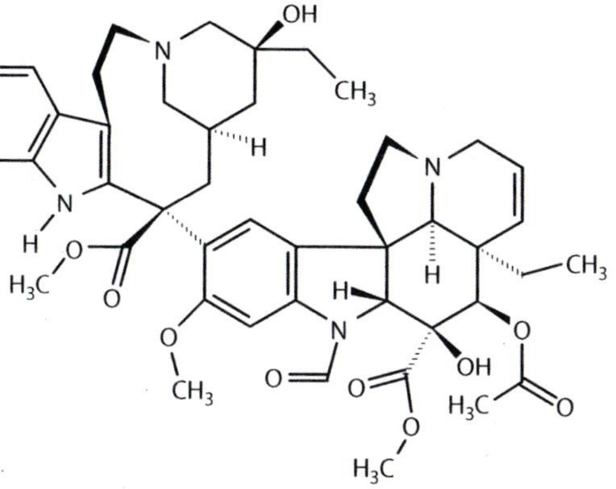

(12.32)

Vinblastine (VLB; Velban®; 12.31) and vincristine (VCR; Oncovin®; 12.32) were the first two plant-derived anticancer drugs to come into use. For hundreds of years, people in different parts of the world (e.g., Madagascar and Jamaica) have used the Madagascar or rose periwinkle, *Catharanthus roseus,* to treat diseases such as diabetes. In 1952, a general practitioner in Toronto named Edward Clark Noble was given some periwinkle leaves from Jamaica and told that Jamaicans used them to make a tea for treating diabetics. He gave them to his brother, Robert Laing Noble, the associate director of the Collip Medical Research Laboratory at the University of Western Ontario, who found that they had a sharp effect on bone marrow and white blood cell counts.

Together with an organic chemist named Charles T. Beer, Noble isolated vinblastine. The first clinical trials were undertaken in 1959 and it was marketed in 1961 by Eli Lilly; research in its laboratories led by James G. Armstrong resulted in the isolation of more alkaloids in 1963, including Vincristine, which Eli Lilly also marketed. These molecules have very similar structures but have different applications as drugs. Vincristine is applied to leukaemias, lymphomas, and sarcomas, with a notable success in improving survival rates in children suffering from acute leukaemia from 10% to 90%.

Vinblastine, derived from vincristine formally by oxidation of a methyl to a formyl group, finds a wide range of applications and is often used in conjunction with other drugs, such as cisplatin and BMN (bleomycin)—notably for advanced testicular cancer and breast cancer, Kaposi's syndrome, Hodgkin's disease, and lymphomas. Subsequently, semisynthetic analogues of these agents came into use, such as vindesine (VDS) and vinorelbine (VRLB; 12.33). The latter is a product of Pierre Potier and his team at CNRS and is presently used to treat non-small-cell lung cancer, sometimes in combination with cisplatin. Recently, a fluorinated compound called vinflunine has come on the scene.

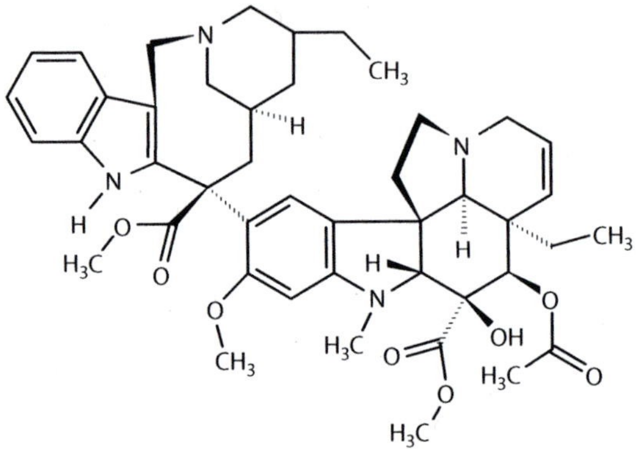

(12.33)

Vinca alkaloids work by interfering with fast-growing cancer cells, which are more vulnerable than healthy cells. This involves binding to the tubulin monomers and stopping them from forming microtubules, thus preventing alignment and separation of chromosomes so that cell division is unable to take place. (In contrast, taxol prevents the microtubules from being broken down.) Vinca alkaloids also affect healthy cells, which is why these drugs have side effects.

These molecules are very active against herbivores and bacteria, so the periwinkle only needs to make very small amounts of them; vinblastine makes up 0.002% of the plant's dry weight. The periwinkle synthesises these molecules from tryptamine and a C10 terpenoid, secologanin; a key intermediate in the synthesis is strictosidine, which is also an intermediate in the biosynthesis of over a hundred other alkaloids, including quinine and strychnine. Sarah O'Connor's research group at MIT has engineered the enzyme strictosidine synthase and expressed it in *Catharanthus roseus;* they have found that, when it was fed substituted tryptamines, unnatural alkaloids resulted. Laboratory synthesis of vinblastine by direct coupling of catharanthine and vindoline has been achieved, notably by the Boger group at the Scripps Research Institute in California, but this is substantially more expensive as a way of making these alkaloids. It has been found that although the plant produces vindoline and catharanthine in leaves, it keeps them apart—the former in plant cells and the latter in the waxy leaf exudates, where its toxicity to insect pests and to fungi can be brought into play.

13 Man-Made Healers

ASPIRIN

Aspirin is the most widely used medication in the world and a perfect example of a medicine that bridges the divide between natural and man-made medicines. Over 2,000 years ago, Hippocrates (440–377 BC) said that the ancient Greeks used powdered willow bark to reduce fever. Subsequently, this knowledge slipped into disuse, until it was relaunched by the Rev Edward Stone (1702–1768), a country clergyman of Chipping Norton in Oxfordshire. He experimented with willow bark in relieving ague fever from malaria and distempers of some 50 people; in 1763, he communicated his findings to the Royal Society. He was influenced by an application of the so-called "doctrine of signatories": Willow grew in damp places and, because agues and fevers were associated with damp places, too, willow might produce a medicine to treat them. In the early nineteenth century, it was found that the glycoside salicin (13.1), extracted from willow bark, relieved pain. Salicin is hydrolysed upon ingestion into glucose and salicyl alcohol, metabolised to salicylic acid. By the 1870s, salicylic acid (13.2) was being used successfully to treat fevers and pain, but gastrointestinal problems were a side effect.

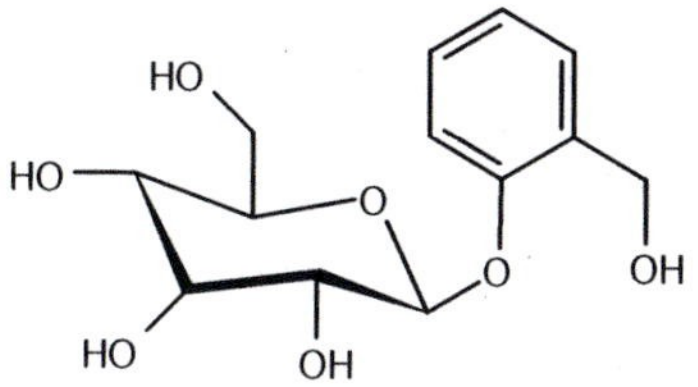

(13.1)

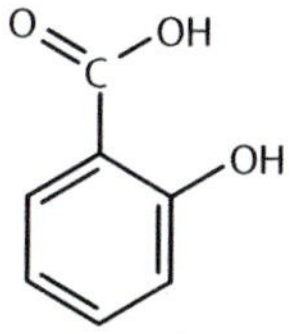

(13.2)

   In 1853, Charles Frederic Gerhardt made an impure sample of acetylsalicylic acid (13.3) from sodium salicylate and acetyl chloride, but did not take this discovery further; in 1897, Felix Hoffmann (1868–1946) made acetylsalicylic acid from the reaction between acetic anhydride and salicylic acid. It is said that Hoffmann's father was rheumatic; because sodium salicylate upset his stomach, his son had the bright idea of making an acetylated version of salicylic acid to try on his father. Or so the story goes. In 1949, after Hoffmann's death, his former colleague Arthur Eichengrün claimed that the synthesis of acetylsalicylic acid had been carried out under his supervision, and in 2000 an analysis of archival and published material by Walter Sneader supported this claim.

(13.3)

Whatever the true story, there was already an adequate supply of salicylic acid, thanks to Kolbe's synthesis of it from sodium phenoxide and carbon dioxide (1856); aspirin began to be sold in 1899 and rapidly became the drug of choice as an analgesic (painkiller), anti-inflammatory, and antipyretic (reduction of temperature). Cautionary notes have to be sounded: Aspirin taken during viral illness can produce fatalities in children suffering from Reye's syndrome. Aspirin was widely used in treating victims of the 1918–1919 influenza pandemic, and it has been suggested that high doses of aspirin may have contributed to the fatalities.

As time went on, there were suggestions that aspirin was a wonder drug with anticlotting properties that could help prevent heart attacks and could limit brain damage in some strokes. It was also thought to be active against colonic cancer, but no one knew how aspirin worked. In 1971, Sir John Vane and his co-workers showed that aspirin inhibits the production of prostaglandins, a discovery for which Vane shared the Nobel Prize in Physiology or Medicine in 1982. Prostaglandins are hormones produced locally in cell membranes in the body in response to infection or tissue damage. They increase the flow of blood to that part of the body as they import more disease-fighting white blood cells, and this leads to swelling and inflammation. Moreover, prostaglandin E_2 (PGE_2) affects temperature regulation in the hypothalamus, causing the body's temperature to rise and leading to fever.

Aspirin blocks prostaglandin synthesis by being an irreversible inhibitor of the cyclo-oxygenase enzymes. It does this by transferring the acetyl group to a serine residue (number 530) in the enzyme (Figure 13.1). This blocks entry of the arachadonic acid substrate, used to make the prostaglandins, to the active site of the enzyme.

In fact, there are two very similar cyclo-oxygenase enzymes: COX-1 and COX-2. Aspirin inhibits both. COX-1 is found in most mammalian cells; COX-2 is normally undetectable, but is generated at the sites of inflammation and cell damage.

COX-2 is involved in production of the prostaglandin PGH_2, which in turn is converted by other enzymes to several other prostaglandins (which mediate pain, fever, and inflammation). Inhibiting this enzyme does reduce pain, fever, and inflammation.

COX-1 makes prostaglandin PGI_2 (prostacyclin), which protects the stomach lining against irritation as it reduces the amount of gastric acid produced, and is also involved in producing protective mucus. Thus, inhibiting COX-1 causes stomach irritation and there is more chance of acid damaging the stomach lining and producing peptic ulcers; a further contributory factor is the acidic nature of aspirin itself. In addition, COX-1 is involved in stimulating platelet aggregation in the blood, leading to clotting; by inhibiting COX-1 activity in platelets, aspirin reduces the formation of

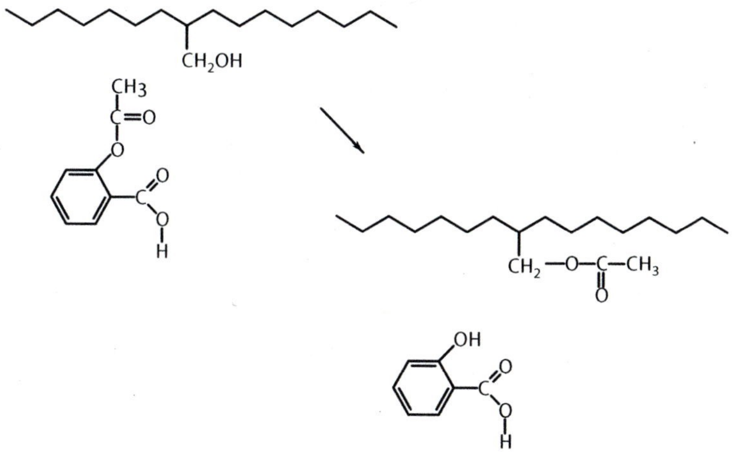

FIGURE 13.1 Aspirin blocks prostaglandin synthesis by transferring the acetyl group to a serine residue (number 530) in the enzyme.

thromboxane A2, which induces platelet aggregation and clotting, leading to thrombosis; therefore, low doses of aspirin are prescribed as cardioprotective.

Thus, the idea was mooted to find a "superaspirin"—a molecule that would inhibit COX-2 and thus stem inflammation and pain, whilst not affecting COX-1 and thus allowing the prostaglandins produced to protect the stomach lining. The best known of these drugs are **celecoxib** (Celebrex; 13.4) and **rofecoxib** (Vioxx; 13.5).

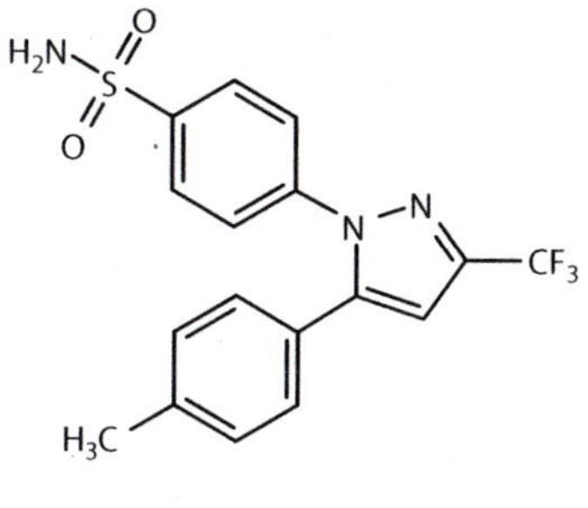

(13.4)

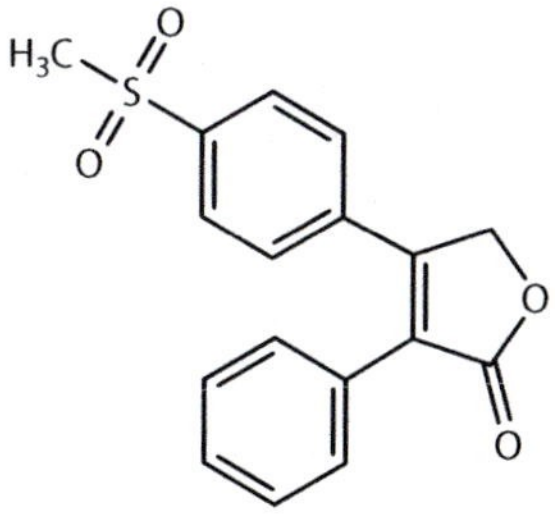

(13.5)

Celecoxib (Celebrex) is currently used to reduce pain and inflammation in patients suffering from osteoarthritis, rheumatoid arthritis, menstruation pains, and acute pain.

Rofecoxib (Vioxx) received FDA approval on 20 May 1999 and rapidly became a medication of choice to treat people with conditions involving chronic or acute pain, such as arthritis. In the year 2003–2004, this drug generated sales of $2.5 billion for the manufacturer, Merck. On 30 September 2004, Merck voluntarily withdrew rofecoxib from the market because it had increased the risk of heart attack for some of its patients—people who had taken it for a long period of time. It was also claimed that Merck had not made public side effects seen in trials. Another COX-2 inhibitor, valdecoxib (Bextra), was also withdrawn from the market, whilst etoricoxib (Arcoxia) is also not marketed in the United States, though it is on sale in over 70 countries worldwide.

Because there are links between cardiovascular risk and exposure to COX-2 inhibitors, it is probably advisable to use the lowest effective dose for the shortest possible period of treatment. COXs act as homodimers made of identical subunits, but just one subunit at a time operates during catalysis, and usually binding of a NSAID to one site inhibits the activity of the dimer. It has recently (2010) been found that celecoxib and some other COX-2 inhibitors do bind tightly to one monomer of COX-1, but without affecting the ability of the other site to convert arachidonic acid into prostaglandins. They do, however, affect the ability of aspirin to inhibit COX-1, and this has health implications. Celecoxib and similar molecules have cardiovascular benefits and thus are often used in conjunction with low doses of aspirin to prevent thrombosis, but taking this mixture may curtail the cardioprotective benefits of aspirin.

IBUPROFEN

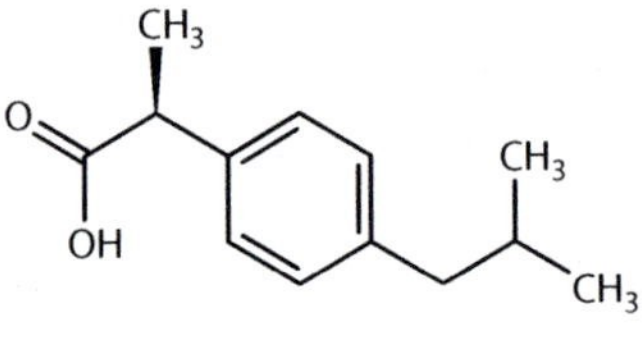

(13.6)

Ibuprofen (13.6) has been marketed as an NSAID by Boots since 1969. It is 20 times more effective than aspirin as a painkiller and can be taken in high doses with no adverse gastric effects. It contains a chiral carbon atom and thus exists as two enantiomers. (S)-(+)-ibuprofen is the isomer active both *in vitro* and *in vivo*, but in solution the inactive (R)-isomer is rapidly converted into the (S)-isomer by an isomerase enzyme; thus, there are no gains in marketing it as a single-isomer drug and it is sold as the racemic mixture. **Naproxen** (13.7) was introduced during the 1970s. An improvement on ibuprofen, it has twice the potency and a longer half-life in the body; a once-daily dose is sufficient.

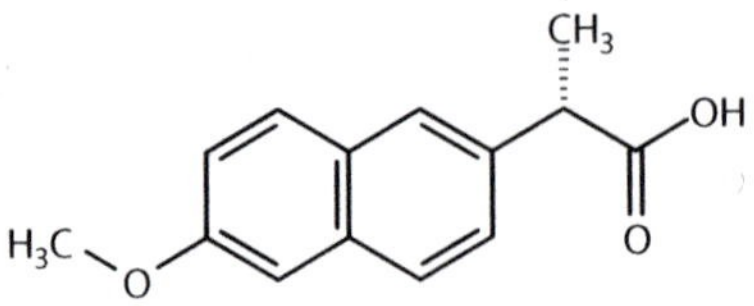

(13.7)

In the late nineteenth century, aspirin was preempted as an analgesic and anti-pyretic drug by some aniline-derived painkillers (of which phenacetin and aceta-nilide were the first); acetaminophen (paracetamol) is the only one of this class in use today.

PHENACETIN

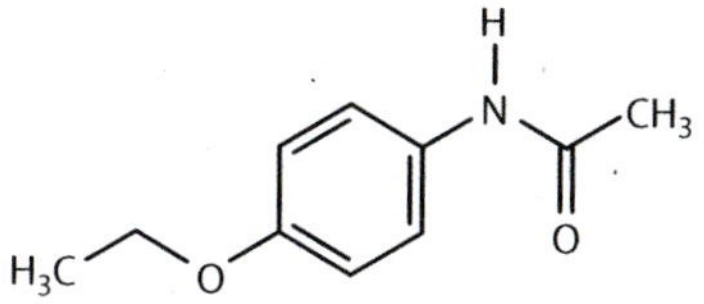

(13.8)

Phenacetin (13.8), introduced in 1887, was a widely used painkiller with antipyretic (but not anti-inflammatory) properties. It was withdrawn because of risks of damage to the kidney, as well as for being a carcinogen.

ACETANILIDE

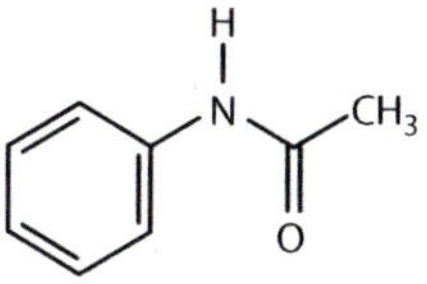

(13.9)

Acetanilide (**Antifebrin;** 13.9) is also an analgesic and antipyretic and was intro-duced into use just before phenacetin. It was withdrawn from use after it was found that its use could lead to high methaemoglobin levels, with resulting reduced abil-ity of the blood to carry oxygen; this is due to aniline formed by hydrolysis of the acetanilide. The analgesic and antipyretic effects were due to another metabolite: paracetamol.

ACETAMINOPHEN

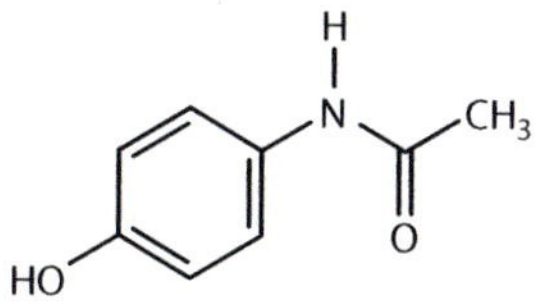

(13.10)

Acetaminophen (**paracetamol** or Tylenol; 13.10) is the one surviving aniline-derived painkiller and febrifuge in use—and very widely used, at that. It supplies these benefits in reduction of pain and fever without any gastric problems, though it does not reduce inflammation or affect clotting. Its mechanism of action is not clear. It is assumed to act upon COX enzymes, but not in the same way as aspirin and other NSAIDs. It is known to inhibit COX-3 (found in the brain), but this may not be related to its action. Paracetamol is toxic; when it is administered at normal therapeutic doses, the body's store of glutathione gets rid of toxic metabolites, but in overdose the body's stock of glutathione can be exhausted, leading to liver damage and possibly liver failure. Its toxicity is greatly increased when it is taken with alcohol. Because paracetamol is used so widely in medications, there is a danger of overdose when paracetamol tablets are combined with other medications such as cough linctus.

ACYCLOVIR

Acyclovir (acycloguanosine, ACV; Zovirax®; 13.11A) is one of several drugs developed by the American biochemist and pharmacologist Gertrude Belle Elion (1918–1999), who won the 1988 Nobel Prize in Physiology or Medicine. It is a modified nucleoside (with a very similar structure to guanosine; 13.11B) and has proved an outstandingly successful antiviral drug for the treatment of infections due to certain specific herpes viruses, such as cold sores, shingles, genital herpes, and chicken pox, because of its activity—particularly against herpes simplex virus types 1 and 2 (HSV-1 and HSV-2).

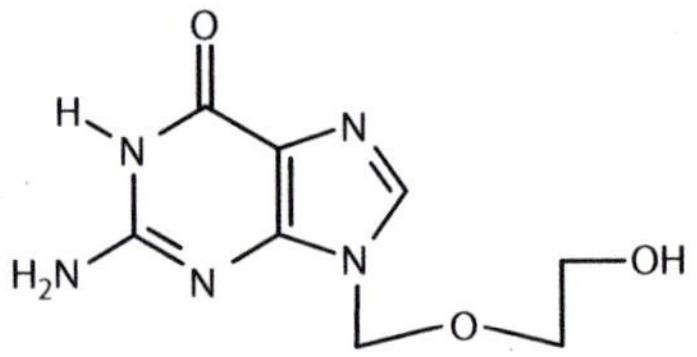

(13.11A)

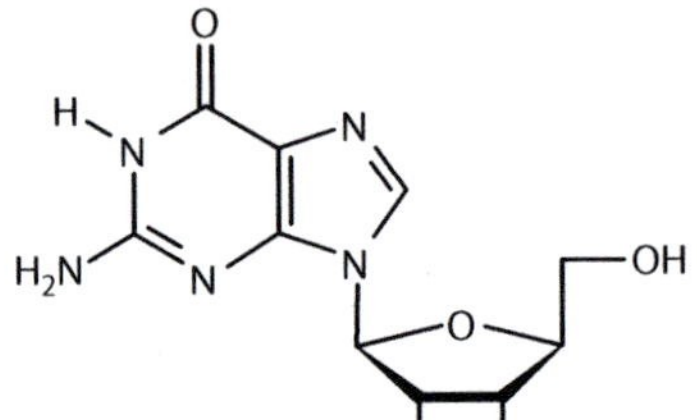

(13.11B)

Acyclovir was approved for use by the FDA in 1982 and came onto the American market that year. Nowadays, the valine ester of acyclovir is used on account of its greater bioavailability, and other antivirals have also come onto the market. Acyclovir acts as a prodrug; it is itself an inactive molecule that, once inside the host, gets phosphorylated to the effective substance acyclovir triphosphate (ACV-TP). This acts as a substrate for a herpes-specific thymidine kinase enzyme, which is present only in infected cells. This enzyme adds on the first phosphate group; other cellular kinases add the other two, forming acyclovir triphosphate. ACV-TP competitively inhibits viral DNA polymerase, for which it has a much greater affinity than for the normal cellular DNA polymerase. It prevents incorporation of deoxyguanosine triphosphate (dGTP) into a growing DNA chain, and when the ACV-TP is inserted into the chain, it causes DNA chain termination and terminates viral replication.

BETA-BLOCKERS AND ATENOLOL

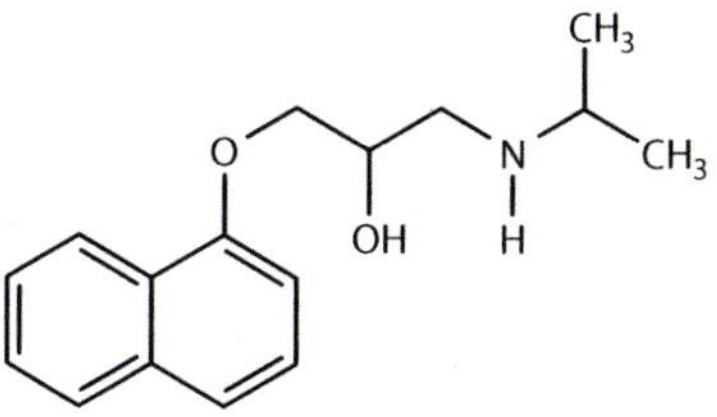

(13.12)

Propranolol (Inderal; 13.12) was the first really successful β-blocker, invented by Sir James Black (who also invented cimetidine; see p. 190) in a project that started in 1958. When the nervous system is stimulated—for example, by shock or fear— adrenaline (epinephrine) is released; like noradrenaline (norepinephrine), it has the effect of stimulating the heart and increasing blood pressure by binding to a β-adrenoreceptor (these are seven-transmembrane proteins). Black had the idea of making molecules that were antagonists for the β-andrenergic receptor (of which the $β_1$ type is found particularly in the heart and arteries), which thus would slow down the heart and reduce blood pressure as well as the demand of the heart muscle for oxygen.

When Black developed propranolol, increased incidence of angina and heart disease was becoming an increasingly serious problem. Angina is caused by the heart receiving too little oxygen; one approach to solving the problem would be to find ways of increasing the oxygen supply to the heart; Black's team followed the reverse strategy of finding ways of reducing the heart's need for oxygen. The team started from the existing β-blocker molecule pronethalol (13.13), which never saw clinical use because it was found to cause cancer in mice.

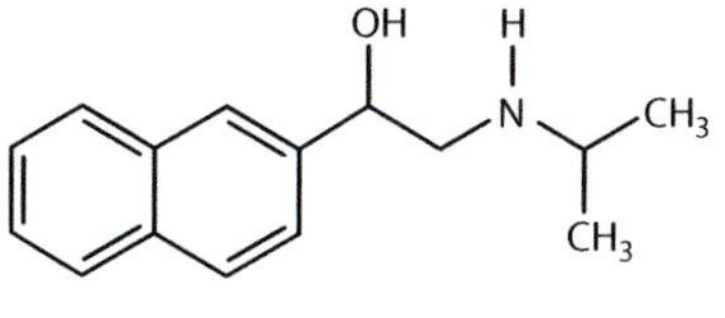

(13.13)

This molecule had established that the presence of the naphthalene rings and the amino ethanol groups was important to activity, so the team decided to study the effect upon drug activity of increasing the distance between these two groupings. As it happened, there was no β-naphthol available, so the synthesis went ahead using α-naphthol instead. The resulting molecule was propranolol, which as a drug delivered all that was desired; when the original target molecule was eventually synthesised, it showed little improvement over pronethalol. Serendipity in drug discovery is a wonderful thing. Propranolol (Inderal) was launched in 1964. It was found to lower blood pressure, reduce the occurrence of heart attacks, and also provide a treatment for angina. It is a chiral molecule; L-propranolol is an adrenoceptor antagonist, but D-propranolol is not.

Atenolol (Tenormin; 13.14) is a selective β_1 receptor agonist with some advantages over earlier β-blockers. It succeeded propranolol as a drug of choice, coming into use in 1976.

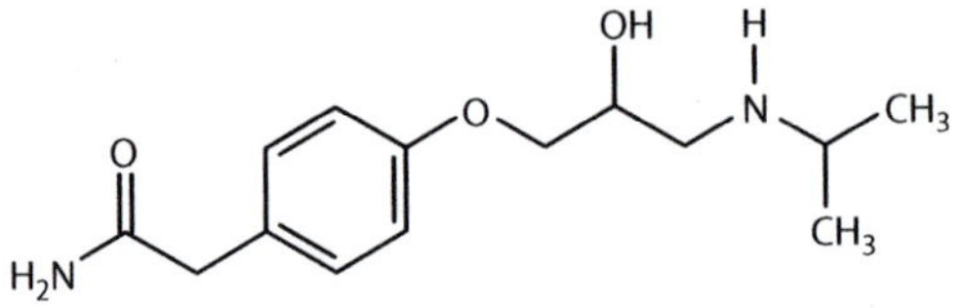

$$(13.14)$$

Because it reduces manifestations of anxiety, like hand tremors, and gives a competitive advantage, β-blocker abuse has occurred at the hands of sportsmen. Beta-blockers like propranolol and atenolol are on the 2008 prohibited list from the World Anti-Doping Agency, which covers many sports. North Korean shooter Jong Su Kim lost silver and bronze medals at the 2008 Beijing Olympics when he tested positive for propranolol. The Canadian snooker star Bill Werbeniuk for many years drank large amounts of lager when competing, in order to control a trembling cue arm caused by a hereditary nervous disorder. Under medical advice to cut out alcohol, he changed to propranolol, only to run into problems when that was placed on the banned list. It has been alleged that a significant number of golfers and archers use β-blockers.

Surgeons have used β-blockers to reduce hand tremors, whilst stage performers like magicians use it to control stage fright; controversy continues over musicians who take it for that reason. The rôle of β-blockers was downgraded in 2006 when it was found that other drugs worked better, especially in the elderly, and that beta blockers, particularly atenolol, carried a risk of provoking type 2 diabetes. British clinical guidelines now advise against their use in treating hypertension.

AZIDOTHYMIDINE

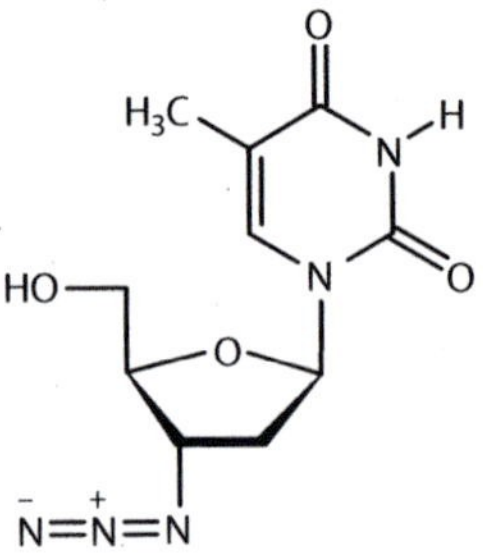

$$(13.15A)$$

$$(13.15B)$$

Azidothymidine (AZT; 13.15A), also known as zidovudine, is historically important because it was the first molecule used for treatment of AIDS (acquired immunodeficiency syndrome), under the names Retrovir and Retrovis. It was first made in 1964 for possible use as a retroviral drug against cancers, but is now known as a reverse transcriptase inhibitor for the HIV virus.

AIDS first came to the attention of the Western world in 1981 and was recognised in 1983 as an immune system disease due to the human immunodeficiency virus (HIV). HIV is itself unable to replicate, but rather turns its RNA genome into DNA and inserts that into the genome of the cell, so the cell replicates the virus. Researchers examined molecules that might be active against HIV; in 1984–1985, researchers showed that AZT was active against the human HIV virus *in vitro*. Clinical trials followed, and the Food and Drug Administration approved AZT for use against AIDS on 20 March 1987.

AZT owes its action to its structural similarity to thymidine (13.15B) and is actually a prodrug. When administered, it is phosphorylated *in vivo* into the triphosphate, and this phosphorylated AZT fools the HIV reverse transcriptase enzyme into inserting it into the DNA it is building, instead of thymine. AZT has an azide group where thymidine has an OH group, so it cannot make a phosphate bridge to the next nucleotide; DNA synthesis is therefore terminated incomplete. AZT is thus an inhibitor of the HIV reverse transcriptase enzyme. AZT does not totally stop the replication of the virus or destroy an HIV infection, but does slow down the development of the disease. It has therefore been employed as part of a cocktail with other reverse transcriptase inhibitors and antiretroviral drugs, as well as HIV protease inhibitors. Better reverse transcriptase inhibitors have followed it.

CISPLATIN AND OTHER PLATINUM ANTICANCER DRUGS

When he felt unwell in early 1996, Lance Armstrong did the male thing. Instead of going to see the doctor, he ignored it. One evening late in September that year he had a headache that did not respond to painkillers; it was gone by the next morning, but then he noticed his vision going blurry. A couple of days later, he coughed up streams of blood. A few days later, on 2 October, when his right testicle was swollen to three times its normal size, he spoke to a friend who was a doctor, and they immediately went to see a urologist. X-rays revealed testicular cancer with large metastasis to the lungs. He had surgery to remove the testicle at 7 a.m. the next day, but within a week a brain MRI scan revealed two spots: The cancer had spread beyond the lungs. Armstrong had stage 3 testicular cancer and faced treatment with a cocktail of chemicals, including cisplatin (13.16). As he said in *It's Not about the Bike,* "I thought, 'Bring it on, give me the platinum'".

$$H_3N \diagdown \diagup Cl$$
$$Pt$$
$$H_3N \diagup \diagdown Cl$$

(13.16)

Armstrong's treatment at the Indiana University medical centre in Indianapolis started with brain surgery to remove the two lesions on the brain; then he began the chemotherapy. This meant a cocktail of vinblastine (pp. 173–174), etoposide, ifosfamide, and cisplatin. He finished chemotherapy on 13 December 1996. He was cured. He entered and won the Tour De France, the world's premier cycle race in 1999. And 2000. And 2001. And 2002. And 2003. And 2004. And 2005. No one had won the race seven times, let alone in consecutive years. If Armstrong had developed his cancer in 1976, rather than 1996, there would have been no hope for him. With a remission rate for testicular cancer of over 90%, why does cisplatin make the difference?

Cisplatin, *cis*-$[Pt(NH_3)_2Cl_2]$, had been discovered by Michel Peyrone in 1845, but it took a chance discovery in 1964 to turn it into a multimillion dollar lifesaver. Barnett Rosenberg of Michigan State University was not trying to make a new drug. Rather, he and his research technician, Loretta Van Camp, were studying the effect of an electric field on cell division. Platinum electrodes were placed in an ammonium chloride buffer containing *Escherichia coli* cells, and an electric current was applied. They found that cells had not divided but instead had grown to 300 times their normal length. It took time to establish that the cause of this was small amounts of a platinum compound formed by reaction between the ammonium chloride buffer and the platinum electrodes. By 1968, they knew that the active compound was *cis*-dichlorodiammineplatinum, which became known as cisplatin (13.16), and that the *trans*-isomer (13.17) was not active.

$$H_3N \diagdown \diagup Cl$$
$$Pt$$
$$Cl \diagup \diagdown NH_3$$

(13.17)

They next tested cisplatin on mice into which a sarcoma-180 tumour had been implanted, finding considerable shrinkage of the tumour; after that, they changed the testing protocol by administering the cisplatin a week after the cancer was implanted. The success rate was nearly 100%, with the mice living to the expected age and dying of natural, age-related causes. Tests continued on dogs and monkeys; the first report on the anticancer activity of cisplatin was published in 1969. In 1972, clinical trials began on cancer patients who had exhausted all alternative treatments, and around 20% of the patients showed whole or partial remission. Further trials revealed high success rates in patients with testicular cancer, and cisplatin received FDA approval in 1978.

Its success was not easily gained because the treatment caused severe nausea and vomiting, and cisplatin also caused kidney toxicity. Drugs have been found to control the nausea and vomiting, whilst hydration therapy counteracts the kidney problem. Years before Armstrong, British jockey Bob Champion was successfully treated with cisplatin, vinblastine, and bleomycin; he was diagnosed with stage 3 testicular cancer in July 1979, and underwent a course of treatment in late 1979 and early 1980. Like Armstrong, he made a successful return to his sport, famously winning the Aintree Grand National on the horse Aldaniti in 1981.

The way in which cisplatin kills cancerous cells is not wholly understood. Cisplatin is normally administered intravenously. The neutral cis-[Pt(NH$_3$)$_2$Cl$_2$] molecules are kept in that form by the high chloride ion concentration in the blood, but when they enter a cell, the much reduced chloride concentration allows hydrolysis, with substitution of chlorides:

$$[Pt(NH_3)_2Cl_2] + H_2O \rightleftharpoons [Pt(NH_3)_2(H_2O)Cl]^+ + Cl^-$$

$$[Pt(NH_3)_2(H_2O)Cl]^+ + H_2O \rightleftharpoons Pt(NH_3)_2(H_2O)_2]^{2+} + Cl^-$$

The ion [Pt(NH$_3$)$_2$H$_2$O(Cl)]$^+$ contains a water molecule that is easily displaced by a DNA base (generally a purine), and hydrolysis of the remaining chloride enables binding to a second base. This usually occurs with the formation of 1,2-intrastrand d(GpG) adducts. It is believed that the formation of intrastrand cross-links cause kinks in the DNA helix and means that the excision repair system cannot function, leading to cell death. Only a small part of the cisplatin actually reacts with the DNA; some is deactivated by binding to protein in the blood, whilst some of that entering the cell reacts with other nucleophiles present.

Traditionally, cisplatin is prepared by a method that takes advantage of the $trans$ effect (Figure 13.2). In this context, this means that in a square planar complex, substitution occurs $trans$ to chloride in preference to substitution $trans$ to ammonia. If ammonia is added to [PtCl$_4$]$^{2-}$, when it comes to substituting [Pt(NH$_3$)Cl$_3$]$^-$, the second ammonia is inserted cis to the first one because chloride has a greater $trans$ effect than ammonia.

Conversely, when chloride is added to [Pt(NH$_3$)$_4$]$^{2+}$, $trans$-[Pt(NH$_3$)$_2$Cl$_2$] is the product.

In practice, if ammonia is added to solutions of [PtCl$_4$]$^{2-}$, a mixture of yellow cis-[Pt(NH$_3$)$_2$Cl$_2$] and green Magnus' salt [Pt(NH$_3$)$_4$]$^{2+}$ [PtCl$_4$]$^{2-}$ is obtained. A superior route involves four steps:

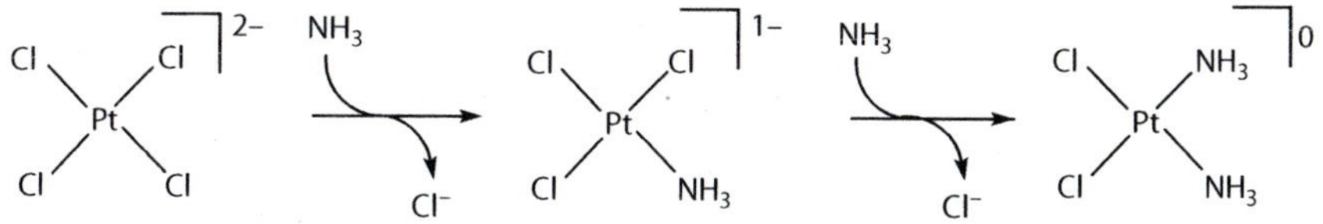

FIGURE 13.2 Cisplatin is prepared by a method that takes advantage of the $trans$ effect.

1. Potassium iodide is added to $K_2[PtCl_4]$, forming $K_2[PtI_4]$.
2. $K_2[PtI_4]$ is reacted with $NH_3(aq)$, which affords a precipitate of slightly soluble cis-$[Pt(NH_3)_2I_2]$.
3. Upon addition of $AgNO_3(aq)$ to a solution of cis-$[Pt(NH_3)_2I_2]$, AgI precipitates; this is filtered off, leaving a solution of cis-$[Pt(NH_3)_2(H_2O)_2]^{2+}$.
4. When KCl(aq) is added to the solution of cis-$[Pt(NH_3)_2(H_2O)_2]^{2+}$, yellow cis-$[Pt(NH_3)_2Cl_2]$ precipitates.

Cisplatin has been a very successful drug, but resistance and side toxicity can present problems, so many other substances have been synthesised as potential anticancer compounds. Only a few of these have seen clinical trials, and just two are in wide use. They both involve bidentate leaving groups and are kinetically more inert.

Carboplatin (13.18) received FDA approval in 1989 for treatment of lung, head, and neck cancers, as well as ovarian carcinoma. It has significantly fewer side effects than cisplatin because it is more stable in the blood and less toxic to the kidneys. **Oxaliplatin** (13.19) has a different activity spectrum to cisplatin and thus circumvents resistance. It has been approved in Europe (1999) and the United States (2002) for secondary treatment of metastatic colorectal cancers.

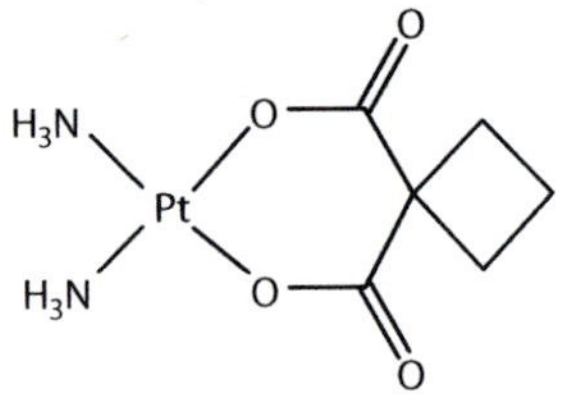

(13.18)

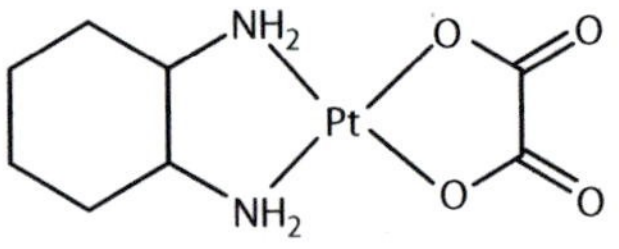

(13.19)

In addition to these, **nedaplatin** (Aqupla®; 13.20) has been used in some parts of the world for cancer chemotherapy, reportedly with fewer side effects.

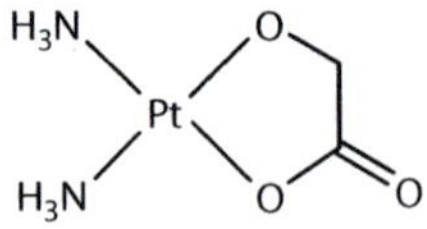

(13.20)

Satraplatin (JM216; 13.21) has been under investigation for some years and underwent phase III trials as a treatment for advanced prostate cancer. It is the first platinum drug to be taken orally; all the others require intravenous administration.

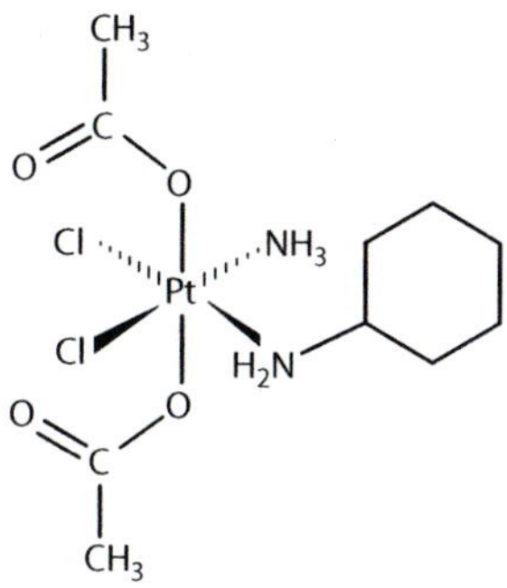

(13.21)

The rather similar **mitaplatin,** c,t,c-[Pt(NH$_3$)$_2$(O$_2$C–CHCl$_2$)$_2$Cl$_2$], has recently been reported but has not yet been in any clinical tests.

Imatinib (Gleevec, Glivec, STI 571; 13.22) is a relatively new drug for the treatment of certain cancers; it belongs to a type known as signal transduction inhibitors (STIs). Unlike many conventional cancer therapies, its mode of action means that it is targeted on cancerous cells by inhibiting a particular enzyme. It specifically deactivates uncontrolled growth of white blood cells by closing down the enzyme responsible for this. In 1960, scientists in Philadelphia discovered that chronic myeloid leukaemia (CML) patients had an unusual chromosome. A piece of chromosome 22 (in the middle of the BCR gene) is transferred to chromosome 9, resulting in a chromosome 9 longer than usual and a chromosome 22 shorter than usual (now called the "Philadelphia chromosome"). The Philadelphia chromosome has an unusual gene resulting from fusion of parts of BCR and ABL genes, which undergoes transcription and translation, forming an unusual protein.

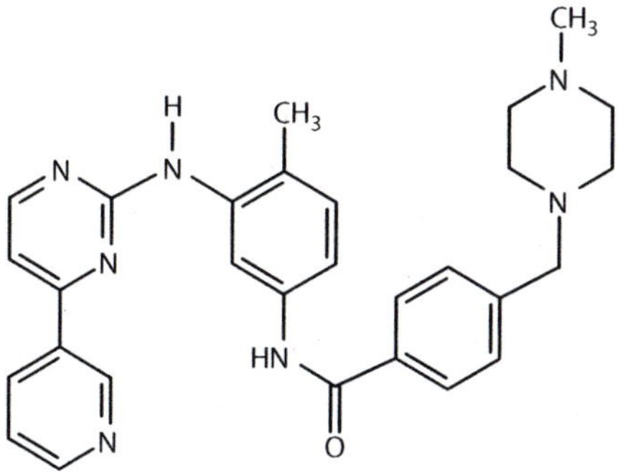

(13.22)

Both the original ABL protein and the "fusion" BCR-ABL protein are tyrosine kinases; they link phosphate groups to other proteins. However, the defective protein lacks the section that tells it when to "switch off". The protein obtains phosphate groups by trapping ATP molecules in a cleft; scientists found that 2-phenylaminopyrimidine was the molecule that inhibited this and then modified the molecule by adding other groups that made Gleevec bind more strongly. Thus, Gleevec fits into this cleft and blocks its ATP binding, stopping the protein from transferring phosphates.

Gleevec was first approved by the US Food and Drug Administration (FDA) on 10 May 2001 for the treatment of CML; in January 2002 it received FDA approval for the treatment of gastrointestinal stromo tumours (GIST), a rare abdominal cancer that can be hard to detect and diagnose in its early stages. Some patients are resistant to Gleevec and are treated with other tyrosine-kinase inhibitors, such as nilotinib (Tasigna) and dasatinib (Sprycel), which may succeed it as frontline treatments.

LINEZOLID

$$(13.23)$$

There was a long period from the 1950s onward when no new type of antibiotic was discovered. New drugs were simply developments of existing compounds; people were wondering if there could be a molecule that could be a backup for vancomycin as a "last resort" drug for cases like MRSA. In the early 1980s, compounds containing the five-membered oxazolidinone ring were found to have antibiotic properties, but they were frequently too toxic for use.

In 1993, Steven Brickner and colleagues at Upjohn Pharmaceuticals found that **linezolid** (13.23) was safe and also very active against Gram-positive bacteria, and it was approved for use by the FDA as Zyvox® in 2000. The antibacterial action of linezolid is quite different from that of other drugs because it blocks protein synthesis at a very early stage. Translation begins with a 50S ribosomal subunit of prokaryotic ribosomes linking up with *N*-formylmethionine-tRNA (fMet-tRNA) and the 30S ribosomal subunit–mRNA complex to generate the 70S assembly, which catalyses protein synthesis. Linezolid binds to the catalytic site of the 50S ribosomal subunit, stopping the subunit joining to the 30S ribosomal subunit and preventing the 70S assembly from forming; thus, fMet-tRNA cannot bind. Because this would be the first amino acid of the peptide chain, protein synthesis in the bacterium is blocked, so bacterial growth and reproduction cannot occur. Bacterial resistance to linezolid is low; however, it has been encountered in some Gram-positive staphylococci, so the endeavour to discover new antibiotics goes on.

ANTIDEPRESSANTS

Depression is usually associated with an imbalance of the neurotransmitters involved in passing on nerve impulses. Serotonin is the molecule that has attracted the most attention because low levels have been associated with people who are depressed, and many antidepressants act to increase serotonin levels. The first type of antidepressant to come into use during the 1950s was monoamine oxidase inhibitors (MAOIs) such as iproniazid. These chemicals inhibit the monoamine oxidase enzyme, which oxidises monoamine neurotransmitters like serotonin, and thus allow serotonin levels to rise. A side effect of MAOIs is that they inhibit the breakdown of other amines—notably tyramine, which is found in various foods like cheese ("cheese syndrome")—so these amines are not catabolised, leading to a possible hypertensive crisis.

MAOIs were followed in the late 1950s by the so-called tricyclic antidepressants (TCAs) that inhibit neurotransmitter reuptake at the synapse so that serotonin levels are raised. The best known of these molecules is probably **dizaepam** (Valium; 13.24); released in 1963, it was the top-selling pharmaceutical in the United States from 1969 to 1982. Some have referred to it as "mother's little helper", though the Rolling Stones' song of that title (1966) was largely about barbiturates.

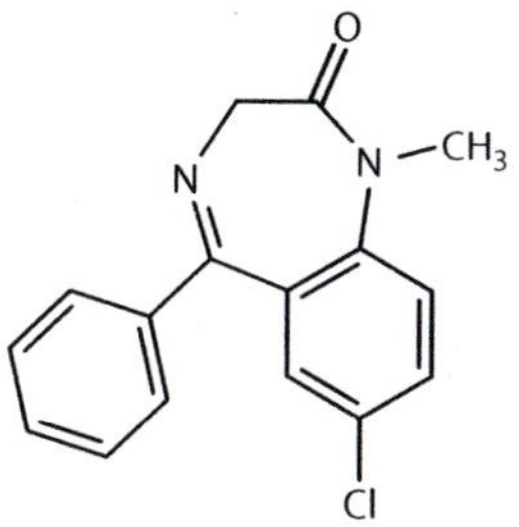

(13.24)

TCAs were followed by selective serotonin reuptake Inhibitors (SSRIs), which have a better margin between a clinical and a dangerous dose. The best-known SSRI is fluoxetine (Prozac), the result of a programme begun in 1970 at Eli Lilly and Co. As often with the best research ideas, the team, led by the Scottish chemist Bryan Molloy, started with a known molecule, diphenhydramine (2-(diphenylmethoxy)-N,N-dimethylethanamine). This was an antihistamine known to have antidepressant properties; the team made many derivatives of this molecule and in 1972 identified fluoxetine as the most strong and selective inhibitor of serotonin reuptake of those tested. Fluoxetine works by inhibiting action of the serotonin uptake pump, which removes serotonin from the synapse, leading to higher levels of serotonin in the synapse.

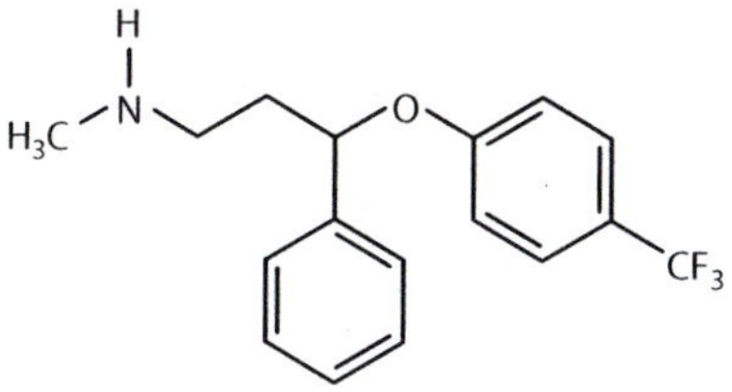

(13.25)

Prozac (fluoxetine; 13.25) was launched in 1986 in Belgium, in the United States in December 1987, and in the UK in 1989; it is prescribed for depression, obsessive-compulsive disorder, bulimia, agoraphobia, and premenstrual dysphoric disorder (PMDD). After patent protection of Prozac expired in 2001, it was given the name of Safarem and marketed particularly for the treatment of PMDD, a severe form of premenstrual syndrome afflicting 3%–8% of women. Prozac proved to be a blockbuster drug for Eli Lilly, with big sales through the 1990s, peaking in 2000 at \$2.5 billion. Even with newer antidepressants on the market, fluoxetine remains extremely popular.

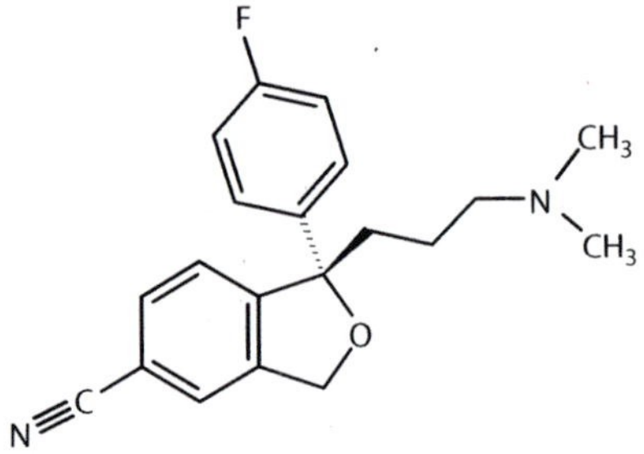

(13.26)

A second generation of SSRIs includes sertraline (Zoloft™) and citalopram (13.26; Citalopram, Celexa, Cipramil). **Citalopram** is a chiral molecule and the drug had been supplied as the racemate; only one isomer, (S)-(+)-citalopram (13.26), has antidepressant effects. When patent rights to citalopram expired in 2003, the Lundbeck pharmaceutical company took out a patent on the single-isomer (S-) form escitalopram (Cipralex, Lexapro™). SSRIs have been followed in their turn by SNRIs (serotonin-norepinephrine reuptake inhibitors), such as duloxetine (Cymbalta™), which appear to have slightly greater efficacy and fewer side effects.

Controversy has surrounded fluoxetine for much of its time on the market. One concern is that SSRIs like Prozac can increase the likelihood of suicidal behaviour; the evidence for this is mixed, but the FDA does require that SSRIs and other antidepressant medications carry a warning box concerning an increased risk of suicidal behaviour in patients younger than 24. The actual benefits of SSRIs have come under scrutiny. Two separate meta-analyses have indicated that, for patients with mild or moderate symptoms of depression, the benefits may be minimal or nonexistent and that the benefits of SSRIs are clinically significant only for patients having severe depression.

ANTIULCER DRUGS

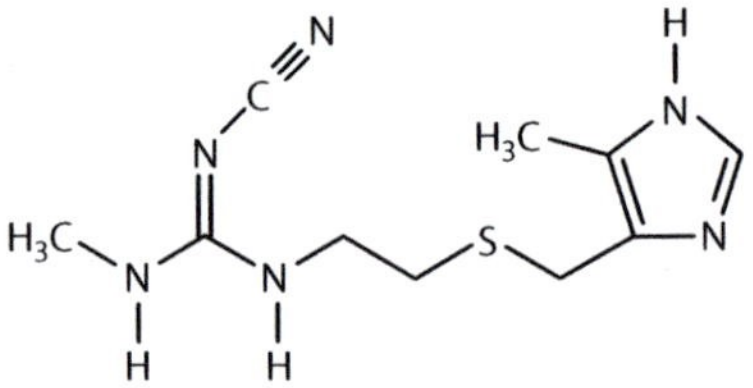

(13.27)

Cimetidine (Tagamet™; 13.27) combines being the molecule that first brought relief to stomach ulcer sufferers from pain, suffering, and the likely threat of death with the tag of being the first drug to reach sales of US\$1 billion—the first "blockbuster drug". In most cases, peptic ulcers are thought to be caused by infection with the *Helicobacter pylori* bacterium, though excess levels of stomach acids (e.g., hydrochloric acid) also contribute. Until recently, the options for treating this complaint relied very much upon alkaline antacids or other drugs and a very boring diet, with surgery as the last port of call.

Cimetidine was a planned molecule. Starting in 1964, a team of British scientists led by Sir James W. Black, at the company Smith, Kline and French (now part of GlaxoSmithKline), looked at what was known about the causes of ulcers. They knew that histamine (13.28) stimulated the release of stomach acid when attaching itself to a receptor in the stomach lining, and they also knew that traditional antihistamine medicines did not affect acid production. Therefore, there had to be two histamine receptors: one that produces allergy and hypersensitivity and another responsible for secreting HCl(aq) into the stomach. The latter one is unaffected by antihistamines (and it has subsequently been found that there are four different histamine receptors: H_1, H_2, H_3, and H_4). In the stomach, H_2 stimulates gastric acid secretion. The team studied molecules with structures based in histamine, but with a longer side chain.

$$(13.28)$$

After 6 years' research, the team found an antagonist (known as burimamide) that worked, and it proved the existence of the H_2 receptor; however, this did not work as an oral drug. Two years later, they had a compound—**metiamide** (13.29)—that worked on oral ingestion and performed very well in clinical trials; however, upon further testing, it was found that a few patients showed a decreased white blood cell count.

$$(13.29)$$

It was thought that the C=S (thiourea) part of the molecule was responsible for the side effects, but when C=S was replaced with C=O or C=NH groups, activity was reduced. In the end, an electron-withdrawing CN group was substituted for the hydrogen in the C=NH group, reducing its basicity so that it was not protonated in the stomach. The product, **cimetidine**, was tested in 1973 and proved completely satisfactory; it went onto the market as Tagamet® in the UK in November 1976 and in the United States the following August. (The name was a fusion of an**tag**onist and ci**met**idine). It is now available in many places as an over-the-counter (OTC) medication.

After the success of cimetidine (Tagamet), chemists at Glaxo looked to produce a similarly valuable histamine H_2 inhibitor. Improvements in modelling led them to produce ranitidine (Zantac), which was an improvement on cimetidine; it had fewer adverse reactions and higher activity, as well as a more prolonged action. **Ranitidine** (Zantac; 13.30) was introduced in 1981 and in less than a decade succeeded cimetidine as the best-selling prescription drug, and it is likewise now widely available as an OTC medication.

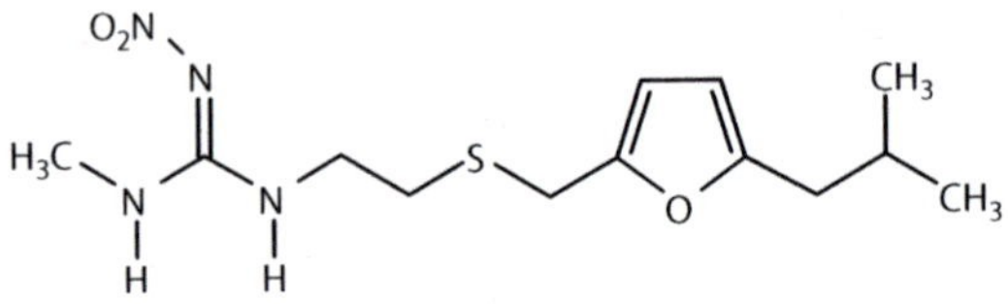

(13.30)

Proton pump inhibitors were the next step in drugs to reduce acid secretion in the stomach and are believed to be the most effective route. They work by inhibiting the H^+-K^+-ATPase enzyme, which pumps the acid into the stomach. They are usually substituted benzimidazoles and bind to the cysteine 813 of the enzyme, forming a covalent disulfide link and thus inhibiting acid secretion. Omeprazole, developed by AstraZeneca, is the best known of these molecules. Known by names including Prilosec, it was introduced to the American market in 1989. It contains a chiral sulfur atom and therefore exists as two enantiomers, but the (R)-form is inactive, though it converts into the active S-enantiomer in vivo. AstaZeneca now markets the single-isomer drug, the pure S-isomer, as **Nexium** (esomeprazole; 13.31). Tests have shown that treatment with PPIs gave faster healing than treatment with either ranitidine or placebo.

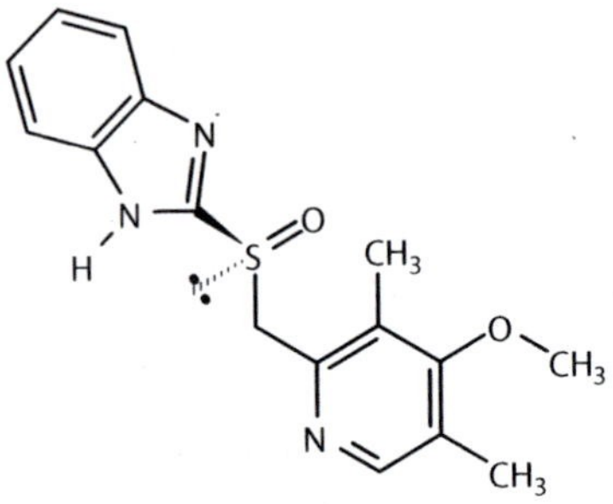

(13.31)

SULFANILAMIDE AND THE SULFONAMIDES

It is easy to forget that until recently people could contract fatal infections in quite trivial ways, such as the vulnerability of women after childbirth to puerperal sepsis ("childbed fever"). In 1924, 16-year-old Calvin Coolidge, Jr., son of the president of the United States, contracted a blister on the toe of his right foot while playing tennis. Septicaemia set in; he was dead in a week. Similarly, in November 1930, W. W. ("Dodger") Whysall, a Nottinghamshire and England Test cricketer, slipped on a dance floor, grazed his elbow, and died of septicaemia a fortnight later. Yet when another president's son, 22-year-old Franklin Roosevelt, Jr., developed a septic sore throat in December 1936, he was injected with a new drug called Prontosil and made a full recovery. Prontosil also sharply reduced the death rate from puerperal sepsis among women who had just given birth.

$$(13.32)$$

Prontosil's action was discovered by a German chemist named Gerhard Domagk (1895–1964), whose experiences with a medical unit in World War I led him to devote his studies to preventing infections. Working for I. G. Farbenindustrie, he methodically set about testing the effect of thousands of dyestuffs on bacteria. After 5 years of work, in late 1932, he found that KI-730, known later as Prontosil (13.32), worked well against streptococcal infections in laboratory mice (*in vivo*), even at very low doses, though it did not work in test-tube tests against the bacterium (*in vitro*). Tests on humans with streptococcal infections— for which there was no other treatment— went well; the drug was patented in December 1934 and came into use.

In early December 1935, 6-year-old Hildegarde Domagk pricked her finger with a needle and a streptococcal infection spread through her arm; it looked as if amputation of the arm was her only chance. Gerhard Domagk treated his daughter with Prontosil and the infection cleared up. Just before then, on 6 November 1935, Daniel Bovet's team at the Pasteur Institute in Paris discovered that sulfanilamide, one of the component molecules of Prontosil, worked just as well. Sulfanilamide (13.33) itself was active both *in vitro* and *in vivo*. Intestinal enzymes in the human body convert Prontosil into sulfanilamide, the active molecule; there were no enzymes in the bacterial cultures upon which Prontosil had been tested. Sulfanilamide was much cheaper than Prontosil and had the further advantage of not turning patients' skins bright red.

$$(13.33)$$

Sulfanilamide was a great success, except for a tragic episode in 1937. A scientist created an "elixir of sulfanilamide" by dissolving the drug in diethylene glycol and adding a dye and raspberry fragrance; this was marketed as a treatment for streptococcal infections of the throat before any animal tests or checks on the literature were conducted. Diethylene glycol was sweet but toxic, and 107 people, mainly children, died as a result. Public outcry resulted in the 1938 Food, Drug & Cosmetic Act, which created safeguards over the introduction of new drugs. Previously, a manufacturer did not have to show that its products were safe, but under the act, officials had to review test results and had the power to ask for extra tests to be carried out. Ultimately, they had the power to prevent a drug being introduced. A 23-year-old PhD student, Frances Oldham Kelsey, conducted the animal tests, which showed that it was the solvent, not the sulfanilamide that was responsible. She remembered the lessons when, in 1960, she was working for the Food and Drug Administration in Washington and was asked to evaluate thalidomide (see Chapter 9).

How sulfanilamide worked was discovered in 1940 by Donald Woods, working in Oxford. He found that para-aminobenzoic acid (PABA; 13.34) stopped sulfanilamide's antibacterial effects and noted that these two molecules had very similar structures.

PABA is used by bacteria in the synthesis of dihydrofolic acid, which is used to make other molecules like the purine bases necessary to construct DNA, so it is absolutely vital to the bacteria. One stage of the synthesis of dihydrofolic acid is catalysed by the enzyme dihydropteroate synthetase; sulfanilamide binds at the active site, and blocks the site to PABA so that the bacterium cannot multiply and grow. Thus, the immune system has the opportunity to clear up the infection.

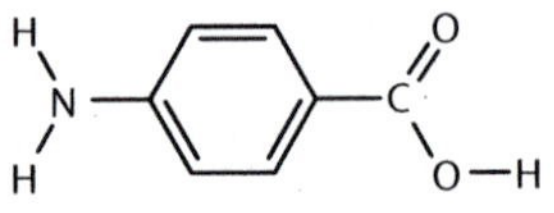

(13.34)

Bacteria make dihydrofolic acid. Humans do not, so they take in folic acid as a vitamin (e.g., from green vegetables); they do not have any dihydropteroate synthetase, which is why sulfanilamide is safe for humans. The 69-year-old British Prime Minister Winston Churchill had two bouts of pneumonia in 1943, the second one just after his meeting with Roosevelt and Stalin at the Teheran conference. He was treated with one of the new sulfa drugs from May and Baker and made a full recovery; this was either M&B 693—sulfapyridine (13.35)—or M&B 760—sulfathiazole (13.36).

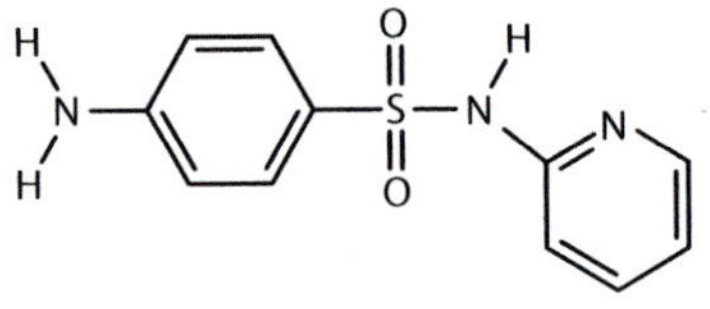

(13.35)

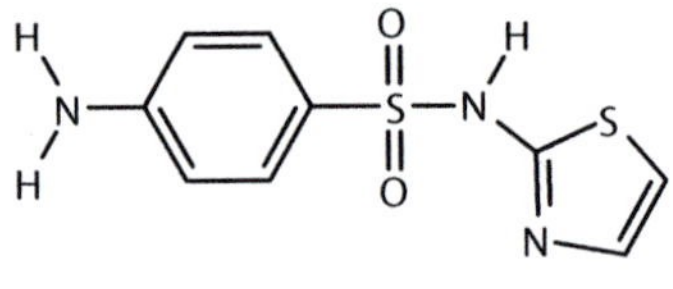

(13.36)

Sulfanilamide hugely reduced the mortality rate of soldiers in World War II. Every American soldier had a first-aid pouch containing sulfa powder and a dressing bandage. They were taught to sprinkle the powder onto any open wound to eliminate infection. Medics in combat carried sulfa powder and sulfa tablets.

Sulfonamides are not used as much now as they were 50 years ago because some bacteria have acquired immunity to them; they can also cause damage to the liver and other side effects if used too frequently. Nevertheless, they were the first widely used and successful examples of chemotherapy, using synthetic chemicals to fight diseases. They were also the first example of the use of a competitive enzyme inhibitor as a drug.

However, they still find applications. New, more durable, and longer lasting sulfonamides are available now. Main uses of sulfonamides nowadays include for urinary

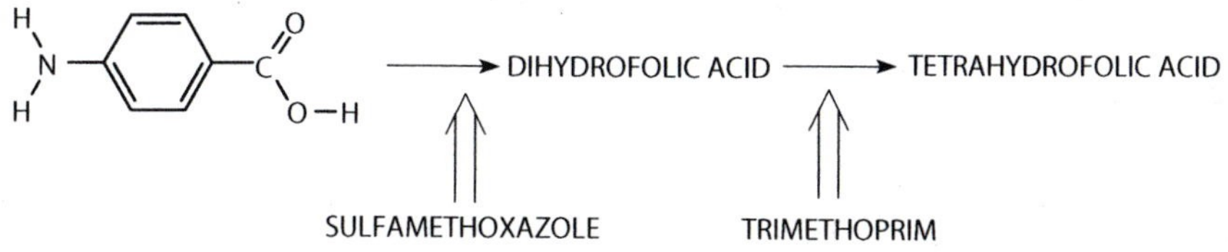

FIGURE 13.3 How co-trimoxazole acts.

tract infections, some sexually transmitted infections such as chlamydia, and certain specific respiratory problems. Co-trimoxazole (Septra®, Septrim®, Bactra®, TMP-SMX), a mixture of sulfamethoxazole and trimethoprim, is used to treat potentially fatal AIDS-linked infections such as pneumonia when immune systems are weakened because of HIV. It is an effective "broad spectrum" antibiotic against a variety of illnesses. Its effectiveness is due to the fact that sulfamethoxazole (13.37) and trimethoprim (13.38) target successive steps in the biosynthesis of tetrahydrofolic acid, so lower doses of each molecule can be used in the drug. It is a kind of double-whammy treatment, but concerns about side effects mean that it sees limited use (Figure 13.3).

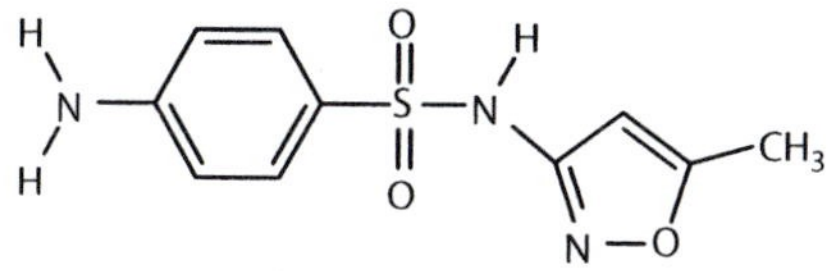

$$(13.37)$$

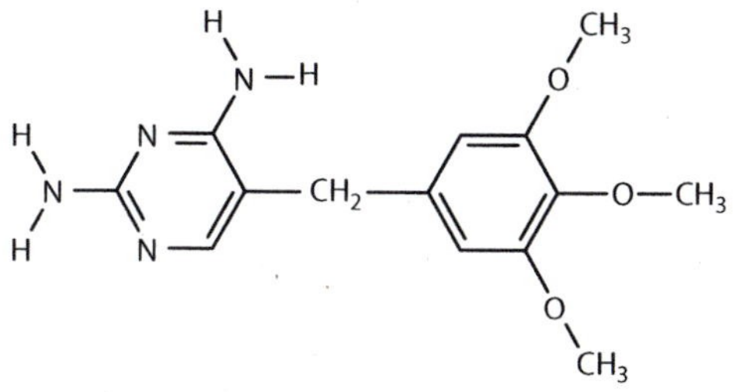

$$(13.38)$$

Sulfa drugs were first used in malaria treatment in the 1930s, but were superseded by other agents like chloroquine, which in its turn has become less effective. A combination of sulfadoxime and pyrimethamine (Fansidar®) is used to treat chloroquine-resistant *Plasmodium falciparium* malaria. This similarly targets the two enzymes in the synthesis of tetrahydrofolic acid. It is only employed for severe cases due to side effects.

ANTIFLU DRUGS

The twentieth century saw three influenza pandemics: 1918–1919 (H1N1; "Spanish flu"), 1957 (H2N2; "Asian flu"), and 1968 (H3N2; "Hong Kong flu"), which killed 40 million, 2 million, and 1 million people, respectively. In recent years, we have anticipated "avian flu" (H5N1) and "swine flu" (H1N1) pandemics. Influenza is caused by a virus that infects host cells, producing copies of itself that go on to attack more cells. The virus binds to cells through a glycoprotein antigen called haemagglutinin, which attaches it to sialic acid receptors on the cell surfaces (in the body, these are cells in the respiratory system), whereupon the membranes of the host cell and the virus fuse. The virus can move into the cell, taking over the cell nucleus and making copies of itself. These copies move out of the cell, taking up a membrane coating. This coating retains sialic acid receptors from the original host cell, and the job of the neuraminidase enzyme is to destroy these receptors (sialic acid acts as the substrate to the enzyme) so that the virus can dissolve a pathway through the cell membranes and escape from the host cell.

Tamiflu® and other antiflu drugs interfere with this process. Influenza viruses change each year, with alterations in the amino acid sequence. However, it is found that the amino acid sequence around the cleft of the catalytic active site in the neuramidinidase enzyme is substantially unchanged, and scientists realised that an inhibitor that fitted this cleft would work on the neuramidinidases of all influenza viruses and stop new virus particles from being released. Thus, Tamiflu and related molecules are "plug drugs".

Discovered in 1996, oseltamivir (Tamiflu®; 13.39) is the best known of these antiviral agents. Although it does not cure flu, it does stop the influenza virus from spreading in the body, but only if it is taken within about 36 hours of the onset of symptoms. By slowing down viral replication, it makes it easier for the body's immune system to deal with the virus, thus limiting how severe the illness is and how long it lasts. In fact, oseltamivir is relatively inactive against the virus, but is rapidly metabolised in the body to the active molecule (known as GS4071; 13.40).

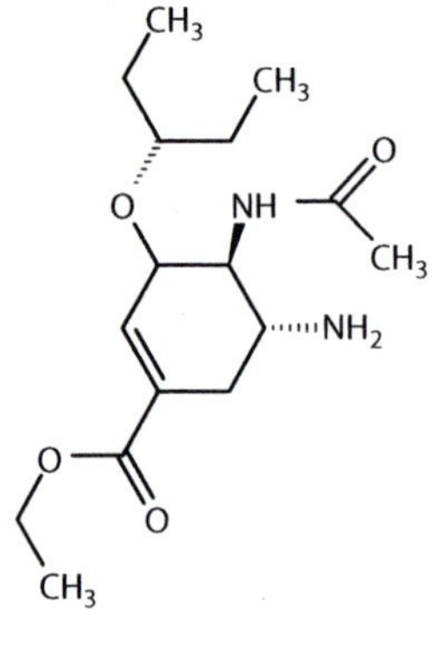

(13.39)

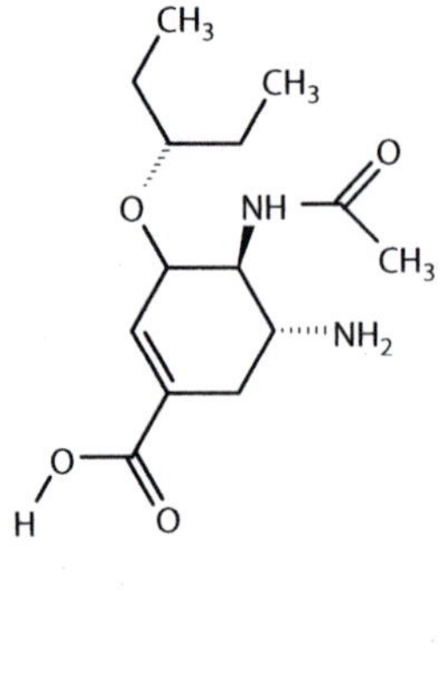

(13.40)

Industrially, oseltamivir is generally made starting from shikimic acid, which is present in the pods of star anise (*Illicium verum*), a star-shaped spice found in the pods of an evergreen tree that grows in China; because of immense demand for oseltamivir, bacterial fermentation methods using engineered *E. coli* are now being employed. Many of the synthetic routes for oseltamivir involve azides, which can present an explosive hazard, so alternatives are being explored. There are alternative treatments to oseltamivir. Zanamivir (Relenza; 13.41) was actually discovered first (1989) but beat Tamiflu to the marketplace by less than a year. It has the disadvantage of having to be administered as a dry powder by an inhaler, rather like an asthma inhaler; many people would rather take pills, like Tamiflu. However, it is reported that no flu virus has any resistance to zanamivir, unlike oseltamivir.

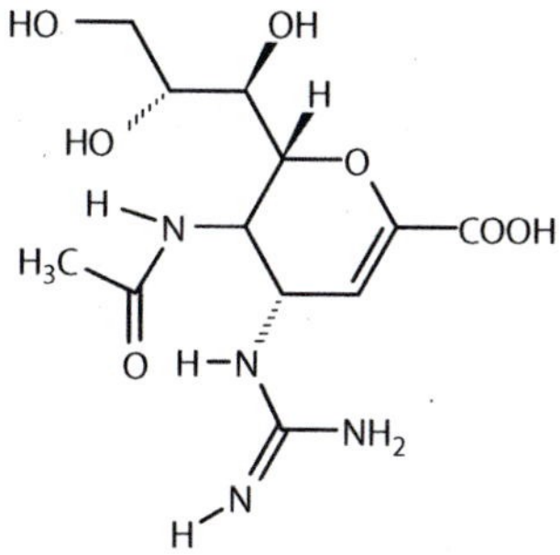

(13.41)

A third drug, peramivir (13.42), has more recently come on the scene, so less is known about it. It has a broad spectrum of availability, is said to be more potent than the others, and has a longer lifetime in the human body. Unlike the others, it is an injectible drug and is used if neither oseltamivir nor zanamivir is suitable.

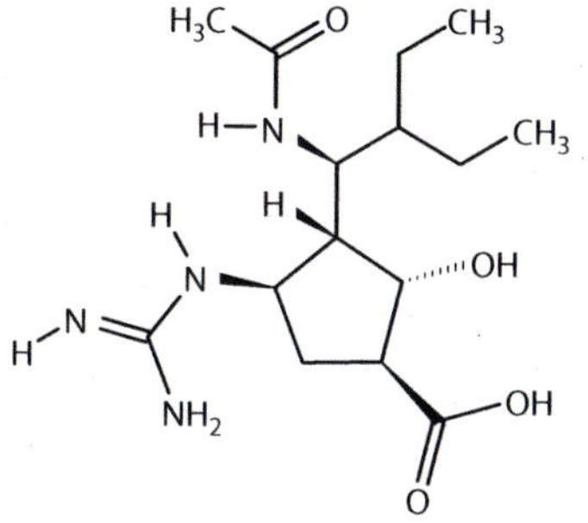

(13.42)

Other antiflu drugs that have been used include the adamantane derivative amantadine (1-aminoadamantane; Symmetrel, 13.43). This is an antiviral drug that was approved in 1976 for treatment of influenza A; rimantadine (1994) is another such molecule (13.44). However, it has been reported that Chinese chicken farmers have been giving amantidine to their birds to ward off avian flu, and it can scarcely be coincidence that many influenza A strains (and virtually all H1N1 "swine flu" strains) are amantadine resistant. Therefore, it is no longer (2009) recommended for treatment of influenza, and neither is rimantidine. The mechanism of action of these molecules is not clear.

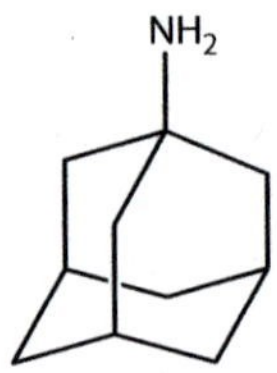

(13.43)

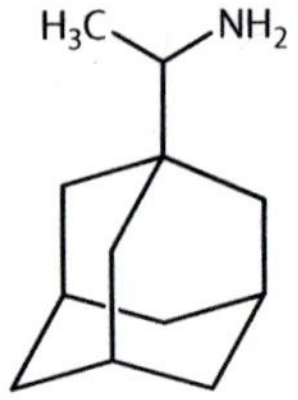

(13.44)

14 Synthetic Polymers

Natural materials like wool, silk, leather, and cotton were all that was available to people making clothing until well into the twentieth century. Synthetic organic chemistry was developing, but when, in 1920, Hermann Staudinger proposed that natural substances like proteins, cellulose, and carbohydrates were made of large numbers of small molecular units joined together, he was pretty much in a minority of one. Yet, in 1953, Staudinger was awarded the Nobel Prize for Chemistry "for his discoveries in the field of macromolecular chemistry", as chemists had made the first synthetic polymers, both deliberately and accidentally.

POLYTHENE (OR POLYETHYLENE)

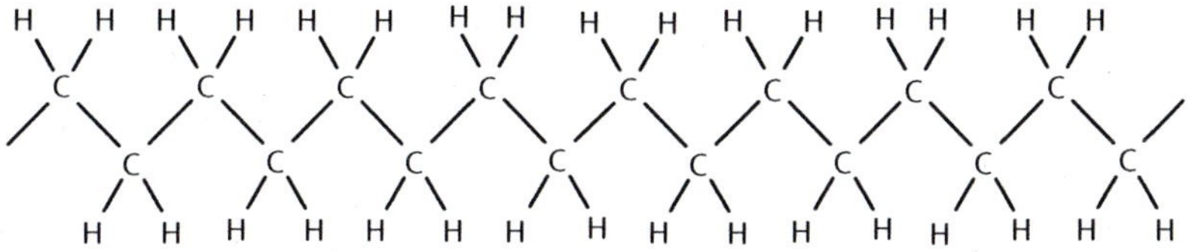

$$(14.1)$$

Polythene (14.1) was first made in 1898 by the German chemist Hans von Pechmann, who heated CH_2N_2, diazomethane, obtaining a white, waxy solid. It was not until Monday, 27 March 1933, that the material was obtained again—and, again, accidentally. This time, two chemists working for ICI, Reginald Gibson and Eric Fawcett, had reacted ethene with benzaldehyde, at 2,000 times atmospheric pressure and 180°C. They were hoping to make ethyl phenyl ketone, $CH_3CH_2COC_6H_5$, by inserting ethene into the C–H bond of the aldehyde group; again, they made a white waxy solid. The synthesis was hard to reproduce, until they realised that traces of oxygen were catalysing the polymerisation.

Subsequently, other catalysts have been developed; the best known is the Ziegler–Natta type based on a mixture of titanium halides and aluminium alkyls, whilst other types have come into prominence, especially metallocene-based systems. The polymerisation of ethene can be carried out using radical and ionic (both anionic and cationic) catalysts. Several different types of polythene are obtained whose properties depend upon the chain length and branching, with between 2,000 and 20,000 ethene units joined together (though a form of polythene is known with 100,000 or more). Upon polymerisation, the double bond is turned into a single bond because two electrons are needed to form the bond to the next two-carbon unit (Figure 14.1). Substituted alkenes are linked in this way in the other addition polymers.

Polythene is effectively a giant alkane; this explains why the discoverers compared it to paraffin wax. Polythene is the most widely used plastic; its introduction

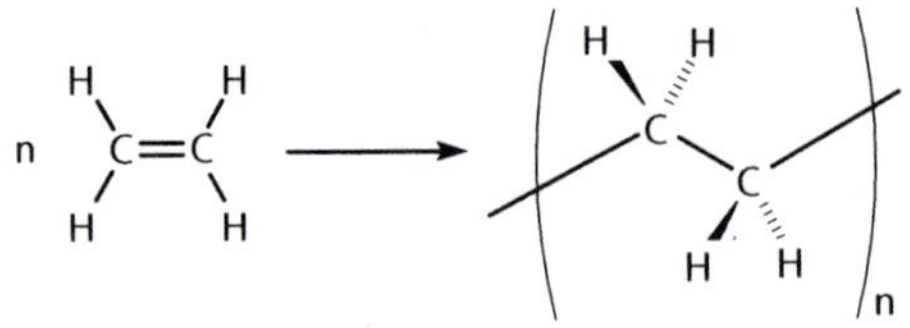

FIGURE 14.1   Formation of polythene.

revolutionised packaging and wrapping, the now iconic "plastic bag" superseding paper bags, with a big advantage in its wet strength (and the downside of its lack of biodegradability). Other familiar uses include washing up bowls and milk bottles, and, in 1958, hula-hoops made of polythene proved a runaway best seller.

POLYVINYL CHLORIDE, POLY(CHLOROETHENE)

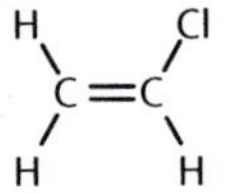

(14.2)

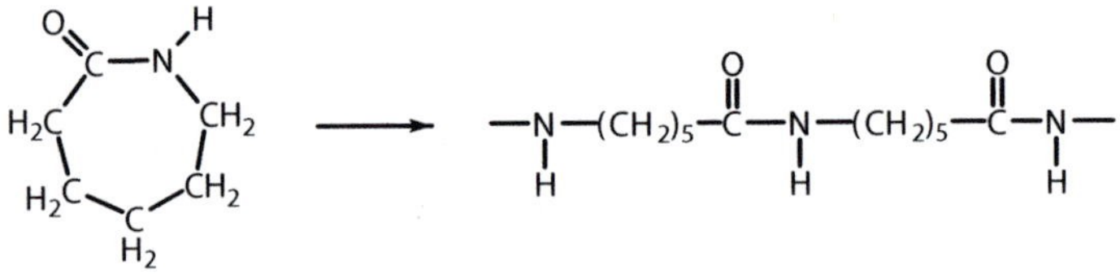

(14.3)

Polymerisation of chloroethene (vinyl chloride; 14.2) leads to poly(chloroethene), PVC (14.3). This reaction was first observed by nineteenth century chemists who found that sunlight brought about the polymerisation, yielding a white solid. For many years, PVC was a laboratory curiosity as a hard, brittle solid; then, chemists found that the right additives ("plasticisers") created a softer and more flexible material. It has many desirable properties: it is very unreactive and resistant to both weather and corrosion, a good insulator, and possesses much higher fire resistance than many polymers. PVC is extensively used in a wide range of applications, including piping, guttering, flooring, membranes in roofing, insulation for electrical cables, and clingfilm, as well as clothing such as raincoats and imitation-leather wear. PVC has been used to make gramophone records ("vinyl") since the 1930s, and it is also the material used to make the credit card that you may have used to buy this book (thank you).

In recent years, questions have been raised over possible health dangers of the additives to PVC—particularly the phthalate plasticizers—because there is evidence that they are hormone mimics and endocrine disruptors. Since PVC has been widely used to make soft toys for very small children, people have been concerned about additives being released from these toys when they are placed in children's mouths. The use of phthalates in PVC is now regulated throughout the EU. It has also been alleged that dioxin is formed as a waste product of both making and incinerating PVC.

Polymerising vinyl acetate (14.4) leads to **poly(ethenyl ethanoate)** (14.5), also known as **polyvinyl acetate** (PVA or PVAc). This is often carried out in emulsions.

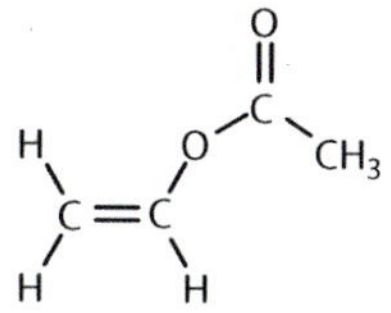

(14.4)

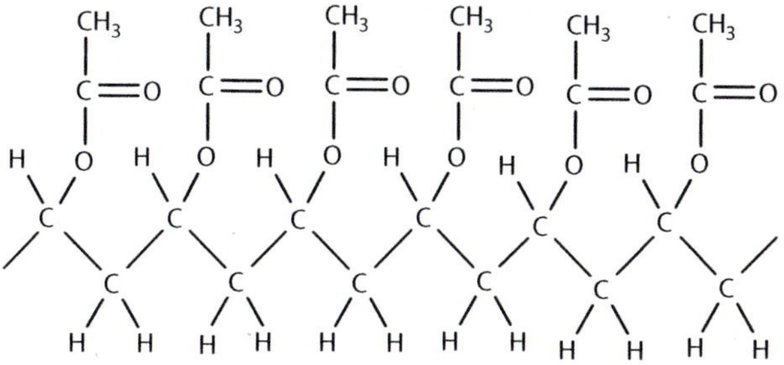

(14.5)

It is very widely used in paints and adhesives, especially as an emulsion in water; particular applications include its use as wood glue and also in bookbinding. Hydrolysis of PVA leads to polyvinyl alcohol (also sometimes confusingly known as PVA), which reacts with borax, cross-linking between the polymer chains to make "polymer slime".

TEFLON, POLY(TETRAFLUOROETHENE), PTFE

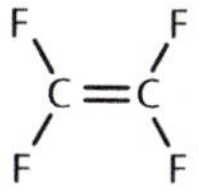

(14.6)

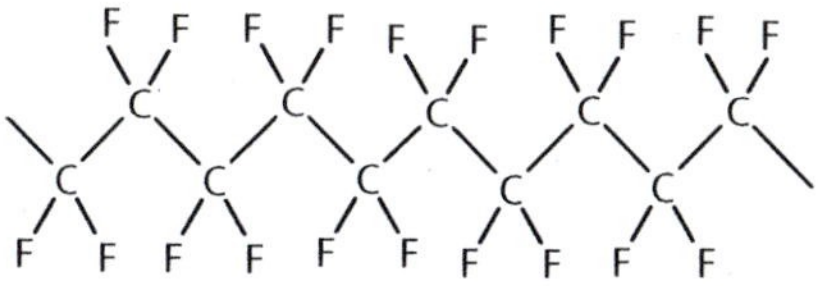

(14.7)

Teflon was discovered in 1938 by American chemist Roy Plunkett; trying to make a CFC refrigerant, he heated tetrafluoroethene (14.6) in an iron container under pressure, obtaining a white solid. It transpired that the iron had catalysed the polymerisation, leading to poly(tetrafluoroethene) (14.7). The strong C–F and C–C bonds cause it to be chemically very inert to other chemicals, making it of almost immediate use in the Manhattan project for the construction of the first atom bomb. The isotopes ^{235}U and ^{238}U were separated by gaseous diffusion of volatile and very reactive UF_6; Teflon was vital for making unreactive and leak-proof valves and seals in the plant.

Today Teflon-coated magnetic stirrer bars are widely used in laboratories. Teflon has an extremely low coefficient of friction and is widely used for a range of non-stick applications—famously, Teflon-coated, nonstick cookware as well as coatings for machinery parts (gears, bearings) and windscreen wiper blades. Teflon taps in burettes prevent sticking and require no lubrication. Teflon tape is used as thread-sealing tape by plumbers, whilst Teflon coatings are used on some armour-piecing bullets (though it is not the Teflon that gives them the armour-piecing property).

POLYSTYRENE (POLY(PHENYLETHENE))

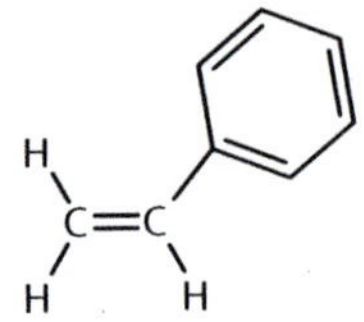

$$(14.8)$$

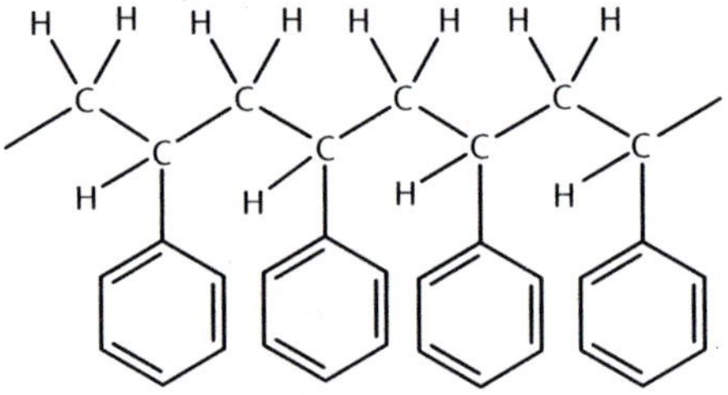

$$(14.9)$$

Phenylethene (styrene; 14.8) polymerises to afford **polystyrene,** poly(phenylethene) (14.9). Although not recognised at the time, polystyrene was probably first made in 1839 by storing the distilled resin of the Oriental sweet gum tree; however, it was not manufactured commercially for almost a further century. It is a hard, rigid material that can produce detailed mouldings, best exemplified in plastic model construction kits. Chemically inert, it is used for many types of containers, such as CD cases and petri dishes in science laboratories. Expanded polystyrene is manufactured from polystyrene beads, pentane, and HCFC blowing agent. Upon heating with steam, the pentane vapourises and causes the beads to expand dramatically, whereupon they are injected into a vacuum mould for further expansion. This is the familiar material used for insulation of buildings and in packaging. The nastiest use is that of a polystyrene–benzene mixture as a very unpleasant formulation for napalm.

POLY(METHYL CYANOACRYLATE)

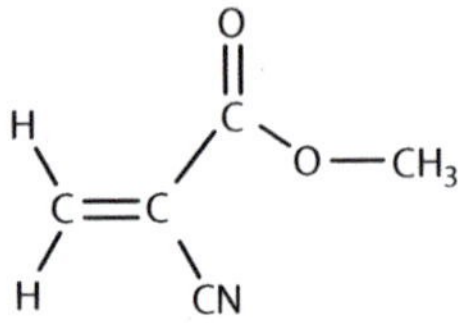

$$(14.10)$$

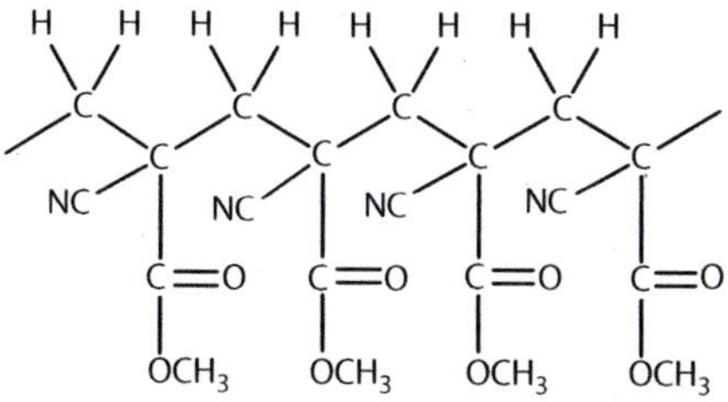

(14.11)

Methyl cyanoacrylate (14.10) contains two electron-withdrawing substituents at the same end of the double bond; this polarises the molecule so much that it is easily polymerised by even traces of water, forming poly(methyl cyanoacrylate) (14.11). This has earned it the name of "superglue" (in practice, most commercial superglues now use ethyl cyanoacrylate). It was invented during World War II, but came into more general use in the 1960s and 1970s. Although it has a low shear strength, methyl cyanoacrylate has many advantageous properties, including its speed of setting and its water resistance. It bonds to most things, notably skin and body tissue, and has been used in surgery to hold tissue together in place of stitches (sutures). One special application is to detect fingerprints on nonporous surfaces. The vapour from warmed cyanoacrylate reacts with traces of moisture in the places where fingers have touched the surface and rapidly polymerises to display the ridges of the prints. There are sufficient free OH^- ions in water to ensure rapid anionic polymerisation.

POLY(METHYL METHACRYLATE) (PMMA)

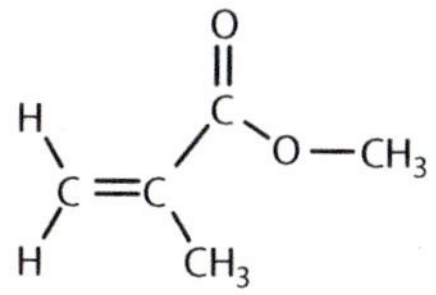

(14.12)

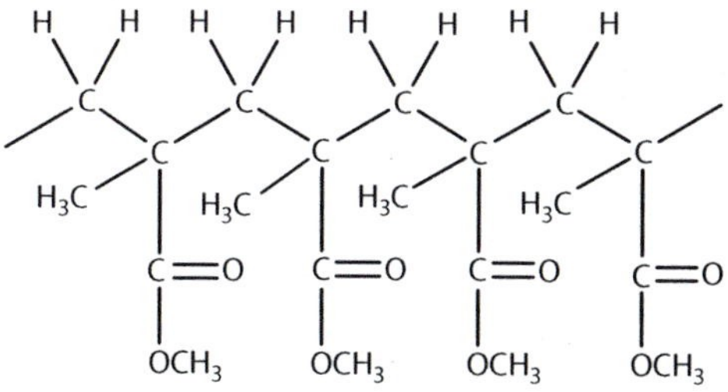

(14.13)

As with many other alkenes, methyl methacrylate (14.12) was probably first polymerised to make poly(methyl methacrylate) (14.13) in the nineteenth century, but the polymer only came into commercial production in the 1930s. Also known as Plexiglas or **Perspex,** it has been widely used as a glass substitute. It has many advantages over glass, including a much lower density, better impact strength, and its ease of joining together using Superglue, by heat, or with solvents such as dichloromethane or chloroform. Although it is easily scratched, such scratches can be polished away. An early application was one-piece "bubble" cockpit canopies and gun turrets on some WWII aircraft, notably the Supermarine Spitfire; related applications include aircraft windows, windshields, helmet visors, aquariums, car headlights, and optical fibres. It is not rejected by human tissue, so it can be used in replacement lenses in the human eye. A further advantage of Perspex is that it can simply be depolymerised to the monomer just by heating it.

NYLON

When Wallace Carothers (1896–1937) moved from Harvard to set up the organic chemistry laboratory at DuPont in Wilmington, Delaware, his objective was to make new polymers. Now if a molecule has only one functional group, at one end, it can only react at that end, and reaction stops there. Carothers hit upon the idea of using monomers with two functional groups, one at each end, so that each end of the monomer could react and form part of a chain. On 28 February 1935, Dr. Gerard Berchet, a member of Carothers's team, reacted hexamethylene diamine (1,6-diaminohexane), $H_2N-(CH_2)_6-NH_2$, with adipic acid, $HOOC-(CH_2)_4-COOH$ (hexane-1,6-dioic acid), and succeeded in making Nylon 6-6 (which means that it is a copolymer of a 6C diamine and a 6C diacid). By that autumn, the first nylon threads had been spun. The acid molecule loses an OH group from each end and the amine loses one hydrogen from each end; an amide link is formed between each reacting monomer, and a molecule of water is eliminated from each amide group (14.14) formed.

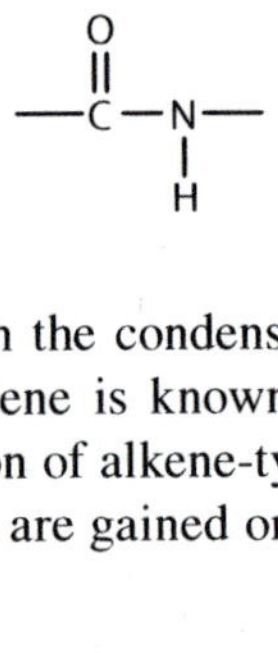

$$-\overset{\overset{\displaystyle O}{\|}}{C}-\overset{\overset{\displaystyle}{}}{\underset{\underset{\displaystyle H}{|}}{N}}- \tag{14.14}$$

The loss of a small molecule in the condensation process (Figure 14.2) during the formation of Nylon and Terylene is known as *condensation polymerisation*. This is in contrast to polymerisation of alkene-type monomers like ethene, styrene, and vinyl chloride, when no atoms are gained or lost in the process called *addition polymerisation*.

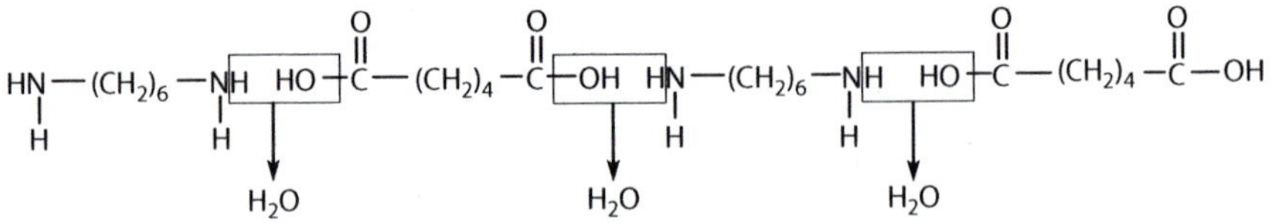

FIGURE 14.2 Condensation polymerisation leading to Nylon 6-6.

When this is repeated at both ends of each monomer, a chain results (14.15).

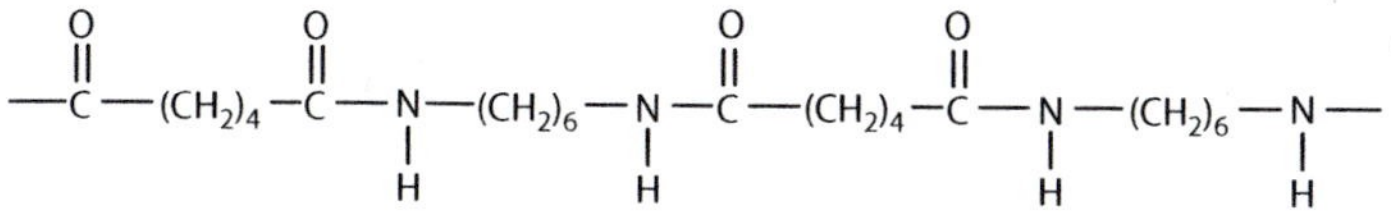

$$-\overset{\overset{\displaystyle O}{\displaystyle \|}}{C}-(CH_2)_4-\overset{\overset{\displaystyle O}{\displaystyle \|}}{C}-\underset{\underset{\displaystyle H}{|}}{N}-(CH_2)_6-\underset{\underset{\displaystyle H}{|}}{N}-\overset{\overset{\displaystyle O}{\displaystyle \|}}{C}-(CH_2)_4-\overset{\overset{\displaystyle O}{\displaystyle \|}}{C}-\underset{\underset{\displaystyle H}{|}}{N}-(CH_2)_6-\underset{\underset{\displaystyle H}{|}}{N}- \quad (14.15)$$

Overall,

$$(n+1)\ H_2N-(CH_2)_6-NH_2 + (n+1)\ HOOC-(CH_2)_4-COOH$$
$$\rightarrow H_2N-(CH_2)_6-NH[CO(CH_2)_4-COHN-(CH_2)_6-NH]_nOC-$$
$$-(CH_2)_4-COOH + (n+1)\ H_2O$$

Sadly, Carothers, who had suffered from mental instability and depression for years, committed suicide on 29 April 1937, shortly after applying for a patent on Nylon. US Patent 2,130,948, "synthetic fibers", was issued on 20 September 1938.

A second type of nylon results from a single monomer, caprolactam, which effectively has two different functional groups. Although it exists as a cyclic lactam, the ring opens on polymerisation, and the monomers (6-aminohexanoic acid) link up to form a single chain of Nylon-6 (Figure 14.3), first made by IG Farben in Germany (1938).

Nylon was a tremendous gift to industry: very tough and nonbiodegradeable, resistant to most chemicals, flexible, and with a low density. It was first used to replace pig's bristles in toothbrushes, but it was soon used to make nylon stockings; they were launched across America on "N-Day", 15 May 1940, a day when 5 million pairs were sold, at \$1.15 a pair. As soon as America entered World War II on 7 December 1941, Nylon was exclusively used for the war effort, and black market prices for stockings went through the roof. Nylon replaced the unobtainable silk for parachutes and found many other uses, including ropes, tyre cords, life rafts, and machinery parts like gears. It was not until the war's end that nylon stockings became available again, in such short supply that huge queues formed; when they sold out, dissatisfied customers rioted, the so-called "Nylon riots". Other, newer polymers have partly taken the place of Nylon in clothing, but it still has many uses, notably in carpets.

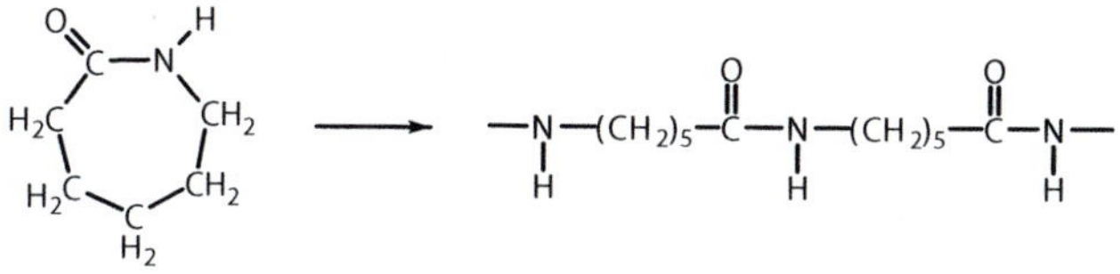

FIGURE 14.3 Nylon-6.

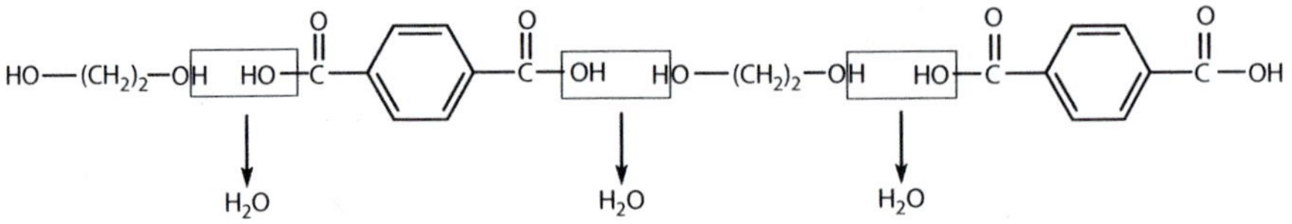

FIGURE 14.4 Synthesis of terylene.

TERYLENE

Carothers's team had made the first polyesters before they hit on Nylon, but the success of the polyamide saw polyesters sidelined. Thus, it was J. T. Dickson and J. R. Whinfield, working in a laboratory at Accrington, in Lancashire, in 1941, who first reacted (Figure 14.4) terephthalic acid (benzene-1,4-dicarboxylic acid) and ethylene glycol (ethane-1,2-diol) and made the first polyester fibres of Terylene.

Terylene (14.16) was marketed by ICI in Britain and by DuPont in the United States, where it was known as Dacron. Analogously to Nylon, a diacid and a diol provide ester linkages at each end of the monomer units used as reactants. Such polyesters have several desirable properties: They do not shrink or stretch, and the ability of Terylene to be "heat-set" at high temperatures so that it can be mistreated without creasing has led to its use in low-maintenance, non-iron clothing like drip-dry shirts ("wash and go"). It mixes well with other fibres, such as wool, and it is also inert to many chemicals. Its low cost was a factor in its becoming unfashionable for a while, but it is increasingly used in smart polyester blends and polyester microfibre materials.

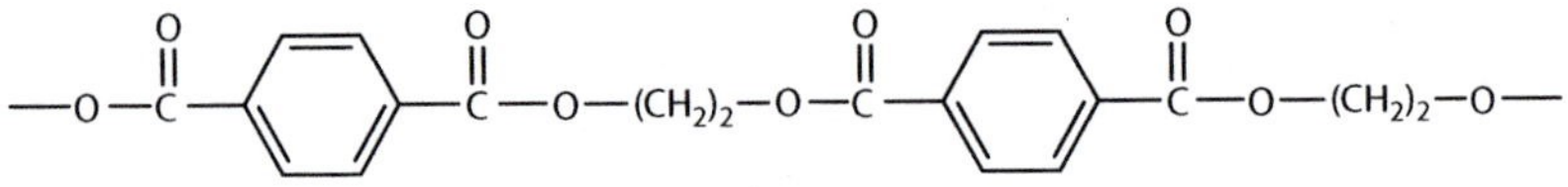

KEVLAR

Like Nylon, Kevlar is a poly(amide) condensation polymer, but with rather different properties. Just as Nylon was the creation of a man, so Kevlar was a woman's discovery (1965). Working for DuPont, Stephanie Kwolek was looking for a strong, lightweight fibre to replace Nylon in car tyres. Kevlar is made by reaction of 1,4-diaminobenzene with benzene-1,4-dicarbonyl chloride (terephthaloyl chloride) (14.17).

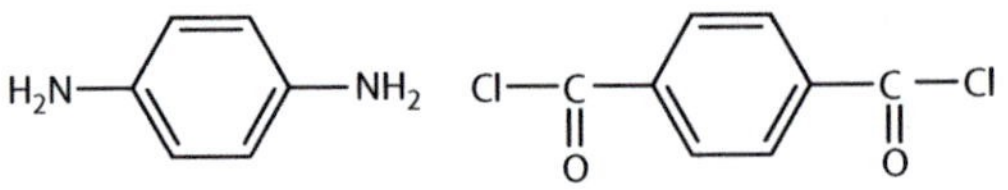

A section of an individual Kevlar chain looks like structure 14.18.

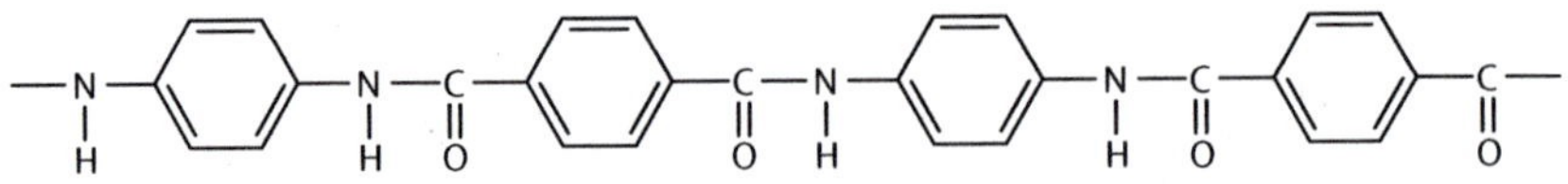

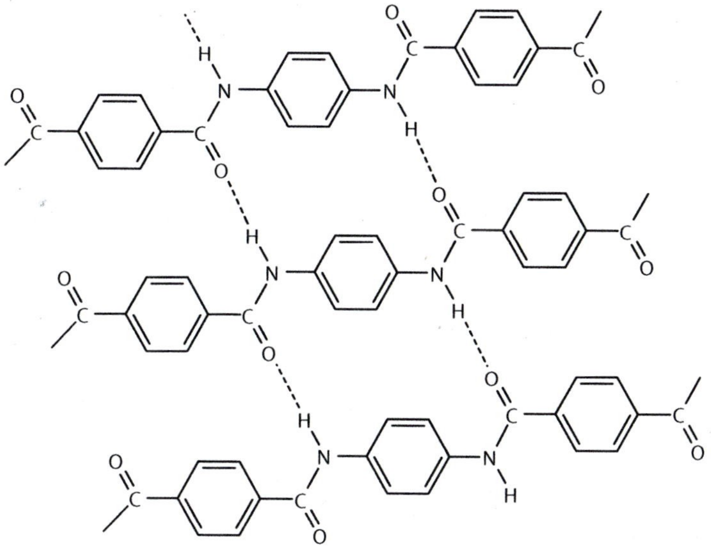

FIGURE 14.5 Alignment of the chains in Kevlar.

Kwolek soon found that Kevlar—unlike Nylon—formed unbreakable fibres when spun. It represented a new class of polymer known as *aramid,* after "aromatic poly-amide". The presence of so many benzene rings leads to rather rigid, oriented chains, and hydrogen bonding between neighbouring chains causes the formation of strong sheets; π–π interactions between the benzene rings in neighbouring sheets in turn causes them to stack and reinforce the strength of the fibre. It is also stable up to above 400°C and is chemically unreactive (the fibres are made by dissolving Kevlar in conc.H_2SO_4, and extruding it into water) (Figure 14.5).

Kevlar's immense strength-to-weight ratio leads to many applications, such as skis, hockey sticks and tennis rackets, aircraft wings, yacht sails, and brake pads, as well as reinforcing tyres. It is best known as the substance used to make body armour and bullet-proof vests because it dissipates the energy of the bullet much faster than do alternative materials and also spreads it over a wider area. The 24-layer Kevlar is supposed to stop Magnum bullets.

Polylactide (also known as **polylactic acid**) is an example of a biodegradeable polyester that has been known for a century. It is now both fashionable and feasible to produce it industrially.

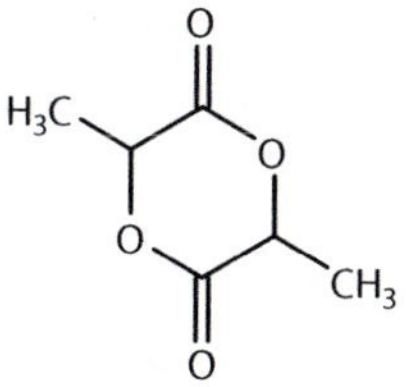

$$(14.19)$$

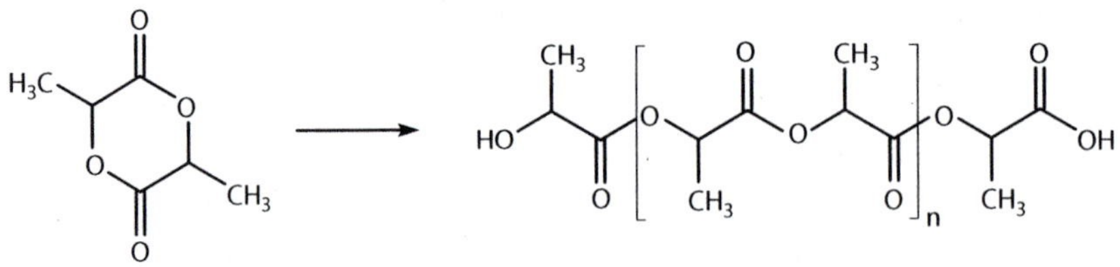

FIGURE 14.6   Poly(lactide).

Sugarcane (Europe) or maize (United States) can be converted into lactic acid by bacterial fermentation; this is then oligomerised and converted into a cyclic lactide monomer (14.19) that, upon polymerisation, gives poly(lactide) (Figure 14.6). Though not yet commercial, a process has been reported (2010) that produces poly(lactide) in one step from glucose, using engineered *Escherichia coli* that has had genes from certain other bacteria spliced into it. Apart from being biodegradable and compostable, poly(lactide) has many advantages; it has a low toxicity, so it can be used in food packaging and containers, disposable cups for cold drinks, and the lining in cups for hot drinks. Disadvantages include brittleness and a low softening temperature; thus, it cannot be used for hot drinks, though using composite materials can offset those problems.

POLYURETHANE

Urethane linkages are generated when an alcohol reacts with an isocyanate group (Figure 14.7). Using a polyisocyanate and a polyol, a polymer chain can be generated. Superficially, this process resembles the condensation reactions used to make polyamides and polyesters, but an addition reaction is used to make polyurethanes (Figure 14.8). Typical raw materials include polyetherpolyols and 2,4- and 2,6-diisocyanatotoluene.

During the polymerisation process, water can react with some of the isocyanate groups to form CO_2 gas, which can supply the gas to make the polyurethane foam, getting trapped in the polymer as it forms. Blowing agents (formerly CFCs and HCFCs, but nowadays more friendly molecules such as pentane or 1,1,1,2-tetrafluorethane) can also be supplied to achieve this. By varying the factors controlling the polymerisation process, a whole range of polyurethanes with different properties can be obtained.

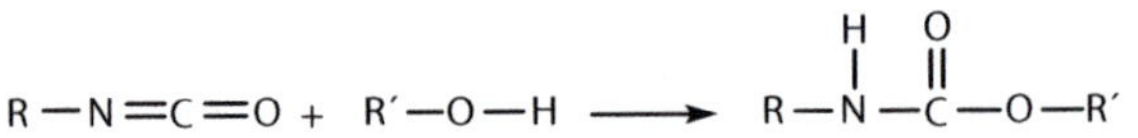

FIGURE 14.7   Formation of urethane group.

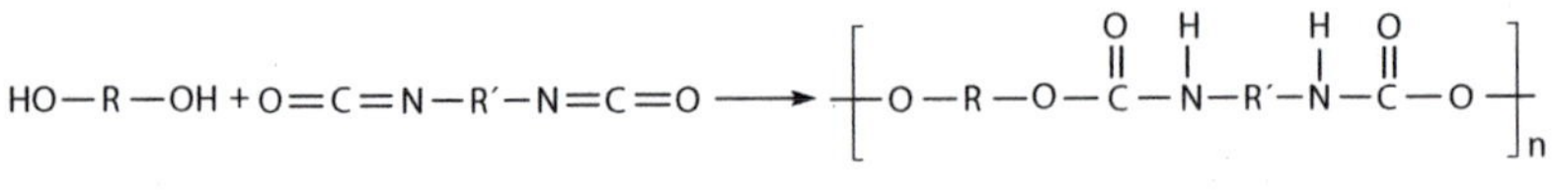

FIGURE 14.8   Formation of a polyurethane group.

Mention "polyurethane" and most people will think of the kind of soft and flexible foam widely used to fill mattresses as well as soft furniture like sofas and car seats. Rigid foams, on the other hand, are employed as thermal insulation in appliances like refrigerators and in buildings, as well as in refrigerated transport. Lightweight composites with rigid polyurethanes supplying the filling can be used in floors, walls, and roofs. Polyurethane elastomers are found in the soles of footwear, and microcellular foams line sports helmets. Polyurethanes are used to make polymer fibres for clothing, and polyurethane adhesives find a wide range and scale of applications.

Bibliography

The General Bibliography includes many books which chemists should consider purchasing for their own library, especially the six listed in the "interesting molecules" section. Many of these books have been referred to in the text on a number of occasions, so condensed references have been used to avoid lengthy repetitions.

GENERAL BIBLIOGRAPHY

BOOKS ABOUT INTERESTING MOLECULES

P. W. Atkins, *Molecules,* New York, W. H. Freeman, 1987; later edition is *Atkins' Molecules,* Cambridge, Cambridge University Press, 2003.

J. Emsley, *Molecules at an Exhibition,* Oxford, Oxford University Press, 1998. (Emsley 1998)

P. Le Couteur and J. Burreson, *Napoleon's Buttons: How 17 Molecules Changed History,* New York, Tarcher, 2003. (Couteur and Burreson 2003)

R. J. Giguere (Ed.), *Molecules That Matter,* Philadelphia, Chemical Heritage Foundation, 2008.

R. L. Myers, *The 100 Most Important Chemical Compounds: A Reference Guide,* Westport, CT, Greenwood Press, 2007. (Myers 2007)

B. Proust, *Petite Geometrie Des Parfums,* Paris, Editions du Seuil, 2006.

This bibliography is primarily concerned with printed resources, but a place of honour should be reserved for the American Chemical Society's "What's That Stuff?" site (http://pubs.acs.org/cen/whatstuff/), which covers many chemicals found in consumer goods.

PERIODIC TABLE

J. Emsley, *Nature's Building Blocks,* Oxford, Oxford University Press, 2001.

CONSUMER CHEMISTRY

J. Emsley, *The Consumer's Good Chemical Guide,* New York, W. H. Freeman, 1994. (Emsley 1994)

J. Emsley, *The Shocking History of Phosphorus,* London, Macmillan, 2000.

J. Emsley, *Vanity, Vitality and Virility, the Science behind the Products You Love to Buy,* Oxford, Oxford University Press, 2004. (Emsley 2004)

J. Emsley, *The Elements of Murder; the History of Poisons,* Oxford, Oxford University Press, 2005.

J. Emsley, *Better Looking, Better Living, Better Loving: How Chemistry Can Help You Achieve Life's Goals,* Weinheim, Germany, Wiley-VCH, 2007. (Emsley 2007)

J. Emsley, *Molecules of Murder: Criminal Molecules and Classic Murders,* Cambridge, Royal Society of Chemistry, 2008.

J. Emsley, *Healthy, Wealthy, Sustainable World,* Cambridge, Royal Society of Chemistry, 2010.

S. Q. Field, *Why There's Antifreeze in Your Toothpaste: The Chemistry of Household Ingredients,* Chicago, Chicago Review Press, 2007. (Field 2007)

K. K. Karukstis and G. R. Van Hecke, *Chemistry Connections: The Chemical Basis of Everyday Phenomena,* San Diego, Academic Press, 2003.

S. Pointer, *The Artifice of Beauty,* Stroud, England, The History Press, 2005. (Pointer 2005)

B. Selinger, *Chemistry in the Market Place,* Sydney, Harcourt Brace, 5th ed., 1998. (Selinger 1998)

R. Winter, *A Consumer's Dictionary of Cosmetic Ingredients,* New York, Crown Press, 1984. (Winter 1984)

MEDICINAL CHEMISTRY AND DRUGS

S. H. Barondes, *Molecules and Mental Illness,* New York, Scientific American Library, 1993.

E. J. Corey, B. Czako, and N. Kürti, *Molecules and Medicine,* Hoboken, NJ, 2007. (Corey, Czako, and Kürti 2007)

J. J. Li, *Laughing Gas, Viagra, and Lipitor: The Human Stories behind the Drugs We Use,* New York, Oxford University Press USA, 2006. (Li 2006)

J. Mann, *The Elusive Magic Bullet,* Oxford, Oxford University Press, 1999; second edition is *Life Saving Drugs: The Elusive Magic Bullet,* Cambridge, Royal Society of Chemistry, 2004.

J. Mann, *Turn On and Tune In. Psychedelics, Narcotics and Euphoriants,* Cambridge, Royal Society of Chemistry, 2009. (Mann 2009)

J. S. Meyer and L. F. Quenzer, *Psychopharmacology: Drugs, the Brain, and Behavior,* Sunderland, MA, Sinauer Associates, 2004.

M. A. Myers, *Happy Accidents: Serendipity in Modern Medical Breakthroughs,* New York, Arcade Publishing, 2007.

K. C. Nicolaou and T. Montagnon, *Molecules That Changed the World,* Weinheim, Germany, Wiley VCH, 2008. (Nicolaou and Montagnon 2008)

G. L. Patrick, *An Introduction to Medicinal Chemistry,* 2nd edition, Oxford, Oxford University Press, 2001. (Patrick 2001)

D. M. Perrine, *The Chemistry of Mind-Altering Drugs,* Washington, DC, American Chemical Society, 1996. (Perrine 1996)

S. H. Snyder, *Drugs and the Brain,* New York, Scientific American Library, 1986.

COOKING

P. Barham, *The Science of Cooking,* Berlin, Springer, 2000.

H. McGee, *On Food and Cooking,* London, Harper Collins, 1986; 2nd (enlarged) ed., 2004.

J. Schwarcz, *An Apple a Day: The Myths, Misconceptions, and Truths about the Foods We Eat,* New York, Other Press, LLC, 2009.

TOXICOLOGY

J. Timbrell, *The Poison Paradox,* Oxford, Oxford University Press, 2005. (Timbrell 2005)

R. W. Waring, G. B. Steventon, and S. C. Mitchell, *Molecules of Death,* London, Imperial College Press, 2nd ed., 2007. (Waring, Steventon, and Mitchell 2007)

CHEMICAL AND BIOLOGICAL WARFARE

G. I. Brown, *The Big Bang,* Stroud, England, Sutton, 1998.

E. Croddy, *Chemical and Biological Warfare,* New York, Copernicus Books, 2002.

R. Harris and J. Paxman, *A Higher Form of Killing*, London, Chatto and Windus, 1982 (new paperback ed., Arrow Books, 2002).

E. M. Spiers, *A History of Chemical and Biological Weapons*, London, Reaktion Books, 2010.

Pheromones and Chemistry in Living Things, and "Drugs from Rain Forests" Theme

W. C. Agosta, *Chemical Communication*, New York, Scientific American Library, 1992.

W. C. Agosta, *Bombardier Beetles and Fever Trees*, New York, Perseus Press, 1995.

W. C. Agosta, *Thieves, Deceivers, and Killers*, Princeton, NJ, Princeton University Press, 2001.

C. Joyce, *Earthly Goods*, Boston, Little Brown, 1994.

M. J. Plotkin, *Medicine Quest*, New York, Viking, 2000.

T. D. Wyatt, *Pheromones and Animal Behaviour: Communication by Smell and Taste*, Cambridge, Cambridge University Press, 2003.

CHAPTER 1

Atmosphere

D. S. Brock and G. J. Schrobilgen, *Journal of the American Chemical Society*, 2011, 133, 6265–6569. (XeO_2)

S. Cagin and P. Dray, *Between Earth and Sky*, New York, Pantheon Books, 2003.

D. Fisher, *Much Ado about (Practically) Nothing: A History of the Noble Gases*, Oxford, Oxford University Press, 2010.

J. Jackson, *A World on Fire*, New York, Viking, 2005.

N. Lane, *Oxygen. The Molecule That Made the World*, Oxford, Oxford University Press, 2002.

L. F. Lundegaard, G. Weck, M. I. McMahon, S. Desgreniers, and P. Loubeyre, *Nature*, 2006, 443, 201–204. (O_8)

P. J. Michaels and R. C. Balling, *The Satanic Gases: Clearing the Air about Global Warming*, Washington, DC, Cato Institute, 2000.

E. A. Parson, *Protecting the Ozone Layer*, Oxford, Oxford University Press, 2003.

P. Pyykkö, *Science*, 2000, 290, 64–65 ("Noblesse Oblige")

E. Roston, *The Carbon Age: How Life's Core Element Has Become Civilization's Greatest Threat*, New York, Walker & Company, 2008.

C. W. Seibel, *Helium: Child of the Sun*, Lawrence, University Press of Kansas, 1968.

T. Volk, *CO2 Rising*, Cambridge, MA, MIT Press, 2008.

G. Walker, *An Ocean of Air*, London, Bloomsbury, 2007.

R. P. Wayne, *Chemistry of Atmospheres*, 2nd ed., Oxford, Oxford University Press, 1991.

Water and Hydrogen Peroxide

P. Ball, *H2O. A Biography of Water*, London, Weidenfeld & Nicolson, 1999.

P. Coffey, *Cathedrals of Science*, Oxford, Oxford University Press, 2008, pp. 208–223 (D_2O).

P. F. Dahl, *Heavy Water and the Wartime Race for Nuclear Energy*, Bristol, Institute of Physics Publishing, 1999.

T. Eisner, *For Love of Insects*, Cambridge, MA, Belknap Press, 2003, pp. 9–43. (bombardier beetle)

T. Eisner, M. Eisner, and M. Siegler, *Secret Weapons: Defenses of Insects, Spiders, Scorpions, and Other Many-Legged Creatures*, Cambridge, MA, Belknap Press, 2005, pp. 157–162. (bombardier beetle)

F. Franks, *Polywater,* Cambridge, MA, MIT Press, 1981.

F. Franks, *Water,* London, Royal Society of Chemistry, 1983.

N. N. Greenwood and A. Earnshaw, *Chemistry of the Elements,* 2nd ed., London, Butterworth-Heinemann, 1997, p. 633 ff. (H_2O_2)

B. Johnson, *The Secret War,* London, BBC, 1978, pp. 278–282. (H_2O_2 and weaponry)

R. Mears, *The Real Heroes of Telemark,* London, Hodder and Stoughton, 2003. (D_2O)

R. Moore, *A Time to Die: The Kursk Disaster,* London, Doubleday, 2002. (H_2O_2)

S. Solomon, *Water,* New York, Harper Collins, 2010.

A. K. Soper and C. J. Benmore, Quantum Differences between Heavy and Light Water, *Physical Review Letters,* 2008, 101, 065502, 1–4.

P. Truscott, *Kursk: Russia's Lost Pride,* London, Simon & Schuster, 2002. (H_2O_2)

R. Wright, *Take Me to the Source. In Search of Water,* London, Harvill Secker, 2008.

Activation energies for the decomposition of H_2O_2 adapted from E. A. Moelwyn-Hughes, *The Kinetics of Reactions in Solution,* 2nd ed., Oxford, Oxford University Press, 1947, p. 299 and J. G. Stark and H. G. Wallace, *Chemistry Data Book,* London, John Murray, 1975, p. 85.

CHAPTER 2

R. Cohen, *Sweet and Low,* New York, Farrar, Straus and Giroux, 2006.

E. Davies, *Chemistry World,* June 2010, 46–49. (artificial sweeteners)

C. de la Pena, *Empty Pleasures: The Story of Artificial Sweeteners from Saccharin to Splenda,* Chapel Hill, University of North Carolina Press, 2010.

W. P. Edwards, *The Science of Sugar Confectionery,* Cambridge, Royal Society of Chemistry, 2000.

S. C. Eggers, T. E. Acree, and R. S. Shallenberger, *Food Chemistry,* 2000, 68, 45–49. (sweetness theory)

L. Hough and S. P. Phadnis, *Nature,* 1976, 263, 800. (sucralose discovery)

S. W. Mintz, *Sweetness and Power: Place of Sugar in Modern History,* New York, Viking, 1985.

S. O'Connell, *Sugar: The Grass That Changed the World,* London, Virgin Books, 2004.

A. B. Richards, S. Krakowka, L. B. Dexter, H. Schmid, A. P. M. Wolterbeek, D. H. Waalkens-Berendsen, A. Shigoyuki, and M. Kurimoto, *Food Chemistry & Toxicology,* 2002, 40, 871–898. (trehalose review)

Z. I. Sabry and N. A. Atallah, *Nature,* 1961, 190, 915–916. (manna of Northern Iraq)

R. V. Stick, *Carbohydrates: The Sweet Molecules of Life,* San Diego, Academic Press, 2001.

N. Teramoto, N. D. Sachinvala, and M. Shibata, *Molecules,* 2008, 13, 1773–1816. (trehalose)

CHAPTER 3

A-G. Bagneres, M. C. Lorenzi, G. Dusticier, S. Turillazzi, and J.-L. Clément, *Science,* 1996, 272, 889–892. (cuticular hydrocarbons)

A. T. Balaban, P. v. R. Schleyer, and H. S. Rzepa, *Chemical Reviews,* 2005, 105, 3436–3447. (benzene and aromaticity)

E. Breitmeier, *Terpenes: Flavor, Fragrances, Pharmaca, Pheromones,* Weinheim, Germany, Wiley-VCH, 2006.

W. J. Broad, *The Oracle: The Lost Secrets and Hidden Message of Ancient Delphi,* Penguin Press, 2006. (ethene)

C. Callow, *Power from the Sea,* London, Victor Gollancz, 1973. (natural gas)

J. Darley, *High Noon for Natural Gas: The New Energy Crisis,* Vermont, Chelsea Green Publishing Company, 2004.

K. S. Deffeyes, *Hubbert's Peak,* Princeton, NJ, Princeton University Press, 2001.

K. S. Deffeyes, *Beyond Oil,* New York, Hill and Wang, 2005.

T. Gold, *The Deep Hot Biosphere,* New York, Springer, 1999.

J. R. Hale, J. Z. de Boer, J. P. Chanton, and H. A. Spiller, *Scientific American,* August 2003, 289, 66–73 (ethene and the Delphic oracle); for a contrary view, see J. Foster and D. Lehoux, *Clinical Toxicology,* 2007, 45, 85–89.

Methane Generation from Human, Animal, and Agricultural Wastes, New York, National Academy of Sciences, 1981.

C. M. Noe and A. Bader, *Chemistry in Britain,* Feb 1993, 126–128. (Loschmidt, Kekulé, and benzene)

F. Paulot, J. D. Crounse, H. G. Kjaergaard, A. Kürten, J. M. St. Clair, J. H. Seinfeld, and P. O. Wennberg, *Science,* 2009, 325, 730–733. (isoprene in the atmosphere)

C. F. Rawnsley and R. Wright, *Night Fighter,* London, Elmfield Press, 1976. (and earlier editions)

P. Roberts, *The End of Oil,* London, Bloomsbury, 2004.

A. J. Rocke, *Image and Reality: Kekule, Kopp, and the Scientific Imagination,* Chicago, University of Chicago Press, 2010.

C. S. Sell, *A Fragrant Introduction to Terpenoid Chemistry,* Cambridge, Royal Society of Chemistry, 2003.

S. Shah, *Crude: The Story of Oil,* New York, Seven Stories Press, 2004.

W. J. Wiswesser, *Aldrichim Acta,* 1989, 22, 17–19. (Loschmidt, Kekulé, and benzene)

M. Yeomans, *Oil: Anatomy of Industry,* New York, New Press, 2004.

CHAPTER 4

D. Asthagiri, L. R. Pratt, J. D. Kress, and M. A. Gomez, *Proceedings of National Academy of Sciences,* 2004, 101, 7229–7233. (hydration of OH⁻(aq))

D. Charles, *Between Genius and Genocide: The Tragedy of Fritz Haber, Father of Chemical Warfare,* London, Jonathan Cape, 2005. (ammonia)

L. K. Gardner and G. D. Lawrence, *Journal of Agriculture & Food Chemistry,* 1993, 41, 693–695. (benzene from benzoic acid)

T. Hager, *The Alchemy of Air: A Jewish Genius, a Doomed Tycoon, and the Scientific Discovery That Fed the World but Fueled the Rise of Hitler,* New York, Harmony, 2008. (ammonia)

R. Hazan, A. Levine, and H. Abeliovich, *Applied & Environmental Microbiology,* 2004, 70, 4449–4457. (benzoic acid as food preservative)

B. G. J. Knols, J. J. van Loon, A. Cork, R. D. Robinson, W. Adam, J. Meijerink, R. de Jong, and W. Takken, *Bulletin of Entomological Research,* 1997, 87, 151–159. (butyric acid as a mosquito attractant)

G. J. Leigh, *The World's Greatest Fix; a History of Nitrogen and Agriculture,* Oxford, Oxford University Press, 2004. (ammonia)

(Myers, 2007), pp. 1–3 (acetic acid); 27–29 (ammonia); 40–42 (benzoic acid); 52–54 (butyric acid and other RCOOH); 62–64 (calcium oxide); 132–133 (formic acid); 141–143 (hydrochloric acid); 194–196 (nitric acid); 257–259 (sodium hydroxide); 271–274 (sulfuric acid).

O. Poizat and G. Buntinx, *Molecules,* 2010, 15, 3366–3377. (OH⁻ ion in solution)

C. A. Reed, *Chemical Communications,* 2005, 1669–1677. (carborane acids and hydronium ions)

R. A. Robergs, F. Ghiasvand, and D. Parker, *American Journal of Physiology—Regulatory Integrative and Comparative Physiology,* 2004, 287, R502–R516. (lactic acid)

E. S. Stoyanov, I. V. Stoyanova, and C. A. Reed, *Journal of the American Chemical Society,* 2010, 132, 1484–1485; *Chemical Science,* 2011, 2, 462–472. (hydrogen ion in water)

K.-J. Tielrooij, R. Timmer, H. Bakker, and M. Bonn, *Physics Review Letters,* 2009, 102, 198303, 1–4. (structure dynamics of the proton in liquid water)

CHAPTER 5

R. Aguilera, C. K. Hatton, and D. H. Catlin, *Clinical Chemistry,* 2002, 48, 629–636. (epitestosterone test)

E. Baulieu, *Science,* 1989, 245, 1351–1357. (RU-486)

J. M. Berg, J. L. Tymoczko, and L. Stryer, *Biochemistry,* 5th ed., New York, W. H. Freeman, 2002, pp. 722–726.

M. S. Brown and J. L. Goldstein, *Cell,* 1997, 89, 331–340. (cholesterol metabolism)

F. Cadepond, A. Ulmann, and E.-E. Baulieu, *Annual Review of Medicine,* 1997, 48, 129–156. (RU-486)

D. H. Catlin, B. D. Ahrens, and Y. Kucherova, *Rapid Communications in Mass Spectrometry,* 2002, 16, 1273–1275. (norbolethone, discovery, synthesis, detection)

D. H. Catlin, M. H. Sekera, B. D. Ahrens, B. Starcevic, Y-C. Chang, and C. K. Hatton, *Rapid Communications in Mass Spectrometry,* 2004, 18, 1245–1249. (THG, synthesis, MS, detection)

S. A. Cotton, *Education in Chemistry,* 2010, 47, 44–48. (steroid abuse in sport)

K. P. de Jésus-Tran, P.-L. Côté, L. Cantin, J. Blanchet, F. Labrie, and R. Breton, *Protein Science,* 2006, 15, 987–999. (structures of androgen receptor with THG and other agonists).

C. Djerassi, *Steroids Made It Possible,* Washington, DC, American Chemical Society, 1990.

M. Fainaru-Wada and L. Williams, *Game of Shadows: Barry Bonds, Balco, and the Steroids Scandal That Rocked Professional Sports,* New York, Gotham Books, 2006.

W. W. Francke and B. Berendonk, *Clinical Chemistry,* 1997, 43, 1262–1279. (East German steroid abuse)

D. Ghosh, J. Griswold, M. Erman, and W. Pangborn, *Nature,* 2009, 457, 219–223. (structure of aromatase in complex with androstenedione)

J. L. Goldstein and M. S. Brown, *Nature,* 1990, 343, 425–430. (cholesterol biosynthesis)

A. Gringauz, *Introduction to Medicinal Chemistry: How Drugs Act and Why,* 2nd ed., New York, Wiley-Blackwell, 1996, pp. 665–681.

D. Lednicer, *Steroid Chemistry at a Glance,* Chichester, England, Wiley-Blackwell, 2010.

L. Marks, *Sexual Chemistry: A History of the Contraceptive Pill,* New Haven: Yale University Press, 2001.

J. M. Riddle, *Eve's Herbs: A History of Contraception and Abortion in the West,* Cambridge, MA, Harvard University Press, 1997, pp. 44–46.

W. Schänzer and M. Donike, *Analytica Chimica Acta,* 1993, 275, 23–48. (use of Me_3Si derivatives in identifying steroids by GC-MS)

M. H. Sekera, B. D. Ahrens, Y-C. Chang, B. Starcevic, and D. H. Catlin, *Rapid Communications in Mass Spectrometry,* 2005, 19, 781–784. ("Madol", discovery, synthesis, and detection in urine)

L. Silvestre, C. Dubois, M. Renault, Y. Rezvani, E. E. Baulieu, and A. Ulmann, *New England Journal of Medicine,* 1990, 322, 645–648. (RU-486)

S. Ungerleider, *Faust's Gold: Inside the East German Doping Machine,* New York, Thomas Dunne Books, 2001.

T. C. Werner and C. K. Hatton, *Journal of Chemical Education,* 2011, 88, 34–40. (drug testing in sport)

OXYTOCIN

T. Baumgartner, M. Heinrichs, A. Vonlanthen, U. Fischbacher, and E. Fehr, *Neuron,* 2008, 58, 639–650. (oxytocin and trust)

G. Froböse and R. Froböse, *Lust and Love: Is It More Than Chemistry?* Cambridge, Royal Society of Chemistry, 2006, pp. 58, 68–77. (oxytocin)

M. Gamera, B. Zurowskia, and C. Büchel, *Proceedings of National Academy of Sciences,* 2010, 107, 9400–9405. (oxytocin and the eye region)

I. Gordon, O. Zagoory-Sharon, J. F. Leckman, and R. Feldman, *Biological Psychiatry,* 2010, 68, 377–382. (oxytocin and the development of parenting)

A. G. Ophir, S. M. Phelps, A. B. Sorin, and J. O. Wolff, *Animal Behavior,* 2008, 75, 1143–1154. (oxytocin and monogamy)

NO AND VIAGRA

A. Butler and R. Nicholson, *Life, Death, and Nitric Oxide,* Cambridge, Royal Society of Chemistry, 2003.

L. Katzenstein, *Viagra: The Remarkable Story of the Discovery and Launch,* New York, Medical Information Press, 2001.

J. J. Li, *Laughing Gas, Viagra, and Lipitor: The Human Stories behind the Drugs We Use,* Oxford, Oxford University Press, 2006, pp. 109–113.

M. Loe, *The Rise of Viagra: How the Little Blue Pill Changed Sex in America,* New York, New York University Press, 2004.

J. Rajfer, W. J. Aronson, P. A. Bush, F. J. Dorey, and L. J. Ignarro, *New England Journal of Medicine,* 1992, 326, 90–94. (NO as a mediator of relaxation of the corpus cavernosum)

CHAPTER 6

VISION AND COLOUR

M. Archetti, T. F. Döring, S. B. Hagen, N. M. Hughes, S. R. Leather, D. W. Lee, S. Lev-Yadun, Y. Manetas, H. J. Ougham, P. G. Schaberg, and H. Thomas, *Trends in Ecology and Evolution,* 2009, 24, 166–173. (leaf colour change)

P. H. M. Bovee-Geurts, I. F. Fernandez, R. S. H. Liu, R. A. Mathies, J. Lugtenburg, and W. J. DeGrip, *Journal of the American Chemical Society,* 2009, 131, 17933–17942. (rhodopsin activation)

M. Ding, R. Feng, S. Y. Wang, L. Bowman, Y. Lu, Y. Qian, V. Castranova, B.-H. Jiang, and X. Shi, *Journal of Biological Chemistry,* 2006, 281, 17359–17368. (chemopreventive and chemotherapeutic activity of cyanidin-3-glucoside)

T. S. Field, D. W. Lee, and N. M. Holbrook, *Plant Physiology,* 2001, 127, 566–574. (leaf colour change)

J. Golley, *John "Cat's-Eyes" Cunningham: The Aviation Legend,* Shrewsbury, England, Airlife Press, 1999.

K. S. Gould and D. W. Lee, *American Scientist,* 2002, 90, 524–531. (leaf colour change)

K. Palczewski, T. Kumasaka, T. Hori, C. A. Behnke, H. Motoshima, B. A. Fox, I. Le Trong, D. C. Teller, T. Okada, R. E. Stenkamp, M. Yamamoto, and M. Miyano, *Science,* 2000, 289, 739–745. (rhodopsin crystal structure)

J. H. Park, P. Scheerer, K. P. Hofmann, H.-W. Choe, and O. P. Ernst, *Nature,* 2008, 454, 183–187. (crystal structure of the ligand-free G-protein-coupled receptor opsin)

C. F. Rawnsley and R. Wright, *Night Fighter,* Morley, England, Elmfield Press, 1976 (and earlier editions)

K. Rodahl and T. Moore, *Biochemical Journal,* 1943, 37, 166–168. (bear and seal liver toxicity)

R. E. Stenkamp, D. C. Teller, and K. Palczewski, *ChemBioChem,* 2002, 3, 963–967. (rhodopsin review)

K. Yoshida, M. Mori, and T. Kondo, *Natural Product Reports,* 2009, 26, 884–915. (anthocyanins and flower colour)

Taste

T. E. Acree, R. S. Shallenberger, and S. Ebeling, *Developments in Food Science*, 1998, 40, 1–13. (the AH–B theory)

J. Ahmed, S. Preissner, M. Dunkel, C. L. Worth, A. Eckert, and R. Preissner, *Nucleic Acids Research*, 2011, 39 (database issue) D377–D382. (sweetening agents)

F. M. Assadi-Porter, M. Tonelli, E. L. Maillet, J. L. Markley, and M. Max, *Biochimica et Biophysica Acta*, 2010, 1798, 82–86. (NMR study of T1R2-T1R3 receptor and sweeteners)

J. Chandrashekar, M. A. Hoon, N. J. P. Ryba, and C. S. Zuker, *Nature*, 2006, 444, 288–294. (taste receptors)

B. Lindemann, *Nature*, 2001, 413, 219–225. (taste receptors)

J. J. López Cascales, S. D. Oliveira Costa, B. L. de Groot, and D. E. Walters, *Biophysical Chemistry*, 2010, 152, 139–144. (binding of glutamate to the umami receptor)

B. Meyers and M. S. Brewer, *Journal of Food Science*, 2008, 73, R81–R90. (sweet taste review)

G. Morini, A. Bassoli, and P. A. Temussi, *Journal of Medicinal Chemistry*, 2005, 48, 5520–5529. (binding sites of the human T1R2/T1R3 receptor)

E. R. D. C. Neta, S. D. Johanningsmeier, and R. McFeeters, *Journal of Food Science*, 2007, 72, R33–R38. (sour taste review)

C. Nofre and J. M. Tinti, *Food Chemistry*, 1996, 56, 263–274. (sweetness theory)

R. K. Palmer, *Molecular Interventions*, 2007, 7, 87–98. (bitter, sweet, and umami taste sensing)

D. V. Smith and R. F. Margolskee, *Scientific American*, March 2001, 26–33. (taste receptors)

P. Temussi, *Advances in Food Nutrition Research*, 2007, 53, 199–239. (sweet taste receptor)

G. Q. Zhao, Y. Zhang, M. A. Hoon, J. Chandrashekar, I. Erlenbach, N. J. P. Ryba, and C. S. Zuker, *Cell*, 2003, 115, 255–266. (sweet and umami receptors)

Odour

J. M. Berg, J. L. Tymoczko, and L. Stryer, *Biochemistry*, 5th ed., New York, W. H. Freeman, 2002, pp. 897–903.

L. B. Buck, *Angewandte Chemie*, international ed., England, 2005, 44, 6128–6140. (Nobel lecture)

L. B. Buck and R. Axel, *Cell*, 1991, 65, 175–187. (genes coding for ORs)

P. Kraft and A. Mannschreck, *Journal of Chemical Education*, 2010, 87, 598–603. (enantioselectivity of odour sensation)

A. B. Malnic, J. Hirono, T. Sato, and L. B. Buck, *Cell*, 1999, 96, 713–723. (combinatorial receptor codes).

C. S. Sell, *Chemistry & Biodiversity*, 2004, 1, 1899–1920. (scent and chirality)

C. S. Sell, *Chemistry in Britain*, March 1997, 39–42; *Angewandte Chemie, International Ed.*, 2006, 45, 6254–6261.

D. A. Wilson and R. J. Stevenson, *Learning to Smell*, Baltimore, MD, John Hopkins University Press, 2006.

F. Xu, N. Liu, I. Kida, D. L. Rothman, F. Hyder, and G. M. Shepherd, *Proceedings of National Academy of Sciences*, 2003, 100, 11029–11034. (aldehydes and olfaction in mice)

Pyrazines

S. Al Abassi, M. A. Birkett, J. Pettersson, J. A. Pickett, and C. M. Woodcock, *Cellular and Molecular Life Science*, 1998, 54, 876–879. (2-isopropyl-3-methoxy-pyrazine as ladybird sexual attractant)

M. S. Allen, M. J. Lacey, and S. Boyd, *Journal of Agricultural and Food Chemistry,* 1995, 43, 769–772. (methoxypyrazines in red wines)

L. Cai, J. A. Koziel, and M. E. O'Neal, *Journal of Chromatography,* 2007, 1147, 66–78. (odourants in Harmonia axyridis)

T. P. Coultate, *Food: The Chemistry of Its Components,* 3rd ed., Cambridge, Royal Society of Chemistry, 1996, pp. 21–26.

D. S. Mottram, in T. H. Parliment, M. J. Morello, and R. J. McGorrin, eds., *Thermally Generated Flavors. Maillard, Microwave, and Extrusion Processes,* ACS Symposium Series 543, 1993, pp. 104–126.

R. Müller and S. Rappert, *Applied Microbiology and Biotechnology,* 2010, 85, 1315–1320. (rev. of pyrazines)

K. E. Murray and F. B. Whitfield, *Journal of Science of Food and Agriculture,* 1975, 26, 973–986. (3-alkyl-2-methoxypyrazines in vegetables)

T. Nawrath, J. S. Dickschat, B. Kunze, and S. Schulz, *Chemistry & Biodiversity,* 2010, 7, 2129–2144 and references therein. (biosynth. of pyrazines)

W. V. Parr, J. A. Green, K. G. White, and R. R. Sherlock, *Food Quality and Preference,* 2007, 18, 849–861. (methoxypyrazines in wines)

R. M. Seifert, R. G. Buttery, D. G. Guadagni, D. R. Black, and J. G. Harris, *Journal of Agricultural and Food Chemistry,* 1970, 18, 246–249. (2-methoxy-3-alkylpyrazines with bell pepper-like odours)

R. Wagner, M. Czerny, J. Bielohradsky, and W. Grosch, *Zeitschrift für Lebensmittel-Untersuchung und -Forschung, A,* 1999, 208, 308–316. (odour thresholds of 80 alkylpyrazines)

J. Weiss, M. Pyrski, E. Jacobi, B. Bufe, V. Willnecker, B. Schick, P. Zizzari, S. J. Gossage, C. A. Greer, T. Leinders-Zufall, et al. *Nature,* 2011, 472, 186–190. (mutations in sodium channel and anosmia)

ANIMAL PHEROMONES

W. C. Agosta, *Chemical Communication,* Scientific American Library, New York, W. H. Freeman, 1993.

A.-G. Bagnères, M. C. Lorenzi, G. Dusticier, S. Turillazzi, and J-L. Clément, *Science,* 1996, 272, 889–892. (parasitic wasp)

D. A. Carlson, M. S. Mayer, D. L. Silhacek, J. D. James, M. Beroza, and B. A. Bierl, *Science,* 1971, 174, 76–78. (housefly pheromone)

D.-H. Choe, J. G. Millar, and M. K. Rust, *Proceedings of National Academy of Sciences,* 2009, 106, 8251–8255. (Argentine ant cues)

S. A. Cotton, *Education in Chemistry,* 2009, 46, 148–152. (bombykol)

G. Dutheuil, M. P. Webster, P. A. Worthington, and V. K. Aggarwal, *Angewandte Chemie, International Ed.,* 2009, 48, 6317–6319. (total synthesis of (+)-faranal)

T. Eisner and J. Meinwald, *Chemical Ecology: The Chemistry of Biotic Interaction,* Washington, DC, National Academy Press, 1995.

T. Eltz, D. W. Roubik, and W. M. Whitten, *Physiological Entomology,* 2003, 28, 251–260. (orchid bees)

T. Eltz, Y. Zimmermann, J. Haftmann, R. Twele, W. Francke, J. J. G. Quezada-Euan, and K. Lunau, *Proceedings of Royal Society B,* 2007, 274, 2843–2848. (orchid bees)

J. B. Free, *Pheromones of Social Bees,* London, Chapman & Hall, 1987.

B. Hölldobler and E. O. Wilson, *The Ants,* Cambridge, MA, Belknap Press, 1990.

C. I. Keeling, K. N. Slessor, H. A. Higo, and M. L. Winston, *Proceedings of National Academy of Sciences,* 2003, 100, 4486–4491. (queen bee)

M. Kobayashi, T. Koyama, K. Ogura, S. Seto, F. J. Ritter, and J. E. M. Brüggemann-Rotgans, *Journal of the American Chemical Society,* 1980, 102, 6602–6604. (faranal)

C. Lautenschlager, W. S. Leal, and J. Clardy, *Structure,* 2007, 15, 1148–1154. (bombykol)

E. D. Morgan, *Biosynthesis in Insects,* Cambridge, Royal Society of Chemistry, 2004.

J. C. Moser, R. C. Brownlee, and R. M. Silverstein, *Journal of Insect Physiology,* 1968, 14, 529–535. (*Atta texana* alarm pheromones)

R. G. Riley and R. M. Silverstein, *Tetrahedron,* 1974, 30, 1171–1174. (*Atta texana* alarm pheromones)

P. K. Visscher, *Animal Behavior,* 1983, 31, 1070–1076. (bee posthumous pheromone)

T. D. Wyatt, *Pheromones and Animal Behaviour: Communication by Smell and Taste,* Cambridge, Cambridge University Press, 2003.

HUMAN PHEROMONES

R. L. Doty, *The Great Pheromone Myth,* Baltimore, MD, Johns Hopkins University Press, 2010.

M. K. McClintock, *Nature,* 1971, 229, 244–245. (menstrual synchrony)

K. Stern and M. K. McClintock, *Nature,* 1998, 392, 177–179. (menstrual synchrony)

B. I. Strassmann, *Human Reproduction,* 1999, 14, 579–580. (nonsynchrony)

H. C. Wilson, *Psychoneuroendocrinology,* 1992, 17, 565–591. (critical review of menstrual synchrony)

T. D. Wyatt, *Pheromones and Animal Behaviour: Communication by Smell and Taste,* Cambridge: Cambridge University Press, 2003.

Z. Yang and J. C. Schank, *Human Nature,* 2006, 17, 434–447. (nonsynchrony).

CHAPTER 7

SHAMPOO

(Winter 1984), pp. 232–233; (Emsley 1998), pp. 90–93; (Emsley 2007), p. 74; (Field 2007), pp. 199–222. (surfactants and conditioners)

HAIR COLOUR AND WAVING

(Emsley 2007), pp. 10–15.
(Selinger 1998), pp. 165–166.

NAIL VARNISH

(Emsley 1998), pp. 121–122.
(Emsley, 2004), p. 16.
(Emsley 2007), pp. 31–35.
(Field, 2007), p. 70.
(Winter 1984), pp. 179–180.

FACE POWDER

P. Begoun, *Don't Go to the Cosmetics Counter without Me,* 7th ed., Washington, DC, Beginning Press, 2008, pp. 63–64.
(Pointer 2005), pp. 136, 180, and 182.
(Winter 1984), pp. 214–215.

Lipstick

(Emsley 2004), pp. 2–6.
D. N. Kayser, A. J. Elliott, and R. Feltman, *European Journal of Social Psychology*, 2010, 40, 901–908. (red and attractiveness)
J. Pallingston, *Lipstick: A Celebration of a Girl's Best Friend*, London, Simon & Schuster, 1999.
(Pointer 2005), especially pp. 148, 172.
M. C. Ragas and K. Kozlowski, *Read My Lips*, San Francisco, Chronicle Books, 1998.
(Selinger 1998), pp. 126–127.
I. D. Stephen, V. Coetzee, M. L. Smith, and D. I. Perrett, *PLoS ONE*, 2009, 4, e5083, pp. 1–7. (skin colour and health)
(Winter 1984), pp. 162–163.

Mascara

(Pointer 2005), pp. 12, 191–192. (kohl)
(Winter 1984), pp. 166–167.

Sunscreens and Tanning

C. Antoniou, M. G. Kosmadaki, A. J. Stratigos, and A. D. Katsambas, *Journal of European Academy of Dermatology and Venereology*, 2008, 22, 1110–1119.
R. Baines and R. Sayer, *Chemistry Review*, 2010, 19 (3), 19–22.
R. Baxter, *ChemMatters*, April 1998, 16, pp. 4–6.
W. A. Bruls, H. Slaper, J. C. van der Leun, and L. Berrens, *Photochemistry and Photobiology*, 1984, 40, 485–494. (transmission of human epidermis and stratum corneum)
M. Burke, *Chemistry World*, October 2004, pp. 48–53.
M. Defranceschi, *La Chimie au Quotidien*, Paris, Ellipses, 2006, pp. 12–21.
(Emsley 2004), pp. 17–27.
(Field 2007), pp. 10–12.
D. R. Kimbrough, *Journal of Chemical Education*, 1997, 74, 51–53. (photochemistry)
(Selinger 1998), pp. 131–139.
(Winter 1984), pp. 257–258.

Toothpaste

(Emsley 2007), pp. 25–31.
(Field 2007), pp. 239–244.
(Selinger 1998), pp. 127–129. (abrasives in toothpaste)

Bad Breath

R. Baxter, *ChemMatters*, Dec. 1996, pp. 6–8. (mouthwash)
L. S. Chiu, *When a Gene Makes You Smell Like a Fish: And Other Tales about the Genes in Your Body*, Oxford, Oxford University Press, 2006, pp. 4–11.
M. Greenberg, P. Urnezis, and M. Tian, *Journal of Agricultural and Food Chemistry*, 2007, 55, 9465–9469. (magnolol and honokiol)
S. C. Mitchell, *Oral Diseases*, 2005, 11 (Suppl 1), 10–13. (fish-odour syndrome and oral malodour)
H. U. Rehman, *Postgraduate Medical Journal*, 1999, 75, 451–452. (fish odour syndrome)

DEODORANTS

(Emsley 1994), pp. 23–25.

(Emsley 2007), pp. 75–81.

R. Emter and A. Natsch, *Journal of Biological Chemistry,* 2008, 283, 20645–20652. (3-methyl-3-sulfanylhexan-1-ol)

A. Natsch, S. Derrer, F. Flachsmann, and J. Schmid, *Chemistry & Biodiversity,* 2006, 3, 1–20. (volatile carboxylic acids from bodily secretions)

(Selinger 1998), pp. 119–121.

A. I. Spielman, X. N. Zeng, J. J. Leyden, and G. Preti, *Experientia,* 1995, 51, 40–47 and references therein. (axillary odour)

(Winter 1984). pp. 89–90. (for sweat)

PERFUMES

(Emsley 1994), pp. 1–23.

A.-D. Fortineau, *Journal of Chemical Education,* 2004, 81, 45–50. (chemistry perfumes your daily life)

D. J. Rowe, *Chemistry and Technology of Flavours and Fragrances,* Oxford, Blackwell and CRC Press, 2004.

C. Sell, ed., *The Chemistry of Fragrances: From Perfumer to Consumer,* 2nd ed., Cambridge, Royal Society of Chemistry, 2006.

L. Turin, *The Secret of Scent,* London, Faber and Faber, 2006.

L. Turin and T. Sanchez, *Perfumes: The Guide,* London, Profile Books, 2008.

ROSES

G. Ohloff, *Perfumer & Flavorist,* 1978, 3 (1), 11–22. (rose oil review)

G. Scalliet, F. Piola, C. J. Douady, S. Réty, O. Raymond, S. Baudino, K. Bordji, M. Bendahmane, C. Dumas, J. M. Cock, and P. Hugueney, *Proceedings of National Academy of Sciences,* 2008, 105, 5927–5932. (tea roses)

T. Yamamoto, H. Matsuda, Y. Utsumi, T. Hagiwara, and T. Kanisawa, *Tetrahedron Letters,* 2002, 43, 9077–9080. (optically active rose oxide)

MUSKS

D. P. Anonis, *Perfumer and Flavorist,* 1997, 22 (1), 43–47. (civetone)

M. G. J. Beets, in *Fragrance Chemistry,* ed., E. T. Theimer, New York, Academic Press, 1982, pp. 77–122.

C. Fehr, N. Chaptal-Gradoz, and J. Galindo, *Chemistry–European Journal,* 2002, 8, 853–858. (Vulcanolide)

S. Fujimoto, K. Yoshikawa, M. Itoh, and T. Kitaharai, *Bioscience, Biotechnology, and Biochemistry,* 2002, 66, 1389–1392. (muscone enantiomers)

German patent 47,599 in 1888; also see A. Baur, *Berichte der Deutschen Chemischen Gesellschaft,* 1891, 24, 2832.

V. P. Kamat, H. Hagiwara, T. Katsumi, T. Hoshi, T. Suzuki, and M. Ando, *Tetrahedron,* 2000, 56, 4397–4403. (muscone)

P. Kraft, Aroma Chemicals IV: Musks, Chapter 7 of D. J. Rowe, ed., *Chemistry and Technology of Flavors and Fragrances,* Oxford, Blackwell and CRC Press, 2004, pp. 143–168.

B. D. Mookherjee and R. A. Wilson, in *Fragrance Chemistry*, ed. E. T. Theimer, New York, Academic Press, 1982, pp. 433–494.

L. Ruzicka et al., *Helvetica Chimica Acta*, 1926, 9, 249–264; 715–729; 1008–1017.

VANILLIN

T. Ecott, *Vanilla: Travels in Search of the Ice Cream Orchid*, New York, Grove/Atlantic, 2004.

L. J. Esposito, K. Formanek, G. Kientz, F. Mauger, V. Maureaux, G. Robert, and F. Truchet, Vanillin, *Kirk-Othmer Encyclopedia of Chemical Technology*, 4th ed., 1997, New York, John Wiley & Sons, vol. 24, pp. 812–825 (see also vol. 11, pp. 20–21 for odour thresholds)

M. B. Hocking, *Journal of Chemical Education*, 1997, 74, 1055–1059. (synth.)

R. Hoffmann, *American Scientist*, 1997, 85, 314–317. (fraudulent vanillin)

O. Negishi, K. Sugiura, and Y. Negishi, *Journal of Agricultural and Food Chemistry*, 2009, 57, 9956–9961. (biosynth.)

P. Rain, *Vanilla: The Cultural History of the World's Favorite Flavor and Fragrance*, New York, Jeremy P. Tarcher, 2004.

A. S. Ranadive, in G. Charalambous, ed., *Spices, Herbs and Edible Fungi (Developments in Food Science)*, Amsterdam, Elsevier, 1994. pp. 517–577.

G. S. Remaud, Y.-L. Martin, G. G. Martin, and G. J. Martin, *Journal of Agricultural and Food Chemistry*, 1997, 45, 859–866. (SNIF-NMR)

C. Sell, ed., *The Chemistry of Fragrances: From Perfumer to Consumer*, Cambridge, Royal Society of Chemistry, 2nd ed., 2006, pp. 44–45, 116–118.

E. J. Tenailleau, P. Lancelin, R. J. Robins, and S. Akoka, *Journal of Agricultural and Food Chemistry*, 2004, 52, 7782–7787. (^{13}C NMR and vanillin analysis)

N. J. Walton, M. J. Mayer, and A. Narbad, *Phytochemistry*, 2003, 63, 505–515. (rev., biosynth)

ALDEHYDES

J. E. Cometto-Muñiz and M. H. Abraham, *Chemical Senses*, 2010, 35, 289. (aldehyde odour thresholds)

G. Darzens, *Comptes Rendus Hebdomadaires des séances de l'Academie des Sciences*, 1904, 139, 1214–1217. (12-MNA synthesis)

D. Enders and H. Dyker, *Liebigs Annalen der Chemie*, 1990, 1107–1110. (enantiomers of 2-methylundecanal)

J. Gibka and M. Gliński, *Flavour and Fragrance Journal*, 2006, 11, 480–483. (odours of aldehydes)

P. Kraft. C. Leard, and P. Goutell, From Rallet No 1 to Chanel No 5 via Mademoiselle Chanel No 1, *Perfumer and Flavorist*, October 2007, 32, 36–48.

D. Krautwurst, K.-W. Yau, and R. R. Reed, *Cell*, 1998, 95, 917–926. (mouse OR-I7)

M. D. Kurland, M. B. Newcomer, Z. Peterlin, K. Ryan, S. Firestein, and V. S. Batista, *Biochemistry*, 2010, 49, 6302–6304. (rat OR-I7)

F. Xu, N. Liu, I. Kida, D. L. Rothman, F. Hyder, and G. M. Shepherd, *Proceedings of National Academy of Sciences*, 2003, 100, 11029–11034. (FMRI of mouse olfactory bulbs)

CHAPTER 8

CURARE

S. Feldman, *Poison Arrows*, London, Metro Books, 2005.

J. B. Traynor, *British Journal of Anaesthesia*, 1998, 81, 69–76.

FROGS

J. W. Daly, *Journal of Natural Products,* 1998, 61, 162–172. (frog alkaloids rev.)

J. W. Daly, H. M. Garraffo, T. F. Spande, M. W. Decker, J. P. Sullivan, and M. Williams, *Natural Product Reports,* 2000, 17, 131–135. (epibatidine discovery and other frog alkaloids rev)

J. P. Dumbacher, A. Wako, S. R. Derrickson, A Samuelson, T. F. Spande, and J. W. Daly, *Proceedings of National Academy of Sciences USA,* 2004, 101, 15857–15860. (batrachotoxin, poison-dart frogs, passerine birds and beetles)

D. A. Evans, K. A. Scheidt, and C. W. Downey, *Organic Letters,* 2001, 3, 3009–3012. (epibatidine synthesis)

R. W. Fitch, H. M. Garraffo, T. F. Spande, H. J. C. Yeh, and J. W. Daly, *Journal of Natural Products,* 2003, 66, 1345–1350. (epibatidine)

R. W. Fitch, T. F. Spande, H. M. Garraffo, H. J. C. Yeh, and J. W. Daly, *Journal of Natural Products,* 2010, 73, 331–337. (phantasmidine: an epibatidine congener)

R. A. Saporito, M. A. Donnelly, R. A. Norton, H. M. Garraffo, T. F. Spande, and J. W. Daly, *Proceedings of National Academy of Sciences USA,* 2007, 104, 8885–8890. (pumiliotoxins)

B. P. Smith, M. J. Tyler, T. Kaneko, H. M. Garraffo, T. F. Spande, and J. W. Daly, *Journal of Natural Products,* 2002, 65, 439–447. (pseudophrynamines and pumiliotoxins)

W. Takada, T. Sakata, S. Shimano, Y. Enami, N. Mori, R. Nishida and Y. Kuwahara, *Journal of Chemical Ecology,* 2005, 31, 2403–2415. (pumiliotoxins and food sources)

TTX

E. D. Brodie, III, C. R. Feldman, C. T. Hanifin, J. E. Motychak, D. G. Mulcahy, B. L. Williams, and E. D. Brodie, Jr., *Journal of Chemical Ecology,* 2005, 31, 343–356. (TTX arms races)

U. Koert, *Angewandte Chemie, International Ed.,* 2004, 43, 5572–5576. (synth.)

J. W. Moore and T. Narahashi, *Federal Proceedings,* 1967, 26, 1655–1663.

H. S. Mosher, Tetrodotoxin, Saxitoxin and the Molecular Biology of the Sodium Channel, *Annals of New York Academy of Science,* 1986, 479, 32–43.

M. Wheatley, in (Waring, Steventon, and Mitchell 2007), pp. 387–414.

CONOTOXINS

D. Alonso, Z. Khalil, N. Satkunanthan, and B. G. Livett, *Mini Reviews in Medicinal Chemistry,* 2003, 3, 785–787.

S. Becker and H. Terlau, *Applied Microbiology and Biotechnology,* 2008, 79, 1–9. (conotoxins)

N. L. Daly and D. J. Craik, *Life,* 2009, 61, 144–150. (conotoxin structures)

J. G. McGivern, *Neuropsychiatric Disease and Treatment,* 2007, 3, 69–85. (ziconotide)

C. I. Schroeder and R. J. Lewis, *Marine Drugs,* 2006, 4, 193–214. (ω-conotoxins)

SPIDERS

M. R. Brown, D. D. Sheumack, M. I. Tyler, and M. E. H. Howden, *Biochemical Journal,* 1988, 250, 401–405. (funnel web spider)

P. Escoubas, *Molecular Diversity,* 2006, 10, 545–554. (spider venoms)

P. Escoubas, S. Diochot, and G. Corzo, *Biochimie,* 2000, 82, 893–907. (spider venom neurotoxins)

P. Escoubas and L. Rash, *Toxicon,* 2004, 43, 555–574. (review of tarantula venoms)

G. Estrada, E. Villegas, and G. Corzo, *Natural Product Reports,* 2007, 24, 145–161. (spider venoms as a source of acylpolyamines and peptides)

S. Lee and K. R. Lynch, *Biochemical Journal,* 2005, 391, 317–323. (fiddleback spider)

E. V. Orlova, M. A. Rahman, B. Gowen, K. E. Volynski, A. C. Ashton, C. Manser, M. van Heel, and Y. A. Ushkaryov, *Nature Structural Biology,* 2000, 7, 48–53. (latrotoxin)

P. K. Pallaghy, D. Alewood, P. F. Alewood, and R. S. Norton, *FEBS Letters,* 1997, 419, 191–196. (funnel web spider)

J. Siemens, S. Zhou, R. Piskorowski, T. Nikai, E. A. Lumpkin, A. I. Basbaum, D. King, and D. Julius, *Nature,* 2006, 444, 208–212. (vanillatoxins)

A. A. Vassilevski, S. A. Kozlov, and E. V. Grishin, *Biochemistry (Moscow),* 2009, 74, 1505–1534. (spider venom review)

R. H. Waring, in (Waring, Steventon, and Mitchell 2007), pp. 345–366.

SCORPIONS

W. J. Cook, A. Zell, D. D. Watt, and S. E. Ealick, *Protein Science,* 2002, 11, 479–486. (Arizona bark scorpion)

M. L. Garcia, H. G. Knaus, P. Munujos, R. S. Slaughter, and G. J. Kaczorowski, *American Journal of Physiology—Cell Physiology,* 1995, 269, C1–C10. (charybdotoxin)

R. MacKinnon and C. Miller, *Journal of General Physiology,* 1988, 91, 335–349. (Nobel lecture on potassium channels)

A. N. Mamelak, S. Rosenfeld, R. Bucholz, A. Raubitschek, L. B. Nabors, J. B. Fiveash, S. Shen, M. B. Khazaeli, D. Colcher, A. Liu, et al., *Journal of Clinical Oncology,* 2006, 24, 3644–3650. (chlorotoxin as medicine)

M. Veiseh, P. Gabikian, S-B. Bahrami, O. Veiseh, M. Zhang, R. C. Hackman, A. C. Ravanpay, M. R. Stroud, Y. Kusuma, S. J. Hansen, et al., *Cancer Research,* 2007, 67, 6882–6888. (chlorotoxin as medicine)

SNAKE VENOMS

C. Betzel, G. Lange, G. P. Pal, K. S. Wilson, A. Maelicke, and W. Saenger, *Journal of Biological Chemistry,* 1991, 266, 21530–21536. (α-cobratoxin from *Naja naja siamensis*)

P. W. R. Corfield, T.-J. Lee, and B. W. Low, *Journal of Biological Chemistry,* 1989, 264, 9239–9242. (erabutoxin a)

V. M. Garsky, P. K. Lumma, R. M. Freidinger, S. M. Pitzenberger, W. C. Randall, D. F. Veber, R. J. Gould, and P. A. Friedman, *Proceedings of National Academy of Sciences USA,* 1989, 86, 4022–4026. (echistatin)

J. P. Imredy and R. MacKinnon, *Journal of Molecular Biology,* 2000, 296, 1283–1294. (mamba)

J. L. Lauer-Fields, M. Cudic, S. Wei, F. Mari, G. B. Fields, and K. Brew, *Journal of Biological Chemistry,* 2007, 282, 26948–26955. (sarafotoxin)

M. H. le Du, P. Marchot, P. E. Bougis, J. Navaza, and J. C. Fontecilla-Camps, *Acta Crystallographica,* 1996, D52, 87–92. (fasciculin 2 from the green mamba, *Dendroaspis angusticeps*)

R. A. Love and R. M. Stroud, *Protein Engineering,* 1986, 1, 37–46. (bungarotoxin)

R. G. Mills, G. B. Ralston, and G. F. King, *Journal of Biological Chemistry,* 1994, 269, 23413–23419. (sarafotoxin).

M. A. Ondetti, B. Rubin, and D. W. Cushman, *Science,* 1977, 196, 441–444. (captopril)

M. J. Plotkin, *Medicine Quest,* New York, Penguin Books, 1999.

B. Rees, J. P. Samama, J. C. Thierry, M. Gilbert, J. Fischer, H. Schweitz, M. Lazdunski, and
D. Moras, *Proceedings of National Academy of Sciences USA*, 1987, 84, 3132–3136.
(cobra cardiotoxin V^{II}4)

I. C. Shaw, in (Waring, Steventon and Mitchell, 2007), 329–344.

V. Tsetlin, *European Journal of Biochemistry*, 1999, 264, 281–286. (review of snake venom
α-neurotoxins and other 'three-finger' proteins)

L. Watanabe, J. D. Shannon, R. H. Valente, A. Rucavado, A. Alape-Girón, A. S. Kamaiguti,
R. D. G. Theakston, J. W. Fox, J. M. Gutiérrez, and R. K. Arni, *Protein Science*, 2003,
12, 2273–2281. (BaP1 from *Bothrops asper*)

STRYCHNINE   ·

C. Beemelmanns and H.-U. Reissig, *Angewandte Chemie, International Ed.*, 2010, 49, 8021–
8025. (total synthesis of strychnine)

J. Bonjoch and D. Solé, *Chemical Reviews,* 2000, 100, 3455–3482. (synthesis of strychnine)

J. Buckingham, *Bitter Nemesis: The Intimate History of Strychnine*, Boca Raton, FL, CRC
Press, 2007.

R. M. Harris, in (Waring, Steventon, and Mitchell 2007), pp. 367–386.

Y. Kaburagi, H. Tokuyama, and T. Fukuyama, *Journal of the American Chemical Society*,
2004, 126, 10246–10247. (total synthesis of (–)-strychnine)

S. D. Knight, L. E. Overman, and G. Pairaudeau, *Journal of the American Chemical Society*,
1993, 115, 9293–9294. (enantioselective total synthesis of (–)-strychnine)

T. Ohshima, Y. Xu, R. Takita, and S. Shimizu, *Journal of the American Chemical Society*,
2002, 124, 14546–14547. (enantioselective total synthesis of (–)-strychnine)

G. Sirasani, T. Paul, W. Dougherty, S. Kassel, and R. B. Andrade, *Journal of Organic Chemistry*,
2010, 75, 3529–3532. (total synthesis of (±)-strychnine)

S. Stevens and A. Bannon, *Book of Poisons*, Cincinnati, OH, Writer's Digest Books, 2007,
28–30.

(Timbrell 2005), pp. 154–156, 227–230.

R. B. Woodward, M. P. Cava, W. D. Ollis, A. Hunger, H. U. Daeniker, and K. Schenker,
Tetrahedron, 1963, 19, 247–288. (the total synthesis of strychnine)

CHAPTER 9

DIMETHYLMERCURY

C. U. Eccles and Z. Aman, eds., *The Toxicity of Methylmercury,* Baltimore, MD, John Hopkins
University Press, 1987.

D. W. Nierenberg, R. E. Nordgren, M. B. Chang, R. W. Siegler, M. B. Blayney, F. Hochberg,
T. Y. Toribara, E. Cernichiari, and T. Clarkson, *New England Journal of Medicine*, 1998,
338, 1672–1676. (fatality)

J. Pazderová, A. Jirásek, M. Mráz, and J. Pechan, *International Archives of Arbeitsmedizin*,
1974, 33, 323–328. (fatality)

TETRAETHYLLEAD

C. Cooksey and A. T. Dronsfield, *Education in Chemistry,* 2005, 159–161.

L. Denworth, *Toxic Truth: A Scientist, a Doctor, and the Battle over Lead*, Boston, Beacon
Press, 2009.

S. Hong, J-P. Candelone, C. C. Patterson, and C. F. Boutron, *Science,* 1994, 265, 1841–1843. (lead in Greenland ice)

G. Markowitz and D. Rosner, *Deceit and Denial, the Deadly Politics of Industrial Pollution,* Berkeley, University of California Press, 2002.

D. Seyferth, *Organometallics,* 2003, 22, 2346–2357; 5154–5178. (rise and fall of tetraethyllead)

C. Warren, *Brush with Death. A Social History of Lead Poisoning,* Baltimore, MD, John Hopkins University Press, 2000.

NERVE AGENTS

J. Burdon in (Waring, Steventon, and Mitchell 2007), pp. 209–231.

D. E. C. Corbridge, Phosphorus: *An Outline of Its Chemistry, Biochemistry, and Technology,* 5th ed., Amsterdam, Elsevier, 1995, pp. 574–598.

C. D. Fleming, C. C. Edwards, S. D. Kirby, D. M. Maxwell, P. M. Potter, D. M. Cerasoli, and M. R. Redinbo, *Biochemistry,* 2007, 46, 5063–5071. (hCE1 complexes with soman and tabun)

R. Harris and J. Paxman, *A Higher Form of Killing: The Secret Story of Chemical and Biological Warfare,* New York, Hill and Wang, 1982, pp. 53–67, 138–139.

A. C. Hemmert, T. C. Otto, M. Wierdl, C. C. Edwards, C. D. Fleming, M. MacDonald, J. R. Cashman, P. M. Potter, D. M. Cerasoli, and M. R. Redinbo, *Molecular Pharmacology,* 2010, 77, 508–516. (hCE1 complexes of cyclosarin and sarin)

W. Lange and K. von Krueger, *Berichte der Deutschen Chemischen Gesellschaft,* 1932, 65, 1598–1601.

V. S. Mirzayanov, *State Secrets,* Denver, CO, Outskirts Press, Inc, 2009.

E. O'Connor, N. Fryer, and T. P. Kee, *Education in Chemistry,* 1998, 123–125. (summary)

B. Sanson, F. Nachon, J.-P. Colletier, M.-T. Froment, L. Toker, H. M. Greenblatt, J. L. Sussman, Y. Ashani, P. Masson, I. Silman, and M. Weik, *Journal of Medicinal Chemistry,* 2009, 52, 7593–7603. (structures of AChE with soman before and after aging)

B. C. Saunders, *Some Aspects of the Chemistry and Toxic Action of Organic Compounds Containing Phosphorus and Fluorine,* Cambridge, Cambridge University Press, 1957.

J. B. Tucker, *War of Nerves: Chemical Warfare from World War I to Al-Qaeda,* New York, Pantheon Books, 2006.

CS AND MUSTARD GAS

B. B. Corson and R. W. Stoughton, *Journal of the American Chemical Society,* 1928, 50, 2825–2837. (CS and CN)

F. T. Fraunfelder, *British Medical Journal,* 2000, 320, 458–459; K. Blaho and M. M. Stark, *British Medical Journal,* 2000, 321, 1; J. Gray, *British Medical Journal,* 2000, 321, 1; G. R. N. Jones, *British Medical Journal,* 2000, 321, 2. (toxicity of CS)

P. B. Goodwin, *Keen as Mustard,* St Lucia, University of Queensland Pr, 1998.

S. Jones, *World War I Gas Warfare Tactics and Equipment,* Oxford, Osprey Publishing, 2007. (mustard gas)

Kirk-Othmer Encyclopedia of Chemical Technology, 4th ed., vol. 5, Chichester, England, John Wiley & Sons, 1993, pp. 797–799. (mustard gas)

R. C. Malhotra, K. Ganesan, K. Sugendran, and R. V. Swamy, *Defence Science Journal,* 1999, 49, 97–115. (sulphur mustard)

M. Sartori, *The War Gases,* New York, Van Nostrand, 1939, p. 156 (CN gas); pp. 217 ff (mustard gas)

THALIDOMIDE

D. A. Blake, G. B. Gordon, and S. P. Spielberg, *Teratology*, 1982, 25, 28A–29A.

R. A. Fine, *The Great Drug Deception. The Shocking Story of Mer/29 and the Folks Who Gave You Thalidomide*, New York, Stein and Day, 1972.

J. N. Gordon and P. M. Goggin, *Postgraduate Medical Journal*, 2003, 79, 127–132. (uses)

W. Lenz, *Teratology*, 1992, 46, 417–418. (a personal perspective on the thalidomide tragedy)

W. G. McBride, *Lancet*, 1962, 2, 1358. (alerting letter)

M. T. Miller and K. Strömland, *Teratology*, 1999, 60, 306–321. (review)

T. Paravar and D. J. Lee, *International Review of Immunology*, 2008, 27, 111–135. (mechanisms of action)

A. Raza, *The Biochemist*, February 2002, 21–23. (thalidomide as a drug)

R. von Moos, R. Stolz, T. Cerny, and S. Gillessen, *Swiss Medical Weekly*, 2003, 133, 77–87. (review: thalidomide as a drug)

T. D. Stephens and R. Brynner, *Dark Remedy: The Impact of Thalidomide and Its Revival as a Vital Medicine*, New York, Perseus Publishing, 2001.

T. Stephens, *Birth Defects*, 2009, A 85, 725–731. (thalidomide and chicken embryos)

Sunday Times Insight Team (P. Knightley, H. Evans, E. Potter, and M. Wallace), *Suffer the Children: The Story of Thalidomide*, London, Andre Deutsch, 1979.

C. Therapontos, L. Erskine, E. R. Gardner, W. D. Figg, and N. Vargesson, *Proceedings of National Academy of Sciences*, 2009, 106, 8573–8578. (thalidomide and limb defects)

M. Wallace and M. Robson, *On Giants' Shoulders: The Terry Wiles Story*, London, Times Books, 1976.

CHAPTER 10

EXPLOSIVES

J. Akhavan, *The Chemistry of Explosives*, Cambridge, Royal Society of Chemistry, 1998.

C. Baker, *Education in Chemistry*, 2004, 41, 159–161. (explosives)

S. Bown, *A Most Damnable Invention: Dynamite, Nitrates, and the Making of the Modern World*, New York, Thomas Dunne, 2005.

G. I. Brown, *The Big Bang! History of Explosives*, Alan Sutton Publishing, Gloucestershire, England, 1998.

F. Dubnikova, R. Kosloff, J. Almog, Y. Zeiri, R. Boese, H. Itzhaky, A. Alt, and E. Keinan, *Journal of the American Chemical Society*, 2005, 127, 1146–1159. (TATP)

G. Marsella, *ChemMatters*, Feb. 1997, pp. 4–5. (car air bags)

DETECTING EXPLOSIVES

T. L. Andrew and T. M. Swager, *Journal of the American Chemical Society*, 2007, 129, 7254–7255. (RDX and PETN detection)

T. F. Jenkins, D. C. Leggett, P. H. Miyares, and M. E. Walsh, *Talanta*, 2001, 54, 501–513. (chemical signatures of TNT-filled landmines)

H. Lin and K. S. Suslick, *Journal of the American Chemical Society*, 2010, 132, 15519–15521. (H_2O_2 detection)

G. Mancino, *Chemistry World*, Dec. 2005, 50–53.

J. B. Miller and G. A. Barrall, *American Scientist*, 2005, 93, 50–57. (NQR and explosive detection)

G. Stix, *Scientific American*, Oct. 2005, 56–59.

S. Vos, *ChemMatters*, April 2008, 26, 7–9. (sniffing landmines)

CHAPTER 11

AMPHETAMINES

S. Bernschneider-Reif, F. Oxler, and R. W. Freudenmann, *Pharmazie,* 2006, 61, 966–972. (the origin of MDMA (ecstasy) from the original documents)

S. R. Braswell, *American Meth: A History of the Methamphetamine Epidemic in America,* New York, iUniverse, Inc., 2006.

P-A. Chouvy and J. Meissonier, *Yaa Baa: Production, Traffic, and Consumption of Methamphetamine in Mainland Southeast Asia,* Singapore, Singapore University Press, 2004.

R. De Grandpre, *Ritalin Nation,* W. W. Norton, New York, 1999.

L. Edeleanu, *Berichte der Deutschen Chemischen Gesellschaft,* 1887, 20, 616–622. (synth. of amphetamine)

L. G. French, *Journal of Chemical Education,* 1995, 72, 484–491. (MDMA syntheses)

M. F. Holley, *Crystal Meth: They Call It Ice,* Mustang, OK, Tate Publishing, 2005.

L. Iversen, *Speed, Ecstasy, Ritalin,* Oxford, Oxford University Press, 2006. (the best single source)

H. L. Jin and T. E. Beesley, *Chromatographia,* 1994, 38, 595; D. W. Armstrong, K. L. Rundett, U. B. Nair, and G. L. Reid, *Current Separations,* 1996, 15, 57. (separating isomers of methamphetamine)

M. Joseph, *Agenda: Speed,* London, Carlton Books, 2000.

U. D. McCann, D. F. Wong, F. Yokoi, V. Villemagne, R. F. Dannals, and G. A. Ricaurte, *Journal of Neuroscience,* 1998, 18, 8417–8422. (methamphetamine and neurotransmitter damage)

A. Ogata, *Journal of Pharmaceutical Society of Japan,* 1919, 445, 193–216; *Chemical Abstracts,* 1919, 13, 1709. (synth. of methamphetamine)

F. Owen, *No Speed Limit: The Highs and Lows of Meth,* New York, St. Martin's Press, 2007.

L. Panizzon, *Helvetica Chimica* Acta, 1944, 27, 1748–1756. (Ritalin synthesis)

N. Rasmussen, *On Speed: The Many Lives of Amphetamine,* New York, New York University Press, 2008.

A. Shulgin and A. Shulgin, *Pihkal: A Chemical Love Story,* Berkeley, Transform Press, 1995. (biographical novel; real synthetic details)

O. Snow, *Amphetamine Syntheses: Industrial,* Melbourne, Australia, Thoth Press, 2002.

A. Szabo, E. Billett, and J. Turner, *British Journal of Sports Medicine,* 2001, 35, 342–343. (phenylethylamine and runner's high)

P. M. Thompson, K. M. Hayashi, S. L. Simon, J. A. Geaga, M. S. Hong, Y. Sui, J. Y. Lee, A. W. Toga, W.Ling, and E. D. London, *Journal of Neuroscience,* 2004, 24, 6028–6036. (methamphetamine and brain abnormalities)

N. D. Volkow, L. Chang, G.-J. Wang, J. S. Fowler, D. Franceschi, M. J. Sedler, S. J. Gatley, R. Hitzemann, Y.-S. Ding, C. Wong, and J. Logan, *American Journal of Psychiatry,* 2001, 158, 383–389. (methamphetamine and neurotransmitter damage)

N. D. Volkow, G.-J. Wang, J. S. Fowler, F. Telang, L. Maynard, J. Logan, S. J. Gatley, N. Pappas, C. Wong, P. Vaska, W. Zhu, and J. M. Swanson, *American Journal of Psychiatry,* 2004, 161, 1173–1180. (Ritalin and mathematical task)

CAFFEINE

M. J. Arnaud, *Progress in Drug Research,* 1987, 31, 273–313. (pharmacology of caffeine)

H. Ashihara, H. Sano, and A. Crozier, *Phytochemistry,* 2008, 69, 841–856. (biosynthesis)

M. C. Cornelis et al., *PLoS Genetics,* 2011, 7, e1002033. (genetic factors for caffeine consumption)

M. Czerny and W. Grosch, *Journal of Agricultural and Food Chemistry*, 2000, 48, 868–872. (odourants in raw arabica)

M. Czerny, F. Mayer, and W. Grosch, *Journal of Agricultural and Food Chemistry*, 1999, 47, 695–699. (roasted arabica)

E. Illy, The Complexity of Coffee, *Scientific American*, June 2002, 86–91.

J. A. Nathanson, *Science*, 1984, 226, 184–187. (pesticide activity)

R. Noever, J. Cronise, and R. A. Relwani, *NASA Tech Briefs*, 1995, 19 (4), 82. [Published in *New Scientist* magazine, 27 April 1995]. (using spider-web patterns to determine toxicity)

Y.-S. Kim, H. Uefuji, S. Ogita, and H. Sano, *Transgenic Research*, 2006, 15, 667–672. (caffeine-producing tobacco plants)

P. Mazzafera, T. W. Baumann, M. M. Shimizu, and M. B. Silvarolla, *Tropical Plant Biology*, 2009, 2, 63–76. (coffee from caffeine-free arabica plants)

M. B. Silvarolla, P. Mazzafera, and L. C. Fazuoli, *Nature*, 2004, 429, 826. (naturally decaffeinated arabica coffee)

B. A. Weinberg and B. K. Bealer, *The World of Caffeine*, New York, Routledge, 2001.

S. K. Yadav and P. S. Ahuja, *Plant Foods & Human Nutrition*, 2007, 62, 185–191. (toward generating caffeine-free tea by metabolic engineering)

COCAINE

S. Bencharit, C. L. Morton, Y. Xue, P. M. Potter, and M. R. Redinbo, *Nature Structural Biology*, 2003, 10, 349–356. (human carboxylesterase 1 and metabolism of heroin and cocaine)

M. Bowden, *Killing Pablo: The Hunt for the World's Richest, Most Powerful Criminal in History*, New York, Atlantic Books, 2001.

W. E. Brewer, R. C. Galipo, K. W. Sellers, and S. L. Morgan, *Analytical Chemistry*, 2001, 73, 2371–2376. (cocaine and metabolites in hair)

J. Calatayud and A. González, *Anesthesiology*, 2003, 98, 1503–1508. (cocaine and anaesthesia)

J. F. Casale, J. R. Ehleringer, D. R. Morello and M. J. Lott, *Journal of Forensic Science*, 2005, 50, 1315–1321. (isotopic fractionation of C and N during cocaine processing)

J. F. Casale and R. F. X. Klein, *Forensic Science Review*, 1993, 5, 95–107. (illicit production of cocaine)

A. Dronsfield, P. M. Ellis, and K. Pooley, *Education in Chemistry*, 2007, 44, 183–186. (anaesthesia)

J. Durlacher, *Cocaine (Agenda)*, London, Carlton Books, 2000.

J. R. Ehleringer, J. F. Casale, M. J. Lott, and V. L. Ford, *Nature*, 2000, 408, 311. (isotopic fingerprint of cocaine from the region of growth)

P. Hazarika, S. M. Jickells, K. Wolff, and D. A. Russell, *Angewandte Chemie, International Ed.*, 2008, 47, 10167–10170. (cocaine and metabolites in fingerprints)

(Myers 2007), pp. 88–91.

(Perrine 1996), pp. 188–192.

R. Sleeman, I. F. A. Burton, J. F. Carter, and D. J. Roberts, *Analyst*, 1999, 124, 103–108. (cocaine on banknotes)

D. Streatfeild, *Cocaine: An Unauthorized Biography*, London, Virgin, 2001.

N. D. Volkow and A. C. Swann, eds., *Cocaine in the Brain*, New Brunswick, NJ, Rutgers University Press, 1990.

DESIGNER DRUGS

B. K. Atwood, J. Huffman, A. Straiker, and K. Mackie, *British Journal of Pharmacology*, 2010, 160, 585–593. (JWH-018)

J. F. Buchanan and C. R. Brown, *Medical Toxicology and Adverse Drug Experience*, 1988, 3, 1–17.

P. Goodman and G. Koz, *Designer Drugs (Encyclopedia of Psychoactive Drugs Series 2)*, New York, Chelsea House, 1988.

P. Jenkins, *Synthetic Panics: The Symbolic Politics of Designer Drugs*, New York, University Press, 1999.

J. W. Langston and J. Palfreman, *The Case of the Frozen Addicts*, New York, Pantheon Books, 1995. (MPPP, MPTP).

A. Shulgin and A. Shulgin, *Pihkal: A Chemical Love Story*, Berkeley, Transform Press, 1995. (amphetamines)

R. F. Staack and H. H. Maurer, *Current Drug Metabolism*, 2005, 6, 259–274.

G. L. Sternbach and J. Varon, *Postgraduate Medicine*, 1992, 91, 169–171, 175–176.

M. R. Meyer, J. Wilhelm, F. T. Peters, and H. H. Maurer, *Analytical and Bioanalytical Chemistry*, 2010, 397, 1225–1233. (β-keto amphetamines)

H. L. Weingarten, *Journal of Forensic Science*, 1988, 33, 588–595. (MPPP, MPTP)

A. Williams, *British Medical Journal*, 1984, 289, 1401–1402. (MPPP, MPTP)

N. Uchiyama, M. Kawamura, R. Kikura-Hanajiri, and Y. Goda, *Forensic Toxicology*, 2011, 29, 25–37. (acetylindoles and naphthoylindoles)

ETHANOL

G. Edwards, *Alcohol: The World's Favorite Drug*, New York, Thomas Dunne Books, 2002.

P. K. Gessner and T. Gessner, *Disulfiram and Its Metabolite Diethyldithiocarbamate. Pharmacology and Status in the Treatment of Alcoholism, HIV Infection, AIDS and Heavy Metal Toxicity*, London, Chapman & Hall, 1992. (disulfiram as antialcohol drug)

S. J. Mihic and R. A. Harris, *Alcohol Health and Research World*, 1997, 21, 127–131. (GABA receptor)

(Myers 2007), pp. 120–123.

R. Rudgley, *The Alchemy of Culture: Intoxicants in Society*, London, British Museum Press, 1993.

D. Stephens and R. Dudley, *Natural History*, December 2004–January 2005, 113, 40–44. (drunken monkey hypothesis)

M. L. Wald, *Scientific American*, January 2007, 296 (1) 28–35. (ethanol biofuel)

LSD

D. Black, *Acid: The Secret History of LSD*, London, Vision Paperbacks, 1998.

L. R. Caporael, *Science*, 1976, 192, 21–26. (ergotism and Salem)

J. González-Maeso, N. V. Weisstaub, M. Zhou, P. Chan, L. Ivic, R. Ang, A. Lira, M. Bradley-Moore, Y. Ge, Q. Zhou, S. C. Sealfon, and J. A. Gingrich, *Neuron*, 2007, 53, 439–452 (LSD and signalling pathways)

M. A. Lee, *Acid Dreams: The Complete Social History of LSD*, New York, Grove Press, 2000. (Mann 2009), pp. 1–25.

M. K. Matossian, *Poisons of the Past: Molds, Epidemics, and History*, New Haven, Yale University Press, 1989. (ergotism and history)

A. Roberts, *Albion Dreaming: A Popular History of LSD in Britain*, London, Marshall Cavendish, 2008.

D. E. Smith, *California Medicine*, 1969, 110, 472–476. (LSD and Haight-Ashbury)

J. Stevens, *Storming Heaven: LSD and the American Dream*, London, Heinemann, 1988.

P. W. J. van Dongen and A. N. J. A. de Groot, *European Journal of Obstetrics & Gynecology*, 1995, 60, 109–116. (ergot alkaloids)

Cannabis

V. Berridge, The Lancet, 2010, 375, 798–799. (cannabis and Queen Victoria)

M. Booth, *Cannabis: A History,* Doubleday, London, 2003.

F. A. Campbell, M. R. Tramèr, D. Carroll, D. J. M. Reynolds, R. A. Moore, and H. J. McQuay, *British Medical Journal,* 2001, 323, 16–21. (cannabinoids in pain management)

M. Earleywine, *Understanding Marijuana: A New Look at the Scientific Evidence,* Oxford, Oxford University Press, 2002.

A. M. Galal, D. Slade, W. Gul, A. T. El-Alfy, D. Ferreira, and M. A. Elsohly, *Recent Patents on CNS Drug Discovery,* 2009, 4, 112–136. (natural and synthetic cannabinoids)

J. Green, *Cannabis: The Hip History of Hemp,* London, Pavilion Books, 2002.

R. D. Hosking and J. P. Zajicek, *British Journal of Anaesthesia,* 2008, 101, 59–68. (cannabis in pain medicine)

J. M. Hurley, J. B. West, and J. R. Ehleringer, *International Journal of Drug Policy,* 2010, 21, 222–228. (tracing origin of cannabis)

L. L. Iversen, *The Science of Marijuana,* Oxford, Oxford University Press, 2000.

J. E. Joy, S. J. Watson, and J. A. Benson, eds., *Marijuana and Medicine: Assessing the Science Base,* Washington, DC, National Academies Press, 1999.

(Mann 2009), pp. 77–96.

(Perrine 1996), pp. 338–369.

C. Sherman, A. Smith, and E. Tanner, *Highlights: An Illustrated History of Cannabis,* Berkeley, CA, Ten Speed Press, 1999.

E. K. Shibuya, J. E. S. Sarkis, O. Negrini-Neto, and L. A. Martinelli, *Forensic Science International,* 2007, 167, 8–15. (C and N isotopes indicating origin of marijuana)

Nicotine

A. P. Blum, H. A. Lester, and D. A. Dougherty, *Proceedings of National Academy of Sciences,* 2010, 107, 13206–13211. (nicotinic pharmacophore)

A. Brandt, *Cigarette Century: The Rise, Fall, and Deadly Persistence of the Product That Defined America,* New York, Basic Books, 2007.

J. J. Buccafusco and A. V. Terry, *Life Science,* 2003, 72, 2931–2942. (cotinine and Alzheimer's)

I. Gately, *La Diva Nicotina: The Story of How Tobacco Seduced the World,* London, Simon & Schuster, 2001.

J. Goodman, *Tobacco in History: The Cultures of Dependence,* London, Routledge, 1993.

S. L. Goodman and Zhou Xun, eds., *Smoke: A Global History of Smoking,* London, Reaktion Books, 2004.

J. W. Gorrod and M-C. Tsai, in (Waring, Steventon, and Mitchell 2007), pp. 233–252.

James I, King of England, *A Counter-Blaste to Tobacco,* imprinted at London by R. B., 1604. (London, the Rodale Press, 1954)

D. Kessler, K. Gase, and I. T. Baldwin, *Science,* 2008, 321, 1200–1202. (nicotine in floral scents)

L. Melton, *Chemistry World,* July 2007, pp. 44–48. (antismoking drugs)

N. Minematsu, H. Nakamura, M. Furuuchi, T. Nakajima, S. Takahashi, H. Tateno, and A. Ishizaka, *European Respiratory Journal,* 2006, 27, 289–292. (genetics of CYP2A6 and smoking)

N. Oreskes and E. M. Conway, *Merchants of Doubt: How a Handful of Scientists Obscured the Truth on Issues from Tobacco Smoke to Global Warming,* New York, Bloomsbury Press, 2010.

T. Parker-Pope, *Cigarettes: Anatomy of an Industry from Seed to Smoke,* New York, New Press, 2000.

A. Steppuhn, K. Gase, B. Krock, R. Halitschke, and I. T. Baldwin, *PLoS Biology,* 2004, 2, 1074–1080. (nicotine's defensive function in nature)

K. Stratton, P. Shetty, R. Wallace, and S. Bondurant, eds., *Clearing the Smoke: Assessing the Science Base for Tobacco Harm Reduction*, Washington, DC, National Academies Press, 2001.

X. Xiu, N. L. Puskar, J. A. P. Shanata, H. A. Lester, and D. A. Dougherty, *Nature*, 2009, 458, 534–537. (nicotine binding to brain receptors)

CHAPTER 12

GALANTHAMINE

S. Berkov, J. Bastida, B. Sidjimova, F. Viladomat, and C. Codina, *Chemistry & Biodiversity*, 2011, 8, 115–130.

S. Berkov, M. Cuadrado, E. Osorio, F. Viladomat, C. Codina, and J. Bastida, *Planta Medica*, 2009, 75, 1351–1355. (alkaloids from snowdrops)

S. Berkov, L. Georgieva, V. Kondakova, A. Atanassov, F. Viladomat, J. Bastida, and C. Codina, *Biotechnology & Biotechnological Equipment*, 2009, 23, 1170–1176. (plant sources of galanthamine)

H. M. Greenblatt, G. Kryger, T. Lewis, I. Silman, and J. L. Sussman, *FEBS Letters*, 1999, 463, 321–326. (acetylcholinesterase complexed with (–)-galantamine)

M. Heinrich, in G. A. Cordell, ed., *The Alkaloids*, 2010, 68, 157–165. (review)

P. Magnus, N. Sane, B. P. Fauber, and V. Lynch, *Journal of the American Chemical Society*, 2009, 131, 16045–16047. (synth.)

B. M. Trost, W. Tang, and F. D. Toste, *Journal of the American Chemical Society*, 2005, 127, 14785–14803. (synth.)

OPIUM, MORPHINE, AND HEROIN

K. W. Bentley and D. G. Hardy, *Proceedings of the Chemical Society*, 1963, 220; *Journal of the American Chemical Society*, 1967, 89, 3281–3292. (etorphine)

D. A. Bird, V. R. Franceschi, and P. J. Facchini, *The Plant Cell*, 2003, 15, 2626–2635. (morphine biosynth)

M. Booth, *Opium. A History*, London, Simon and Schuster, 1996.

R. A. Braithwaite, in (Waring, Steventon, and Mitchell 2007), pp. 95–118.

P-A. Chouvy, *Opium: Uncovering the Politics of the Poppy*, London, I B Tauris & Co., 2009.

A. D. Corbett, G. Henderson, A. T. McKnight, and S. J. Paterson, *British Journal of Pharmacology*, 2006, 147, S153–S162. (opioid receptor)

D. S. Goodsell, *The Oncologist*, 2004, 9, 717–719. (opioid receptor)

J. M. Hagel and P. J. Facchini, *Nature Chem. Biol.*, 2010, 6, 273–275. (morphine biosynth.)

B. Hodgson, *Opium: A Portrait of the Heavenly Demon*, London, Souvenir Press, 2000.

B. Hodgson, *In the Arms of Morpheus: The Tragic History of Laudanum, Morphine, and Patent Medicines*, Buffalo, NY, Firefly Books, 2001.

C. Meadway, S. George, and R. Braithwaite, *Forensic Science International*, 1998, 96, 29–38. (heroin testing)

G. L. Patrick, *An Introduction to Medicinal Chemistry*, 2nd ed., Oxford, Oxford University Press, 2001, pp. 511–550. (opioid receptor)

B. D. Paul, E. T. Shimomura, and M. L. Smith, *Clinical Chemistry*, 1999, 45, 510–519. (heroin testing)

(Perrine 1996), pp. 57–58, 68. (opioid receptor)

P. L. Schiff, *American Journal of Pharmaceutical Education*, 2002, 66, 186. (opium and its alkaloids)

G. F. Stegmann, *Journal of South African Veterinary Association* 1999, 70, 164. (etorphine and big game)

D. F. Taber, T. D. Neubert, and A. L. Rheingold, *Journal of the American Chemical Society,* 2002, 124, 12416–12417. (synthesis of (–)-morphine)

J. D. White, P. Hrnciar, and F. Stappenbeck, *Journal of Organic Chemistry,* 1997, 62, 5250–5251. (synthesis of (+)-morphine)

K. Yamaguchi, M. Hayashida, H. Hayakawa, M. Nihira, and Y. Ohno, *Forensic Toxicology,* 2011, 29, 69–71. (heroin testing)

PENICILLIN

R. Bud, *Penicillin: Triumph and Tragedy,* Oxford, Oxford University Press, 2007.

F. W. Diggins, *British Journal of Biomedical Science,* 1999, 56, 83–93; F.W. Diggins, *Biologist (London),* 2000, 47, 115–119. (the discovery of penicillin)

E. Lax, *The Mould in Dr. Florey's Coat: How Penicillin Began the Age of Miracle Cures,* London, Little, Brown, 2004.

J. Mann, *The Elusive Magic Bullet,* Oxford, Oxford University Press, 1999, pp. 39–60.

G. L. Patrick, *An Introduction to Medicinal Chemistry,* 4th ed., Oxford, Oxford University Press, 2009.

S. W. Queener, *Antimicrobial Agents and Chemotherapy,* 1990, 34, 943–948. (penicillin and cephalosporin biosynthesis).

P. L. Roach, I. J. Clifton, C. M. H. Hensgens, N. Shibata, C. J. Schofield, J. Hajdu, and J. E. Baldwin, *Nature,* 1997, 387, 827–830. (isopenicillin *N* synthase and penicillin formation)

V. P. Sandanayaka and A. S. Prashad, *Current Medicinal Chemistry,* 2002, 9, 1145–1165. (resistance to beta-lactam antibiotics)

S. Selwyn, *The Beta-lactam Antibiotics: Penicillins and Cephalosporins in Perspective,* London, Hodder 1980.

J. C. Sheehan, *Enchanted Ring: Untold Story of Penicillin,* Cambridge, MA, MIT Press, 1982.

I. Trehan, F. Morandi, L. C. Blaszczak, and B. K. Shoichet, *Chemistry & Biology,* 2002, 9, 971–980. (structure of amoxycillin bound to β-lactamase)

QUININE AND ANTIMALARIALS

L. J. Bruce-Chwatt, *British Medical Journal,* 1984, 288, 796. (mepacrine and Wingate's toxic psychosis)

A. R. Butler and Y. L. Wu, *Chemical Society Reviews,* 1992, 21, 85–90. (artemisinin).

D. Chaturvedi, A. Goswami, P. P. Saikia, N. C. Barua, and P. G. Rao, *Chemical Society Reviews,* 2010, 39, 435–454. (artemisinin and its derivatives)

(Corey, Czako, and Kürti 2007), pp. 166–169.

A. M. Dondorp et al., *New England Journal of Medicine,* 2009, 361, 5. (artemisinin resistance)

Y. Dong, S. Wittlin, K. Sriraghavan, J. Chollet, S. A. Charman, W. N. Charman, C. Scheurer, H. Urwyler, J. S. Tomas, C. Snyder, et al., *Journal of Medicinal Chemistry,* 2010, 53, 481–491. (OZ-277)

U. Eckstein-Ludwig, R. J. Webb, I. D. A.Van Goethem, J. M. East, A. G. Lee, M. Kimura, P. M. O'Neill, P. G. Bray, S. A. Ward, and S. Krishna, *Nature,* 2003, 424, 957–961. (artemisinin)

M. Honigsbaum, *The Fever Trail: In Search of the Cure for Malaria,* London, Macmillan, 2001.

D. L. Klayman, *Science,* 1985, 228, 1049. (artemisinin)

(Le Couteur and Burreson 2003), pp. 330–350. (quinine)

(Myers 2007), pp. 238–240.

(Nicolaou and Montagnon 2008), pp. 58–66.

P. M. O'Neill, V. E. Barton, and S. A. Ward, *Molecules*, 2010, 15, 1705–1721. (molecular mechanism of action of artemisinin)

P. M. O'Neill and G. H. Posner, *Journal of Medicinal Chemistry*, 2004, 47, 2945–2964. (artemisinin)

G. H. Posner and P. M. O'Neill, *Accounts of Chemical Research*, 2004, 37, 397–404. (artemisinin)

P. Rabe and K. Kindler, *Chemische Berichte*, 1918, 51, 466–467; R. B. Woodward and W. E. Doering, *Journal of the American Chemical Society*, 1944, 66, 849. (quinine synth.)

A. Robert, F. Benoit-Vical, O. Dechy-Cabaret, and B. Meunier, *Pure and Applied Chemistry*, 2001, 73, 1173–1188. (malaria drugs)

F. Rocco, *The Miraculous Fever-Tree: Malaria, Medicine and the Cure That Changed the World*, London, HarperCollins, 2003.

J. I. Seeman, *Angewandte Chemie, International Ed.*, 2007, 46, 1378; A. C. Smith and R. M. Williams, *Angewandte Chemie, International Ed.*, 2008, 47, 1736. (Woodward–Doering synthesis of quinine)

A. Spielman and M. D'Antonio, *Mosquito: The Story of Man's Deadliest Foe*, London, Faber and Faber, 2001.

G. Stork, D. Niu, A. Fujimoto, E. R. Koft, J. M. Balkovec, J. R. Tata, and G. R. Dake, *Journal of the American Chemical Society*, 2001, 123, 3239. (stereoselective total synthesis of quinine)

A. Strecker, *Annalen*, 1854, 91, 155. (quinine)

TAXOL (PACLITAXEL)

L. A. Amos and J. Löwe, *Chemistry & Biology*, 1999, 6, R65–R69. (taxol and microtubule structure)

M. D. Chau, S. Jennewein, K. Walker, and R. Croteau, *Chemical Biology*, 2004, 11, 663–672. (biosynth.)

G. Chauvière, D. Guénard, F. Picot, V. Sénilh, and P. Potier, *Comptes rendus des Séances de l'Académie des Sciences ser II*, 1981, 293, 501–503. (10-deacetylbaccatin III)

J. N. Denis, A. E. Greene, D. Guenard, F. Gueritte-Voegelein, L. Mangatal, and P. Potier, *Journal of the American Chemical Society*, 1988, 110, 5917–5919. (semisynthesis of taxol)

Y. Fu, S. Li, Y. Zu, G. Yang, Z. Yang, M. Luo, S. Jiang, M. Wink, and T. Efferth, *Current Medicinal Chemistry*, 2009, 16, 3966–3985. (review)

D. A. Fuchs and R. K. Johnson, *Cancer Treatment Reports*, 1978, 62, 1219–1222. (trials)

J. Goodman, and V. Walsh, *The Story of Taxol: Nature and Politics in the Pursuit of an Anti-Cancer Drug*. Cambridge, Cambridge University Press, 2001.

D. Guenard, F. Gueritte-Voegelein, and P. Potier, *Accounts of Chemical Research*, 1993, 26, 160–167. (taxol and taxotere rev)

H. Hartzell, *The Yew Tree: A Thousand Whispers*, Eugene, Hulogosi, 1991.

F. A. Holmes, R. S. Walters, R. L. Theriault, A. U. Buzdar, D. K. Frye, G. N. Hortobagyi, A. D. Forman, L. K. Newton, and M. N. Raber, *Journal of National Cancer Institute*, 1991, 83, 1797–1805. (trials against metastatic breast cancer)

R. A. Holton, C. Somoza, H. B. Kim, F. Liang, R. J. Biediger, P. D. Boatman, M. Shindo, C. C. Smith, S. Kim, H. Nadizadeh, Y. Suzuki, C. Tao, P. Vu, S. Tang, P. Zhang, K. K. Murthi, L. N. Gentile, and J. H. Liu, *Journal of the American Chemical Society*, 1994, 116, 1597–1598.

D. G. I. Kingston, *Journal of Natural Products*, 2000, 63, 726–734. (rev.).

D. G. I. Kingston, *Chemical Communications*, 2001, 867–880. (rev.).

W. P. McGuire, E. K. Rowinsky, N. B. Rosenshein, F. C. Grumbine, D. S. Ettinger, D. K. Armstrong, and R. C. Donehower, *Annals of Internal Medicine*, 1989, 111, 273–279. (trials against ovarian cancer)

K. C. Nicolaou, R. K. Guy, and P. Potier, *Sci. Am.*, June 1996, 84–88 (rev.).

K. C. Nicolaou, Z. Yang, J. J. Liu, H. Ueno, P. G. Nantermet, R. K. Guy, C. F. Clairborne, J. Renaud, E. A. Couladouros, K. Paulvannan, and E. J. Sorensen, *Nature*, 1994, 367, 630–634.

P. B. Schiff, J. Fant, and S. B. Horwitz, *Nature*, 1979, 277, 665–667.

P. B. Schiff and S. B. Horwitz, *Proceedings of National Academy of Sciences*, 1980, 77, 1561–1565.

A. L. Wheeler, R. M. Long, R. E. B. Ketchum, C. D. Rithner, R. M. Williams, and R. Croteau, *Archives of Biochemistry and Biophysics*, 2001, 390, 265–278. (biosynthesis of Taxol)

TETRACYCLINE

D. E. Brodersen, W. M. Clemons, A. P. Carter, R. J. Morgan-Warren, B. T. Wimberly, and V. Ramakrishnan, *Cell*, 2000, 103, 1143–1154. (how tetracycline works)

J. Mann, *Life Saving Drugs: The Elusive Magic Bullet*, Cambridge, Royal Society of Chemistry, 2004, pp. 70–71.

M. L. Nelson, A. Dinardo, J. Hochberg, and G. J. Armelagos, *American Journal of Physical Anthropology*, 2010, 143, 151–154. (tetracycline in skeletons from Sudanese Nubia 350–550 CE)

C. Sun, Q. Wang, J. D. Brubaker, P. M. Wright, C. D. Lerner, K. Noson, M. Charest, D. R. Siegel, Y.-M. Wang, and A. G. Myers, *Journal of the American Chemical Society*, 2008, 130, 17913–17927.

VANCOMYCIN

D. A. Evans, M. R. Wood, B. W. Trotter, T. I. Richardson, J. C. Barrow, and J. L. Katz, *Angewandte Chemie, International Ed.*, England, 1998, 37, 2700–2704. (total synth. of vancomycin and eremomycin aglycons)

D. Kahne, C. Leimkuhler, Wei Lu, and C. Walsh, *Chemical Reviews*, 2005, 105, 425–448. (review)

K. C. Nicolaou, C. N. C. Boddy, S. Bräse, and N. Winssinger, *Angewandte Chemie, International Ed.*, 1999, 38, 2096–2152. (glycopeptide antibiotics)

K. C. Nicolaou, H. J. Mitchell, N. F. Jain, N. Winssinger, R. Hughes, and T. Bando, *Angewandte Chemie, International Ed.*, England, 1999, 38, 240–244. (total synth.)

M. Schäfer, T. R Schneider, and G. M. Sheldrick, *Structure*, 1996, 4, 1509–1515. (crystal structure of vancomycin)

M. Shnayerson and M. Plotkin, *The Killers within: The Deadly Rise of Drug-Resistant Bacteria*, Boston, Little Brown and Company, 2002.

D. H. Williams and B. Bardsley, *Angewandte Chemie, International Ed.*, 1999, 38, 1172–1193. (vancomycin antibiotics)

VINCA ALKALOIDS

J. Duffin, *Canadian Bulletin of Medical History*, 2000, 17, 155–192. (discovery of vinblastine)

I. S. Johnson, J. G. Armstrong, M. Gorman, and J. P. Burnett, *Cancer Research*, 1963, 23, 1390–1427. (isolation of vinca alkaloids)

E. McCoy and S. E. O'Connor, *Journal of the American Chemical Society,* 2006, 128, 14276–14277. (biosynthesis in periwinkle)

H. Ishikawa, D. A. Colby, and D. L. Boger, *Journal of the American Chemical Society,* 2008, 130, 420–421. (vinblastine synthesis)

H. Ishikawa, D. A. Colby, S. Seto, P. Va, A. Tam, H. Kakei, T. J. Rayl, I. Hwang, and D. L. Boger, *Journal of the American Chemical Society,* 2009, 131, 4904–4916. (vinca alkaloid synthesis)

J. Roepke, V. Salim, M. Wu, A. M. K. Thamm, J. Murata, K. Ploss, W. Boland, and V. De Luca, *Proceedings of National Academy of Sciences,* 2010, 107, 15287–15292. (vinca components in plant)

W. Runguphan, X. Gu, and S. E. O'Connor, *Nature,* 2010, 468, 461–464. (engineered biosynth.)

W. Runguphan and S. E. O'Connor, *Nature Chemical Biology,* 2009, 5, 151–153.

J. R. Wright, *Canadian Medical Association Journal,* 2002, 167, 1391–1396. (discovery of vinblastine)

A. Y.-S. Yip, E. Y.-Y. Ong, and L. W-C. Chow, *Expert Opinion on Investigational Drugs,* 2008, 17, 583–591. (vinflunine)

CHAPTER 13

ASPIRIN AND OTHER NSAIDs

B. J. Anderson, *Pediatric Anesthesia,* 2008, 18, 915–921. (how paracetamol acts)

(Corey, Kürti, and Czakó 2007), pp. 38, 40–41; Vioxx, p. 43.

S. Houlton, *Chemistry World,* December 2007, 56–59. (Vioxx withdrawal)

D. Jeffreys, *Aspirin, the Story of a Wonder Drug,* London, Bloomsbury, 2004.

R. G. Kurumbail, A. M. Stevens, J. K.Gierse, J. J. McDonald, R. A. Stegeman, J. Y. Pak, D. Gildehaus, J. M. Iyashiro, T. D. Penning, K. Seibert, P. C. Isakson, and W. S. Stallings, *Nature,* 1996, 384, 644–648. (COX-2)

(Li 2006), pp. 219–224.

P. J. Loll, D. Picot, O. Ekabo, and R. M. Garavito, *Biochemistry,* 1996, 35, 7330–7340. (COX-1)

P. J. Loll, D. Picot, and R. M. Garavito, *Nature Structural Biology,* 1995, 2, 637–643. (aspirin activity and crystal structure of inactivated prostaglandin H2 synthase)

C. C. Mann and M. L. Plummer, *The Aspirin Wars,* New York, A. A. Knopf, 1991.

(Myers 2007), pp. 10–12.

T. Nesi, *Poison Pills: The Untold Story of the Vioxx Drug Scandal,* New York, Thomas Dunne Books, 2008.

(Nicolaou and Montagnon 2008), pp. 22–27.

C. Osborn, ed., *Aspirin,* London, Royal Society of Chemistry, 1998.

D. Picot, P. J. Loll, and R. M. Garavito, *Nature,* 1994, 367, 243–249. (COX-1)

K. D. Rainsford, ed., *Aspirin and Related Drugs,* London, Taylor & Francis, 2004.

G. Rimon, R. S. Sidhu, D. A. Lauver, J. Y. Lee, N. P. Sharma, C. Yuan, R. A. Frieler, R. C. Trievel, B. R. Lucchesi, and W. L. Smith, *Proceedings of National Academy of Sciences,* 2010, 107, 28–33. (celecoxib binding to COX-1)

B. S. Selinsky, K. Gupta, C. T. Sharkey, and P. J. Loll, *Biochemistry,* 2001, 40, 5172–5180. (ibuprofen binding to COX-1)

P. Sheldon, *The Fall and Rise of Aspirin—The Wonder Drug,* Studley, England, Brewin Books, 2007.

W. Sneader, *British Medical Journal,* 2000, 321, 1591–1594. (discovery of aspirin: a reappraisal)

K. M. Starko, *Clinical Infectious Diseases,* 2009, 49, 1405–1410. (salicylates and 1981–1999 flu pandemic)

K. M. Starko, C. G. Ray, L. B. Dominguez, W. L. Stromberg, and D. F. Woodall, *Pediatrics,* 1980, 66, 859–864. (Reye's syndrome and salicylate use)

J. R. Vane and R. M. Botting, eds., *Therapeutic Roles of Selective COX-2 Inhibitors,* London, William Harvey Press, 2001.

J. R. Vane and R. M. Botting, *Thrombosis Research,* 2003, 110, 255–258. (mechanism of action of aspirin).

ACYCLOVIR

(Corey, Czako, and Kürti 2007), p. 148.

E. De Clercq and H. J. Field, *British Journal of Pharmacology,* 2006, 147, 1–11. (antiviral prodrugs)

G. B. Elion, Nobel Lecture, 8 December 1988, from Jan Lindsten (ed.), *Nobel Lectures, Physiology or Medicine 1981–1990,* Singapore, World Scientific Publishing Co., 1993, pp. 458–463.

G. B. Elion, P. A. Furman, J. A. Fyfe, P. de Miranda, L. Beauchamp, and H. J. Shaeffer, *Proceedings of National Academy of Sciences USA,* 1977, 74, 5716–5720. (selectivity)

B. Golankiewicz and T. Ostrowski, *Antiviral Research,* 2006, 71, 134–140. (tricyclic nucleoside analogues as antiherpes agents)

A. Gringauz, *Introduction to Medicinal Chemistry,* New York, Wiley-VCH, 1997, pp. 326–327.

(Nicolaou and Montagnon 2008), pp. 306–307.

ATENOLOL AND β-BLOCKERS

A. Barrett and C. Cullum, *British Journal of Pharmacology,* 1968, 34, 43–55. (biological properties of the optical isomers of propranolol)

J. W. Black, *British Journal of Pharmacology,* 1997, 120 (Suppl.), 283–284.

J. W. Black, W. A. M. Duncan, and R. G. Shanks, *British Journal of Pharmacology,* 1965, 25, 577–591. (comparison of some properties of pronethalol and propranolol)

(Corey, Czako, and Kürti 2007), p. 68.

A. Gringauz, *Introduction to Medicinal Chemistry,* New York, Wiley-VCH, 1997, pp. 428–438.

G. L. Patrick, *An Introduction to Medicinal Chemistry,* 1st ed., Oxford, Oxford University Press, 1995, pp. 94–95, 102–103.

R. Silverman, *The Organic Chemistry of Drug Design and Drug Action,* San Diego, Academic Press, 1992, p. 79. (chirality)

M. P. Stapleton, *Texas Heart Institute Journal,* 1997, 24, 336–342. (Sir James Black and propranolol)

K. Stoschitzky, G. Zernig, and W. Lindner, *Journal of Clinical and Basic Cardiology,* 1998, 1, 14. (racemic beta-blockers)

AZT (AZIDOTHYMIDINE)

(Corey, Czako, and Kürti 2007), p. 151.

I. Dyer, J. N. Low, P. Tollin, H. R. Wilson, and R. A. Howie, *Acta Crystallographica,* 1988, C 44, 767–769. (structure of AZT)

M. A. Fischl et al., *New England Journal of Medicine,* 1987, 317, 185–191. (trials)

(Li 2006), pp. 122–128.

H. Mitsuya, K. J. Weinhold, P. A. Furman, M. H. St Clair, S. N. Lehrman, R. C. Gallo, D. Bolognesi, D. W. Barry, and S. Broder, *Proceedings of National Academy of Science USA,* 1985, 82, 7096–7100. (AZT works against HIV in vivo)
(Nicolaou and Montagnon 2008), pp. 307–310.
R. Yarchoan, R. Klecker, K. Weinhold, P. Markham, H. Lyerly, D. Durack, E. Gelmann, S. Lehrman, R. Blum, and D. Barry, *Lancet,* 1986, 327, 575–580. (trials)

Cisplatin

R. A. Alderden, M. D. Hall, and T. W. Hambley, *Journal of Chemical Education,* 2006, 83, 728–734. (cisplatin discovery and development of Pt anticancer drugs)
L. Armstrong, *It's Not about the Bike,* London, Yellow Jersey Press, 2000. (esp. p. 86)
R. Champion and J. Powell, *Champion's Story,* London, Victor Gollancz, 1981.
H. Choy, C. Park, and M. Yao, *Clinical Cancer Research,* 2008, 14, 1633–1638. (satraplatin)
S. Dhar and S. J. Lippard, *Proceedings of National Academy of Sciences,* 2009, 106, 22199–22204. (mitaplatin)
S. C. Dhara, *Indian Journal of Chemistry,* 1970, 8, 193–194. (best synth.)
G. B. Kauffman, R. Pentimalli, S. Dolti, and M. B. Hall, *Platinum Metals Review,* 2010, 54, 250–256. (Peyrone and cisplatin)
L. Kelland, *Nature Reviews Cancer,* 2007, 7, 573–584.
L. R. Kelland, G. Abel, M. J. McKeage, M. Jones, P. M. Goddard, M. Valenti, B. A. Murrer, and K. R. Harrap, *Cancer Research,* 1993, 53, 2581–2586. (satraplatin)
A. Kuwahara, M. Yamamori, K. Nishiguchi, T. Okuno, N. Chayahara, I. Miki, T. Tamura, T. Inokuma, Y. Takemoto, T. Nakamura, K. Kataoka, and T. Sakaeda, *International Journal of Medical Science,* 2009, 6, 305–311. (nedaplatin)
B. Lippert, *Cisplatin: Chemistry and Biochemistry of a Leading Anticancer Drug,* Weinheim, Germany, Wiley-VCH, 1999.
J. Lokich, *Cancer Investigation,* 2001, 19, 756–760. (What is the "best" platinum drug?)
B. Rosenberg, L. Van Camp, and T. Krigas, *Nature,* 1965, 205, 698–699. (products from Pt electrode inhibit *Escherichia coli* division)
R. C. Todd and S. J. Lippard, *Metallomics,* 2009, 1, 280–291. (Pt antitumor compounds)
D. Wang and S. J. Lippard, *Nature Reviews Drug Discovery,* 2005, 4, 307–320.

Gleevec

B. J. Druker and N. B. Lydon, *Journal of Clinical Investigation,* 2000, 105, 3–7.
C. Gambacorti-Passerini, *Lancet Oncology,* 2008, 9, 600.
A. Hochhaus, B. Drucker, C. Sawyers, et al. *Blood,* 2008, 111, 1039–1043.
H. Kantarjian et al., *New England Journal of Medicine,* 2010, 362, 2260–2270. (dasatinib vs. imatinib)
G. Saglio et al., *New England Journal of Medicine,* 2010, 362, 2251–2259. (nilotinib vs. imatinib)
M. Talpaz et al., *New England Journal of Medicine,* 2006, 354, 2531–2541. (dasatinib)
D. Vasella and R. Slater, *Magic Cancer Bullet: How a Tiny Orange Pill May Rewrite Medical History,* New York, HarperBusiness, 2003.

Linezolid

J. A. Ippolito, Z. F. Kanyo, D. Wang, F. J. Franceschi, P. B. Moore, T. A. Steitz, and E. M. Duffy, *Journal of Medicinal Chemistry,* 2008, 51, 3353–3356. (structure of linezolid bound to the 50S ribosomal subunit)

J. B. Locke, J. Finn, M. Hilgers, G. Morales, S. Rahawi, G. C. Kedar, J. J. Picazo, W. Im, K. J. Shaw, and J. L. Stein, *Antimicrobial Agents and Chemotherapy,* 2010, 54, 5337–5343. (structure–activity relationships of oxazolidinones)

D. Shinabarger, *Expert Opinion on Investigational Drugs,* 1999, 8, 1195–1202. (how oxazolidinones work)

D. L. Shinabarger, K. R. Marotti, R. W. Murray, A. H. Lin, E. P. Melchior, S. M. Swaney, D. S. Dunyak, W. F. Demyan, and J. M. Buysse, *Antimicrobial Agents and Chemotherapy,* 1997, 41, 2132–2136.

S. Stefani, D. Bongiorno, G. Mongelli, and F. Campanile, *Pharmaceuticals,* 2010, 3, 1988–2006. (linezolid resistance in staphylococci)

D. L. Stevens, D. Herr, H. Lampiris, J. L. Hunt, D. H. Batts, B. Hafkin, and the Linezolid MRSA Study Group, *Clinical Infectious Diseases,* 2002, 34, 1481–1490. (linezolid vs. vancomycin for treatment of MRSA)

M. H. Wilcox, *Expert Opinion on Pharmacotherapy,* 2005, 6, 2315–2326. (update)

PROZAC AND ANTIDEPRESSANTS

(Corey, Czako, and Kürti 2007), pp. 229–231.

J. C. Fournier, R. J. DeRubeis, S. D. Hollon, S. Dimidjian, J. D. Amsterdam, R. C. Shelton, and J. Fawcett, *Journal of the American Medical Association,* 2010, 303, 47–53. (meta-analysis of SSRIs)

D. Gunnell, J. Saperia, and D. Ashby, *British Medical Journal,* 2005, 330, 385–389. (SSRIs and suicide)

(Li 2006), pp. 137–150.

D. Healy, *Psychotherapy and Psychosomatics,* 2003, 72, 71–79. (SSRIs and suicide)

D. Healy, *Let Them Eat Prozac: The Unhealthy Relationship between the Pharmaceutical Industry and Depression,* New York, NYU Press, 2004.

I. Kirsch, *The Emperor's New Drugs: Exploding the Antidepressant Myth,* New York, Basic Books, 2010.

I. Kirsch, B. J. Deacon, T. B. Huedo-Medina, A. Scoboria, T. J. Moore, and B. T. Johnson, *PLoS Medicine,* 2008, 5, 0260–0268. (meta-analysis of SSRIs)

P. D. Kramer, *Listening to Prozac. A Psychiatrist Explores Antidepressant Drugs and Remaking of the Self,* New York, Viking Press; 1993.

(Myers 2007), pp. 127–129.

A. Tone, *Age of Anxiety: A History of America's Turbulent Affair with Tranquilizers,* New York, Basic Books, 2009.

D. T. Wong, F. P. Bymaster, J. S. Horng, and B. B. Molloy, *Journal of Pharmacology and Experimental Therapeutics,* 1975, 193, 804–811. (in vitro and in vivo study of fluoxetine in rat)

D. T. Wong, J. Horng, F. Bymaster, K. Hauser, and B. Molloy, *Life Sciences,* 1974, 15, 471–479.

D. T. Wong, K. W. Perry, and F. P. Bymaster, *Nature Reviews Drug Discovery,* 2005, 4, 764–774. (discovery of fluoxetine)

E. Wurtzel, *Prozac Nation: Young and Depressed in America—A Memoir,* New York, Houghton Mifflin, 1994.

RANITIDINE (ZANTAC) AND CIMETIDITNE (TAGAMET)

K. Abe, K. Tani, and Y. Fujiyoshi, *Nature Communications,* 2011, 2, 155–161. (conformational rearrangement of gastric H^+,K^+-ATPase)

K. Abe, K. Tani, T. Nishizawa, and Y. Fujiyoshi, *The EMBO Journal,* 2009, 28, 1637–1643. (crystal structure of gastric H+,K+-ATPase)

(Corey, Czako, and Kürti 2007), p. 103.

A. Gringauz, *Introduction to Medicinal Chemistry,* New York, Wiley-VCH, 1997, pp. 634–638.

K. Munson, R. Garcia, and G. Sachs, *Biochemistry,* 2005, 44, 5267–5284. (inhibitor and ion binding sites on gastric H, K-ATPase)

(Nicolaou and Montagnon 2008), pp. 311–313.

G. L. Patrick, *An Introduction to Medicinal Chemistry,* 1st ed., Oxford, Oxford University Press, 1995, pp. 281–312. (history of the development of cimetidine and ranitidine)

V. F. Roche, *American Journal of Pharmaceutical Education,* 2006, 70, 101, 11 pp. (PPI mechanism)

G. Sachs, J. M. Shin, and C. W. Howden, *Alimentary Pharmacology & Therapeutics,* 2006, 23, 2–8.

G. Sachs, J. M. Shin, and R. Hunt, Novel Approaches to Inhibition of Gastric Acid Secretion *Current Gastroenterology Reports,* 2010, 12, 437–447.

M. Salas, A. Ward, and J. Caro, *BMC Gastroenterology,* 2002, 2, 17, 7 pp. (PPI testing)

S. Shi and U. Klotz, *European Journal of Clinical Pharmacology,* 2008, 64, 935–951. (clinical use of PPIs)

J. M. Shin, K. Munson, O. Vagin, and G. Sachs, *Pflugers Archive,* 2009, 457, 609–622. (the gastric HK-ATPase: structure, function, and inhibition)

J. M. Shin and G. Sachs, *Current Gastroenterology Reports,* 2008, 10, 528–534. (pharmacology of PPIs)

R. B. Silverman, *The Organic Chemistry of Drug Design and Drug Action,* San Diego, Academic Press, 1992, p. 88.

R. B. Silverman, *The Organic Chemistry of Drug Design and Drug Action,* 2nd ed., San Diego, Academic Press, 2004, p. 159.

SULFANILAMIDE AND THE SULFONAMIDES

The account of the discovery of Prontosil is based on that given in Chapters 10–15 of Hager's book (previously cited). This is carefully researched from contemporary sources, notably the laboratory notebooks of Domagk and others, together with Domagk's unpublished reminiscences. Many previous accounts imply (following Domagk's description of a single experiment in his initial publication) that it was based on a single experiment. It is similarly sometimes implied that Hildegarde Domagk was the first person to benefit from Prontosil.

C. M. De Costa, *Medical Journal of Australia,* 2002, 177, 668–671. (puerperal sepsis)

G. Domagk, *Deutsche Medizinische Wochenschaft,* 1935, 250–253.

R. J. Franklin, *Medic! How I Fought World War II with Morphine, Sulfa and Iodine Swabs,* Lincoln, University of Nebraska Press, 2006.

I. Galdston, *Behind the Sulfa Drugs,* New York, Appleton-Century Company, 1943.

M. Gilbert, *Winston S. Churchill, Volume VII,* London, Heinemann, 1986, pp. 340, 603–622.

T. Hager, *The Demon under the Microscope: From Battlefield Hospitals to Nazi Labs, One Doctor's Heroic Search for the World's First Miracle Drug,* New York, Harmony, 2006.

K. Halle, *Irrepressible Churchill,* Facts on File, 1985, p. 196.

M. Korsinczky, K. Fischer, N. Chen, J. Baker, K. Rieckmann, and Q. Cheng, *Antimicrobial Agents and Chemotherapy,* 2004, 48, 2214–2222. (antimalarials)

J. E. Lesch, *The Inside Story of Medicines: A Symposium,* G. J. Higby and E. C. Stroud, eds., Madison, WI, American Institute of the History of Pharmacy, 1997, pp. 101–119. (M & B 693)

J. E. Lesch, *The First Miracle Drugs: How the Sulfa Drugs Transformed Medicine,* Oxford, Oxford University Press, 2006.

A. G. N., *Canadian Medical Association Journal,* 1937, 37, 590. (fatal elixir)

D. D. Woods, *British Journal of Experimental Pathology,* 1940, 21, 74–90. (PABA and sulfanilamide)

TAMIFLU

V. Farina and J. D. Brown, *Angewandte Chemie, International Ed.,* 2006, 45, 7330–7334. (Tamiflu supply)

G. Laver, *Future Virology,* 2006, 1, 577–586. (Tamiflu)

G. Laver, *Education in Chemistry,* 2007, 44, 48–52. (flu drugs).

W. Lew, X. Chen, and C. U. Kim, *Current Medicinal Chemistry,* 2000, 7, 663–672. (oseltamivir discovery)

A. Moscona, *New England Journal of Medicine,* 2005, 353, 1363–1373. (neuraminidase inhibitors)

J. C. Rohloff, K. M. Kent, M. J. Postich, M. W. Becker, H. H. Chapman, D. E. Kelly, W. Lew, M. S. Louie, L. R. McGee, E. J. Prisbe, L. M. Schultze, R. H. Yu, and L. Zhang, *Journal of Organic Chemistry,* 1998, 63, 4545–4550. (synthesis of GS-4104)

N. Satoh, T. Akiba, S. Yokoshima, and T. Fukuyama, *Angewandte Chemie, International Ed.,* 2007, 46, 5734–5736. (Tamiflu synthesis)

M. Shibasaki and M. Kanai, *European Journal of Organic Chemistry,* 2008, 1839–1850. (synthetic strategies review)

B. M. Trost and T. Zhang, *Angewandte Chemie, International Ed.,* 2008, 47, 3759–3761. (Tamiflu synthesis)

M. von Itzstein, *Nature Reviews Drug Discovery,* 2007, 6, 967–974. (sialidase inhibitors)

Y.-Y. Yeung, S. Hong, and E. J. Corey, *Journal of the American Chemical Society,* 2006, 128, 6310–6311.

CHAPTER 14

R. Adams, Biographical Memoir of Wallace Hume Carothers 1896–1937, *National Academy of Sciences of the USA,* 1939, Vol. XX.

P. Atkins, *Atkins' Molecules,* 2nd ed., Cambridge, Cambridge University Press, 2003, pp. 93–97.

M. Chisholm, *Chemistry in Britain,* April 1998, 33–36. (Perspex)

S. Fenichell, *Plastic. The Making of a Synthetic Century,* London, HarperCollins, 1996.

S. Freinkel, *Plastic: A Toxic Love Story,* Boston, Houghton Mifflin Harcourt, 2011.

W. Gratzer, *Giant Molecules: From Nylon to Nanotubes,* Oxford, Oxford University Press, 2009, pp. 126–127, 157–158.

S. Handley, *Nylon: The Manmade Fashion Revolution,* London, Bloomsbury Publishing, 1999.

M. Hermes, *Enough for One Lifetime, Wallace Carothers, the Inventor of Nylon,* Washington, DC, Chemical Heritage Foundation, 1996.

(Le Couteur and Burreson 2003), pp. 105–123.

H. F. Mark, *Giant Molecules,* Time-International Nederland, 1968, pp. 105–109.

R. O. MacRae, C. M. Pask, L. K. Burdsall, R. S. Blackburn, C. M. Rayner, and P. C. McGowan, *Angewandte Chemie,* international ed., 2011, 50, 291–294. (synthesis and colouration of poly(lactic acid))

S. B. McGrayne, *Prometheans in the Lab: Chemistry and the Making of the Modern World,* New York, McGraw–Hill, 2001, pp. 106–147.

F. M McMillan, *The Chain Straighteners,* London, Macmillan, 1979.

(Selinger 1998), pp. 226–228 (polyurethane)

T. H. Yang, T. W. Kim, H. O. Kang, S.-H. Lee, E. J. Lee, S.-C. Lim, S. O. Oh, A.-J. Song, S. J. Park, and S. Y. Lee, *Biotechnology and Bioengineering,* 2010, 105, 150–160. (biosynthesis of polylactide)

Index